A Long Voyage to the Moon

Outward Odyssey
A People's History of Spaceflight

Series editor
Colin Burgess

A Long Voyage to the Moon

The Life of Naval Aviator and *Apollo 17* Astronaut Ron Evans

Geoffrey Bowman

Foreword by Jack Lousma

UNIVERSITY OF NEBRASKA PRESS • LINCOLN

Library of Congress Cataloging-in-
Publication Data
Names: Bowman, Geoffrey (Geoffrey
Alexander), author. | Lousma, Jack, 1936–
writer of foreword.
Title: A long voyage to the moon: the life
of naval aviator and Apollo 17 astronaut
Ron Evans / Geoffrey Bowman; foreword
by Jack Lousma.
Description: Lincoln: University of
Nebraska Press, [2021] | Series: Outward
odyssey: a people's history of spaceflight |
Includes bibliographical references and
index.
Identifiers: LCCN 2020051336
ISBN 9781496213198 (hardback)
ISBN 9781496228260 (epub)
ISBN 9781496228277 (pdf)
Subjects: LCSH: Evans, Ronald E., 1933–
1990. | Astronauts—United States—
Biography. | Project Apollo (U.S.)
Classification: LCC TL789.85.E93 B69
2021 | DDC 629.450092 [B]—dc23
LC record available at
https://lccn.loc.gov/2020051336

Set in Garamond Premier by Laura Buis.

For Jan

This is your story, too.

What is it, I wondered, that makes a man willing to sit up on top of an enormous Roman candle, such as a Redstone, Atlas, Titan or Saturn rocket, and wait for someone to light the fuse?

—Tom Wolfe, *The Right Stuff*

Contents

Illustrations

Foreword

Ron Evans was a very special individual. Knowing him as I did during those golden years for space exploration, it was not merely a classic case of right place, right time for him, as he was a man endowed with a tremendous combination of qualities and skills that came together to set him on a steady course to the moon—one of only twenty-four human beings to do so. His career certainly paralleled mine in many ways, and as I reflect on his life and accomplishments for this book, I can easily recall his openness and sincerity, his determination to do the best he could, and his pride in overcoming whatever challenges he faced, be it as a U.S. Navy combat pilot or a NASA astronaut. Through all of our shared years, I also knew him as a devoted husband and father to his beloved wife, Jan, and their children, Jaime and Jon.

In past interviews, I have pointed out that when I was young—and this would apply equally to Ron—there were no such things as astronauts. In fact, that word did not even exist. We grew up in an era of magical childish imagination, fueled by unmissable Saturday movie serials, comic books, and cartoon characters like Alley Oop, the time-traveling caveman, with his dinosaur companion. We also had *fictional* space heroes like Buck Rogers and Flash Gordon bringing us the wonders—and terrors—of space travel. Never in our wildest dreams could we have imagined that one day we, too, would be riding rockets into space. It still feels somewhat surreal that we did those things.

Ron Evans and I were selected in NASA's Group 5 astronaut contingent in April 1966, but our lives were running somewhat parallel before then. In 1963, having served as a jet attack pilot at the U.S. Marine Corps Air Station Cherry Point, North Carolina, and overseas at Iwakuni, Japan, I began two years of study at the U.S. Naval Postgraduate School in Monterey, California, where I had undertaken the aeronautical engineering course. There I met or at least knew of several top-notch pilots who would one day join NASA's astronaut corps, including Gene Cernan, Dick Gordon, Paul Weitz, and Ron Evans. I was a full year behind Ron, and I would later learn that he had already applied to NASA for selection as a Group 3 astronaut, albeit unsuccessfully. The master of science degree he achieved at the postgraduate school, also in aeronautical engineering, surely added an additional, vital qualification that helped him to succeed the next time he applied.

Military officers at Monterey were still on active duty during the postgraduate school's summer (no class) period and had to "do something military-career productive." I now know that Ron took a short summer course in the Russian language at the Monterey-based Defense Language Institute. Some years later, when I feared that the Skylab Program might be canceled, I took a University of Maryland off-campus study course in the Russian language, on my own time, in case the rumored mission with the Soviets might be my only option for a space flight. Neither Ron nor I could have known that a few years later we would actually be training together and speaking Russian with our onetime Cold War enemies for the Apollo-Soyuz Test Project.

It was an exciting time when we were both attending the Naval Postgraduate School. The United States and the Soviet Union were then engaged in a monumental undertaking that had come to be known as the space race to the moon. Already the first astronauts and cosmonauts had begun flying ever longer and increasingly adventurous missions in Earth orbit. How we admired them! Despite this, I had not yet acquired the credentials even to apply to become an astronaut in our nation's space program. Of course, Ron had, and as it turned out, we ended up at NASA together!

After graduating from the postgraduate school, I was assigned to a U.S. Marine Corps Photo Reconnaissance Squadron at Cherry Point. Ron and I both flew the Vought F-8 Crusader but at different times and places. We loved flying that powerful supersonic jet aircraft. It was a single-seater, so you didn't have to fly with anybody else. It had an afterburner, and you could fly high and fast or low and fast. I was flying photo recon in the F-8, again with the Second Marine Air Wing based at Cherry Point. I was also trained to operate off aircraft carriers and made about eighty carrier landings in various airplanes, while Ron was assigned to sea duty aboard the USS *Ticonderoga* (CV-14), flying the F-8 in combat operations over Southeast Asia. During two combat cruises, he completed more than a hundred missions. In the meantime, we had both learned that NASA was seeking a new group of astronauts, and we applied.

In February 1966, following a battery of thorough physical examinations at the School of Aviation Medicine in San Antonio, we faced an intense interview and evaluation process before a panel of NASA astronauts and engineers in Houston. Both of us made it through, and in April 1966, Ron and I joined seventeen other brand-new astronauts, ready to begin training specifically for the Apollo lunar landing missions and for what would become the Earth-orbiting Skylab Program. At the time NASA notified him of his astronaut selection, Ron was the maintenance officer for the U.S. Navy's VF-51 Squadron and still engaged on a combat deployment. He only returned to the States a few days before we all had to report to NASA in Houston. Both of our families purchased houses in the new El Lago subdivision, which is near the Manned Spacecraft Center (known later as the Johnson Space Center) and where many of the astronauts had their homes. My family did not live on the same street as the Evans family, but we lived close by and among other astronauts and engineers from the Mercury, Gemini, and Apollo Programs.

For the first six months or so, we mostly had classroom training in Houston combined with familiarization visits to other NASA space centers, contractors' plants, and other facilities associated with the space program. We also went on geology trips around the country, and while we were busy during these excursions, we also had some free time at home to engage in recreational activities. Through fishing and hunting, Ron and I truly became good friends, especially on the geology training trips to Alaska and Iceland, where summer daylight hours were so long that we could go fishing early in the morning before breakfast or until late at night after our astronaut training work was done for the day. We got in a lot of fishing together.

Ron and I also hunted together in locations around the United States, usually at "celebrity" hunts with several teams of three or four sports figures, politicians, retired military, astronauts, and locals. These one-day hunts culminated in an evening social event benefiting the community. They also enabled a welcome change of pace in a usually jam-packed schedule of mission training, technical briefings, spacecraft testing, and public relations events.

Ron was a good guy, very sociable and so easy to get along with. He also had a wonderful, dry sense of humor. I was thrilled for him when he was assigned to the backup crew for *Apollo 14* along with Gene Cernan and Joe Engle (who had flown the X-15 rocket plane). With the crew rotation system then in place, it looked as if Ron would be on the prime crew of *Apollo 17*. I will leave it to the author to tell this story, but on that mission in 1972 as the command module pilot, Ron performed his lunar-orbiting duties and beyond with utmost competence, skill, and diligence—all alone, by the way—while his buddies were on the moon.

Between 1973 and 1975, Ron and I were teamed up on the backup crew for the history-making Apollo-Soyuz Test Project along with Alan Bean, who had walked on the moon on *Apollo 12*. This "Era of Détente" initiative, one of several, was a particularly visible and impressive technical and political experience on which to work jointly with many Soviet cosmonauts and other personnel in both Moscow and Houston.

Two years later, in March 1977, Ron resigned from NASA. Unlike me, he wasn't prepared to hang around and wait an unknown number of years for the space shuttle program, still under development, to begin. He had done what he came to do with NASA. He had flown to and orbited the moon on the final Apollo mission and had carried out a deep-space EVA (extravehicular activity, or spacewalk) on the way back. Those spectacular experiences are pretty hard to beat. He had accomplished what he had set out to do, and he knew for him and his family it was time to move on.

Meanwhile, shortly after *Apollo 17*, in 1973 I flew as the pilot on the Skylab 3 space station mission for fifty-nine days, which for a short time was NASA's longest-duration spaceflight. It would be another eight years before the first space shuttle orbital flight test mission, STS-1, left the launchpad in April 1981. The following year I flew into space for a second time as the commander on the eight-day STS-3 test mission. Having participated in a broad spectrum of NASA's manned space opportunities, like Ron, I decided it was time to be moving on, so I resigned from NASA in 1983 to engage in other challenges in the private sector.

Ron and I remained in touch after leaving NASA. I visited with him at the aerospace company where he worked in Phoenix. We also participated in numerous special events for the Astronaut Scholarship Foundation and at the Kennedy Space Center in Florida, in Washington DC, and at various locations across the nation.

I was devastated when I learned that my friend and onetime astronaut colleague Capt. Ron Evans (USN) had passed away from a heart attack on 7 April 1990 at the relatively young age of fifty-six. Ron's many accomplishments as a dedicated naval aviator and astronaut will always remain his legacy and be a key element of American and spaceflight history. Even though he left us far too soon and is now in God's tender care, he is still sorely missed by his loving family and all those who knew him well.

I certainly consider myself privileged to have enjoyed the long friendship of Ron Evans and his family, and to have been given the opportunity to reflect on that friendship for this marvelous and respectful biography of his life.

Col. Jack Lousma, USMC (Ret.)
Pilot, Skylab 3
Commander, STS-3

Preface

The history of the United States is rich with examples of individuals who rose from obscure or humble beginnings to lofty positions or high office, but only a very few—twenty-four to date—have risen high enough to be able to hide the whole of our world and all its teeming billions behind the tip of a raised thumb.

A Long Voyage to the Moon is the story of one of those men. Coming from the very heart of the American Midwest, he charted a long course through family tragedy, the perils of aerial combat over the skies of Vietnam, and the rigors of astronaut selection and training before making his voyage to the moon.

In September 1962, speaking at Rice University in Texas, President John F. Kennedy referred to outer space in nautical terms: "We set sail on this new sea because there is new knowledge to be gained and new rights to be won, and they must be won, and used for the progress of all people." Appropriately, many of the early sailors on that "new sea" were officers in the U.S. Navy. Their number included naval aviator and commander (later captain) Ron Evans.

Ron Evans was not the first to visit our nearest neighbor in space, nor will he be the last. But he and his two crewmates on *Apollo 17* were the last of the early pioneers. Their scientific legacy continues to influence our understanding of the moon today.

Acknowledgments

The origins of this book lie in an email exchange with Colin Burgess, the managing editor of the Outward Odyssey book series, in 2017. Knowing that I had recently retired after four decades as a litigation lawyer, Colin referenced my contribution of two chapters to *Footprints in the Dust*, an earlier volume in the series, and wondered if I would be interested in writing "a whole book."

Colin and I shared the opinion that Apollo command module pilots have tended to be overshadowed in the wider story of the exploration of the moon. We agreed that the life of Ron Evans, the command module pilot of *Apollo 17*, was certainly worthy of greater public recognition. It is only scratching the surface of Ron's military and astronaut careers to point out that he survived more than a hundred combat missions in the Vietnam War, that he still holds the record for the longest time in lunar orbit, and that he was the last American astronaut to orbit solo in space.

As this book has progressed from an idea to reality, Colin has been a tower of strength, a constant guide, a wise counsel, and a long-distance friend. In 2018 on a European tour, he and his delightful wife, Pat, actually managed to visit me in Belfast for a very welcome opportunity to mix business with pleasure. Without Colin, this book would not have happened.

Another key individual to whom mere thanks are entirely inadequate is Jan Evans. A widow since Ron's untimely death in 1990, Jan has been a loyal and articulate "keeper of the flame," maintaining contact with other wives and widows from the Apollo era and arranging reunions (coronavirus permitting).

Since I first approached Jan by letter, we have spent well over twenty hours on the telephone, discussing the life and career of naval aviator and astronaut Ron Evans. Jan's phenomenal memory, together with numerous family photographs and Ron's navy records, brought to life an era that I previously only knew from books, newspapers, and TV coverage. My respect and admiration for Jan know no bounds. If this biography fails to capture at least some of the spirit of those heady days of Apollo, the fault is entirely mine.

In August 2019 I had the opportunity to visit Jan at her home in Scottsdale, Arizona. She then drove me to the former Evans family home, now the home of her daughter, Jaime, and her son-in-law, Lamar, where I experienced Arizona hospitality at what must surely be its finest. I want to express my gratitude to Jaime not only for two fascinating telephone interviews but also for spending many hours scanning those family photographs and her father's aviation logbooks—a priceless record of Ron's home life and his military and NASA careers. Thank you, Jaime!

Also present that evening in Scottsdale were Andy and Vicki Chaikin (who would give me a lift to Spacefest in Tucson the next day), and this gives me the opportunity to pay grateful tribute to Andy for agreeing to share information with me that has allowed the authentic voice of Ron Evans to be heard in this biography. Andy visited Scottsdale in 1986 to interview Ron and Jan for his acclaimed account of Project Apollo, *A Man on the Moon*. Given the scope of that book, covering all of the Apollo flights, there was no room left for many hours' worth of biographical material he had amassed, and with Jan's consent, Andy provided me with digitized copies of those interviews. Ron's life story benefits enormously from their inclusion. Thank you, Andy. You are a scholar and a gentleman.

I am indebted to four other acclaimed space historians and writers. Professor John Logsdon is the author of definitive accounts of the space policies of Presidents John F. Kennedy and Richard M. Nixon. Professor Logsdon generously reviewed what I had written about how close *Apollo 17* came to being canceled—unknown to the crew members who were hoping to fly the mission. The full story in *After Apollo? Richard Nixon and the American Space Program* is most unnerving for Apollo aficionados! I would also like to thank Ryan Pettigrew of the Richard Nixon Presidential Library and Museum for providing copies of crucial correspondence, memos, audio recordings, and photographs.

Michael Cassutt helped me clarify certain issues relating to Ron Evans's brief period working on the space shuttle. This included raising points with George Abbey (the subject of Michael's recent biography, *The Astronaut Maker*) for which I am most grateful. Of course, *Deke!*, one of Michael's earlier works, is a mine of information for space writers.

Many thanks also to Francis French, a stalwart of Outward Odyssey and the coauthor, with the late Al Worden, of *Falling to Earth*. I conducted a most enjoyable and informative interview with Al for this book, and following Al's passing, Francis agreed to cast his expert eye over the relevant section. Thank you for that, Francis.

I have known Dave Shayler for many years and have many of his meticulously researched books in my study. Dave provided me with much useful background on NASA's Group 5 astronauts and put me in touch with the late Jerry Carr, who was a wonderfully helpful and engaging interviewee.

Ron Evans was born in the small community of St. Francis, Kansas, in November 1933. Marvin Miller was a friend and neighbor of Ron's in St. Francis, and I am most grateful to Marv for providing firsthand details of Ron's childhood and (with helpful contributions from his wife, Anita) later episodes in their lives. Marv provided me with copies of the *Saint Francis Herald* containing reports he wrote on *Apollo 17* and the joyful return of Ron Evans to his hometown after the mission. I am grateful to Marv, and to Steve Haynes of *Colby Free Press*, for permission to use extracts from those reports.

Jay Evans, who, like his older brother Ron, served his country with great distinction, provided me with additional insights into Ron's life. Jay himself sadly died only a few months after our telephone discussion. His daughter Kelly, immensely proud of her father and her astronaut uncle, provided additional details of Evans family history and several photographs for this book.

The Evans family moved to Topeka just after Ron had started high school. Emily Denney, née Wolverton, was a friend of Ron's in high school and at the University of Kansas, and I am indebted to her for her recollections. Thanks also to Highland Park High School for providing me with extracts from the 1951 yearbook and to Kathy Lafferty and Stephanie Shackelford at the university for helping to clarify details of Ron's graduation.

Ron left no account of his years as a cadet in the Naval Reserve Officers Training Corps, but Tamara Horner of the corps's office at the University of Kansas put me in touch with alumnus Bob Settles, who went through the same training process as Ron and recorded his story in the chapter "Seven Years a Sailor" in his entertaining blog *Stories from the Funny Farm*. Bob's contributions very helpfully fill a potential gap in Ron's story, although some of Ron's anecdotes in Andy Chaikin's interviews tie in very neatly with Bob's memories.

When I began my research for this book, I thought it would prove difficult to learn much about Ron's career as a naval aviator. How wrong I was! I had the good fortune to make contact with Dave Cowles of the Crusader Association, who circulated my email to fellow members. I was astounded by the number of replies I received from former squadron mates of Ron's who flew the F-8 Crusader with him from the USS *Ranger*, from the USS *Oriskany*, in combat from the USS *Ticonderoga*, and in the VF-124 "Crusader College." My grateful thanks to Jack Allen, John Allen, Gail Bailey, Howard Brown, Dick Cavicke, Fred Dale, Gene Geronime, Marlo Holland, John Holm, Ralph James, Craig Kintzel, Tom Klein, Wayne Skaggs, and Jimmie Taylor. I listened in awe as these superb aviators told me about flying their beloved supersonic F-8 Crusaders in peacetime and at war. Sadly, since completing my research, I must record the passing of Gene Geronime, who flew with Ron in 1959.

My thanks also to Phil Morrison for his recollections of Ron Evans and his account of life "below decks" on the USS *Ticonderoga*. He was one of the dedicated support crew who kept the Crusaders flying in the heat of combat.

After the Evans family arrived in El Lago, Houston, one of Ron's neighbors, Ron Ammons, became his closest friend. Ammons has contributed many anecdotes that allow us to glimpse something of the home life of the Apollo astronaut as a husband, a father, and a member of the community. Many thanks to you, Ron. Members of the NASA family who came to know astronaut Ron Evans best also have been delighted to offer their insights: Dee O'Hara tells of the devoted family man whose combat experiences earned him the respect of his peers. Gerry Griffin admired his friend Ron's dedication and team spirit during the Apollo years. Dr. Farouk El-Baz trained Ron to be a keen observer of the moon, and they became good friends. To all of you, my gratitude for your invaluable contributions.

I sincerely wish I could have interviewed the late Gene Cernan for this book, but that was not to be. Happily, Barbara Butler, the former Barbara Cernan, was and remains one of Jan Evans's closest friends from "the old days," and I was delighted that Barbara agreed to speak to me about those days and in particular about Ron, a man she liked enormously. Her observations are insightful, profound, and often very funny.

This book would have been so much the poorer without the contribution of the surviving *Apollo 17* crew member, Dr. Harrison Schmitt. I am most grateful to Dr. Schmitt for talking to me about his shipmate Ron and the spacewalk they both enjoyed so much; for reviewing the chapters on *Apollo 17*; and for his many helpful suggestions.

Six of the surviving members of NASA's Group 5 astronauts also agreed to be interviewed, and they provided a wealth of anecdotes and observations. They were Vance Brand, Jerry Carr, Joe Engle, Fred Haise, Jack Lousma, and Al Worden. Ken Mattingly and Charlie Duke also contributed. The details of the historic missions they all flew are still crystal clear in my memory, and I am proud to record my gratitude to them all. Sadly, since I completed my research, Al Worden and Jerry Carr have joined the list of space pioneers who have passed into history.

Tom Stafford knew Ron Evans as a near neighbor in El Lago, as a friend, and as a colleague on the Apollo-Soyuz Test Project. I was delighted that General Stafford agreed to be interviewed and am grateful for his observations. My thanks also to Max Ary of the Stafford Air and Space Museum for facilitating the interview.

Although Dave Scott never shared the same training cycle with Ron, he has contributed a number of insightful observations and helped me avoid several blind alleys in my research. I greatly appreciate his enthusiastic assistance.

It would be difficult to write an Apollo biography without access to several superb research tools compiled by dedicated and learned experts in the field. First, I refer to the incredible *Apollo Lunar Surface Journal*, edited over many years by Eric Jones and more recently by Ken Glover. It is wonderfully complemented by the *Apollo Flight Journal*, and in the cases of *Apollo 14* and *17*, I thank David Woods, Ben Feist, Ronald Hansen, and Lennox Waugh for their diligent research. Ben Feist also deserves a loud round of applause for his highly impressive internet guide to *Apollo 17* in words and pictures. I would be remiss if I did not add to this list Robert Pearlman's collectSPACE website, truly an Aladdin's cave of space wisdom.

My thanks to Ed Hengeveld for sourcing some of the photographs in this book and to Rich Orloff and David Harland for their welcome advice. Thanks also to Garrett Hutchinson of CMG Worldwide, Inc., acting on behalf of the Purdue Research Foundation, relating to the inclusion in this book of a very pertinent quotation by the late Neil Armstrong.

I have already thanked Colin Burgess, but I am not forgetting my guiding lights in Lincoln, Nebraska. Many thanks to Rob Taylor and Courtney Ochsner for answering my numerous questions and for helping me turn a big bundle of words into a real book. And to Leif Milliken for helping me with the official paperwork to make sure royalties are not taxed by both the

U.S. and UK exchequers! In the last months of the process, my project editor, Sara Springsteen, patiently walked me through the complexities of copyediting software, yet another attribute of my computer of which I had been blissfully unaware. It fell to my copyeditor, the very patient and skilled Vicki Chamlee, to unleash the power of that computer software on my scribblings and to help me protect American readers from certain expressions and spellings that should remain on my side of the Atlantic. Thanks, Vicki!

And, finally, to my long-suffering wife, Sandra, who probably thinks I was actually writing a sequel to *War and Peace*, my love and gratitude for your understanding, and for your encouragement and sympathy on the days when inspiration dried up!

A Long Voyage to the Moon

Prologue

His bulky spacesuit was airtight, the gold-visored helmet securely attached. His similarly attired colleague edged out of the way to allow him room for maneuver. As the last few molecules of the spacecraft's atmosphere fled into the vacuum beyond, the hatch finally opened.

Back on Earth, giant antennae pointed skyward, awaiting a signal from a carefully positioned TV camera. At home in El Lago, Texas, the first astronaut's wife, Jan, and their two children stared intently at their TV set, wondering what to expect.

A live picture flashed onto the screen—a slash of white against a dark background. The astronaut eased through the hatch, taking care not to snag any part of his suit. Then his figure appeared on the screen. As he began to speak, his words triggered the microphone in front of his lips. What would he say? Across the gulf of space, an enthusiastic voice exclaimed, "Hot diggity dog!"

Ron Evans, command module pilot of the *Apollo 17* mission, floated out of the open hatch of spacecraft *America*. He was, quite literally, having his day in the sun after being eclipsed by the more newsworthy activities of his companions during their three days on the moon.

Looking back toward the hatch, Ron spotted his own reflection in the mirrored visor of Dr. Harrison "Jack" Schmitt. Resembling a snail popping out of his metal shell, Jack kept his eye on Ron's air hose to avoid tangles. This spacewalk was not for show. It was a hazardous but essential part of their mission to retrieve film cassettes from the belly of the spacecraft where cameras and other instruments had been scrutinizing the moon. Ron had trained repeatedly for this task. Of course, it hardly compared with that famous televised moonwalk conducted by his former El Lago neighbor Neil Armstrong. For one thing, Ron's audience was only a fraction of the numbers that had watched Neil's "giant leap for mankind."

No one would have expected Neil to have been as exuberant as Ron was on live TV, but Armstrong was doing more than making footprints in the lunar dust. He was making history. Ron Evans knew that *his* spacewalk, in comparison, was almost the final act in the Apollo story.

One feature of the two excursions was entirely comparable—the level of dedication and training required of each man. Ron knew how to conduct his crucial task with perfect precision. He took his job extremely seriously. But nothing in the rules said you couldn't *enjoy* doing your job, and right now Ron Evans—thirty-nine-year-old husband of Jan, father of Jaime and Jon, and proud American from a little town in Kansas—was having the time of his life doing what he would later call "the best job in the world."

1. Kansas Kid

Mid pleasures and palaces though we may roam,
Be it ever so humble, there's no place like home.

—J. H. Payne (1791–1852)

On some days, when the wind blew, the sky would turn as black as night. Writhing tsunamis of desiccated topsoil came sweeping across the midwestern states like one of the plagues of Egypt: from northern Texas and across the Oklahoma panhandle into New Mexico, Colorado, western Kansas, and southern Nebraska. Chilling photographs of the period show black clouds of churning filth engulfing farms, roads, and whole communities, forcing the populations to retreat into their homes and protect their eyes and noses from the choking dust. As the dry waves rolled across the prairies, every window, every table, every chair, and every face bore witness to the wind-borne assault.

These were the "Dirty Thirties," and this was the infamous Dust Bowl adding meteorological insult to the financial injuries inflicted by the Great Depression. For every Wall Street stockbroker who responded to the Crash of 1929 by hurling himself from a high window, there were hundreds of midwestern farmers whose tragedies never made a headline. By 1932 the region was suffering the severest economic downturn in its history. In the worst-hit areas, farm prices fell to all-time lows, and farm incomes plummeted by nearly 60 percent. John Ford's classic film treatment of *The Grapes of Wrath* by John Steinbeck provides a memorable Hollywood portrayal of the plight of the families that were forced to leave their ruined Oklahoma farms to seek a new life in California.

The state of Kansas, lying right at the heart of the continental United States, was perhaps not affected as badly, either by the Great Depression or the Dust Bowl phenomenon, as some neighboring states, but western Kansas was certainly not spared the twin onslaught. The driest parts of the state coincide on the map with the regions most badly affected by the rhythmic tides of dust that ebbed and flowed during the decade. On Black Sunday, 14 April 1935, a massive frontal system packing winds of sixty miles per hour rolled across the Great Plains from the northwest and threatened anyone caught outdoors with suffocation and death. These "black blizzards" even reached as far east as New York City and Washington DC.

Cheyenne County lies in the extreme northwest of Kansas, providing the top-left square on the checkerboard of the state's counties. The county seat is the small town of St. Francis, founded in 1887 and lying along the south fork of the Republican River. The town's 1930 census showed a population of 944 citizens.

On Friday, 10 November 1933, St. Francis had another mouth to feed. Dr. Haddon Peck had called at 321 West Webster, the home of Jimmy and Marie Evans, the previous evening. Marie had been in labor, and around four o'clock in the morning, Ronald Ellwin Evans finally made his appearance. His first home was a very modest wooden structure with just two rooms—a living area and a bathroom. Formerly a garage, Jimmy had converted it into a small house. Years later it would become a garage again before eventually burning down.

But this is not a rags-to-riches story. Like most people in St. Francis, Jimmy and Marie were poor but not poverty-stricken, and although Ron's life would be remarkable, he would never make millions. His riches would be of another kind. St. Francis in the 1930s had dirt roads and mostly wooden homes, but the little town was very clean and well maintained (when the dust wasn't blowing) and had a close-knit community. Marvin Miller was born in 1936, too late to have any personal memories of the Dust Bowl except from his mother's accounts of having to press wet bedsheets along the windows of their basement home to keep out the dust.

Ron and Marv became acquainted as soon as the younger boy was old enough to walk. His mother, Hollis, and Marie Evans were best friends, almost like sisters, with each treating the other's home like her own and each helping out through the other's pregnancies.

In 1935 Jimmy Evans got a better job that paid more. When his second son, Larry Joe, was born, Jimmy realized the family was outgrowing the converted garage and acquired a smart three-bedroom house with a basement across the street from the local hospital. In 1939 with the birth of a third son, Jay Dale, the family was complete.

Growing up in the 1930s and 1940s, Ron was an energetic, hardworking boy. Often, when others were playing, Ron was mowing lawns or helping out at the local creamery. He also spent time swimming and playing with his brothers and the other children of the neighborhood, but he was (according to Marv) "pretty dedicated to working." Jimmy made sure all three of his sons understood the importance of work. At a time of economic malaise, having a job made the difference in keeping your head above water, and keeping a job meant working hard. Born at the end of 1911, Clarence Ellwin Evans was always known as Jim or Jimmy for reasons now lost to history. His own father, Joseph Delmar Evans, known to everyone as "J.D.," was a familiar figure in St. Francis. He had one of the better homes in town and was an electrician by trade. There wasn't much J.D. didn't know about fixing magnetos, rewinding electric motors, and generally keeping the town's tractors and combine harvesters moving. These qualities of hard work and effort seem to have passed down from grandfather to father and to Jimmy's three sons. Jimmy was clever and could turn his hand to just about anything. He worked for the local farming cooperative as a kind of jack-of-all-trades, busying himself in the co-op's service station, John Deere agricultural machinery dealership, grocery store, creamery, and lumberyard. He also worked for the co-op at nearby Bird City, a grandiose name for a settlement that made St. Francis look populous.

If you need a solid work ethic to become an astronaut, Jimmy certainly set a good example for Ron. But Ron always found his father a hard man to get close to. Jimmy was a loner by nature but, as it turned out, a loner with a wandering eye.

By contrast, Marie was warm and outgoing. To Marv Miller, Marie Evans was like a favorite aunt. She was "a wonderful mother and a wonderful person" and would do anything to help her family, neighbors, and friends. Marie was also hardworking; she had to be to carry out her combined roles of housewife, mother, nurse, and—when Hollis Miller was at work—occasional childminder for Marv and his older sister.

In the fall of 1939, as war was breaking out in Europe, Ron began his formal education at the St. Francis grade school. Hard work, although helpful, does not necessarily translate into academic achievement, but Tessie Lawless, Ron's first-grade teacher, soon saw that he was going to be a star pupil. According to his brother Jay, Ron was "intensely intelligent all the way through school" and "always excelled at anything he did." Many years later, writing in the *Saint Francis Herald* during the buildup to the *Apollo 17* mission, Marv Miller reminisced about his old friend's school days: "[Ron] will be remembered as a typical youth growing up in the community. He was active in the school athletic program, as well as achieving high scholastic standing. Not only was he an ambitious Scout, he loved to hunt, fish and camp. Already, at this early age, he showed much determination and initiative by making early morning deliveries of the *Denver Post*. He could also be found working at his father's creamery which was later to be operated by Harold Couse for whom he was also employed."

Both several years younger than Ron, Marv Miller and Jay Evans clearly looked up to the older boy. Jay recalled that "you couldn't ask for a better brother to have around the house." Ron seemed happy to keep an eye on his younger brothers and the Miller boy. The Evans home had a basement where Jimmy set up table tennis. None of the local children could beat Ron at ping-pong, and he passed on some of his skills to Marv, teaching him how to put a spin on the little white ball.

In addition to his schoolwork and assorted chores, Ron was often persuaded to help his grandfather in J.D.'s electrical shop, winding copper wire onto small electric motors. Even for Ron, the pleasure in helping his grandfather occasionally wore off, and on one hot day when the local children were all heading to Keller's flooded sandpit to swim, Ron confided in Marv that he hoped J.D. wouldn't need him as he was "so tired of the wiring jobs."

As a boy, Ron enjoyed making model aircraft. He was also skillful at building kites. The children of St. Francis didn't have money for store-bought kites, but with a couple of sticks, a whittling knife, string, and some scraps of material, Ron could turn out a kite that flew far better than anything the shops sold. He was a victim of his own success, with a line of children all jostling to have their own "Evans kite." As Marv Miller recalls, "When the winds came in March and April, gosh, we had more kites to fly than you could shake a stick at—and Ron made every one of them!"

Ron would later say that he thoroughly enjoyed his childhood in St. Francis, which he called "a neat place, a good place!"

It might seem obvious that this hardworking young maker of model aircraft and kites was yearning for the life of a pilot. After all, the wide-open spaces of the midwestern states have incubated many of the most famous aviators and astronauts of all time. From Ohio, the Wright brothers, John Glenn, and Neil Armstrong. From Oklahoma, Gordon "Gordo" Cooper and Tom Stafford. From Wisconsin, Donald K. "Deke" Slayton. From Kansas itself, Joe

Engle—who would gain his astronaut wings flying the x-15 rocket-plane and the even faster space shuttle—and, of course, one of its most famous daughters, the globe-spanning aviatrix Amelia Earhart.

But here's the thing about the young Kansan Ron Evans, future fighter pilot and Apollo astronaut: no one who knew him ever remembered his expressing the slightest interest in flying. As a child, he never said anything about it to Jay or to Marv Miller. St. Francis, like most small rural communities, had its own airstrip, but unlike the young Neil Armstrong, who was breathing aero-engine fumes from childhood and soloing at sixteen, Ron showed no obvious interest in flying.

As for becoming an astronaut, in Ron's youth there wasn't any such thing—unless you counted fictional guys like Buck Rogers or Flash Gordon. Asked about it in 1986, Ron observed, "You would look at the moon once in a while, never even considering the possibility that man would ever *go* to the moon. Of course, when you grow up in a little town in western Kansas, it's kind of the farthest thing from your mind!"

By 1941, as Ron entered the third grade, the clouds of war were casting their shadow over the United States. It was becoming obvious to Jimmy Evans that he would likely be drafted. He was prepared to answer his country's call but confided in Marv Miller's father that if he had to go to war, his first choice would be the navy and his second choice the air force. He didn't want to be a foot soldier. Marv never knew whether Jimmy was drafted or whether he volunteered, but, one way or the other, he got his wish and served in the U.S. Navy until 1945.

Meanwhile, in St. Francis and at the tender age of eight, Ron became the man of the house, a role he took seriously. The farms surrounding St. Francis all had dairy herds, and the farming co-op ran a creamery where Ron had helped his father test the cream, wash out the cans for customers, and then refill them. After Jimmy went to war, Ron continued to work at the creamery under the new manager, Harold Couse. Harold was a friend of the Miller family and told Marv's parents that Ron knew more about running the creamery than he did. He admitted, "I didn't teach him anything; he taught me a lot!"

In spite of being right in the middle of the country, St. Francis was not spared the effects of the Second World War. The town lost a dozen men. Marv Miller had an uncle who was shot twice, in the leg and the shoulder, and endured over a year as a prisoner of the Japanese. The brother-in-law of Marv's future wife, Anita, died—and was buried—at sea.

Even in Kansas, families experienced wartime rationing of certain commodities, especially gasoline, sugar, and tires. They had to collect stamps in a ration book that could then be exchanged for sugar. The rationing led to fights and family feuds. Two farming brothers stopped speaking when one got a set of tires and the other didn't. Obviously, money had changed hands somewhere along the line, or at least that's how it appeared to the brother who lost out.

When Germany fell, Hiroshima and Nagasaki succumbed to nuclear fire, and Japan surrendered, the American troops came back to a world that had changed forever. Many of the men who returned to St. Francis had memories they were reluctant to share. Marv Miller's uncle would only say, "What you don't know, don't hurt you." But Jimmy Evans seemed unscathed and sufficiently proud of his naval service that he brought back a box of white sailor hats, at least thirty or forty of them, and handed them out to his three sons and to most of the neighborhood children. Ron gave one to Marv, and for a long time any school photograph was sure to contain at least a couple of sailor-hatted youngsters. An Evans family photograph shows Marie, Jimmy in his dark blue navy uniform and hat, and the three boys lined up in their white hats.

While many fathers and uncles were away at war, the armed forces acted as positive role models for the children of America. All too frequently, press pictures, recruiting literature, and wartime newsreels showed the heroic young servicemen puffing away on cigarettes, so it was hardly surprising that impressionable youngsters were encouraged to experiment with this most manly of practices. Unfortunately, Ron seems to have developed a taste for cigarettes at a very young age. Marv would tag along with Ron, Larry, and their friends Stan and Jack on swimming trips to the banks of the Republican River, where they would swing on ropes like Tarzan. They would also pool their money and cough their way through packs of Wings cigarettes, a popular economy brand introduced in 1929 as a product of the Depression. Ron always seemed to cover his tracks, but Marv came home a couple of times smelling of cigarettes and got his "little butt tanned."

Photographs of the three Evans boys showed an interesting progression. At first, Ron was obviously the tallest, followed by Larry, with the youngest brother, Dale (as Jay was then called), a distant third. Then Dale overtook Larry. Then he overtook Ron. By the time he reached university, Dale was a superb footballer. Six feet five inches of brawn with arms like Popeye's, he would go on to play football professionally before serving his country with great distinction as a much-decorated U.S. Marine. But another Dale Evans had become well known to cinema and TV audiences in the forties and fifties. One day while Dale was playing football, someone who must have been either very brave or very stupid shouted out, "Hey, look! Roy Rogers's wife scored that touchdown!" The fate of the heckler is unknown, but as a result of the incident, Dale henceforth preferred to be known by his initials, "J.D.," like his grandfather. Later, in the marines, most people called him by his first name, Jay, but even many years later, Ron's children would still refer to their uncle as Uncle Dale.

Although Ron and Jay would go on to great things, the middle boy, Larry, seemed to lack their vitality. Jay always remembered Larry as having poor health. By the time Larry entered the fifth grade at St. Francis School, his health was noticeably deteriorating. One of the neighbors had a boxing ring set up in the basement, and the local children all took turns putting on boxing gloves and sparring with each other. Larry didn't lack gumption and took part with the others, but he always warned Marv Miller, "Marv, please don't hit me in the stomach!" Marv's mother took it further, warning Marv never to lay a hand on Larry.

By then Jimmy was back from the navy, and he and Marie sought medical advice. Some accounts suggest Larry had developed leukemia, but Marv was in the house the day Marie came to Hollis Miller to break the news: tests at the local hospital had revealed liver cancer. Haddon Peck, the Evans family's doctor, knew a specialist in the state capital, Topeka, who wasn't actually an oncologist but worked with cancer patients. Peck's advice to Marie had been unequivocal: "You need to get Larry to somewhere like Topeka or even Denver where there are doctors who can treat him, because his liver is completely engulfed."

At the best of times, uprooting a family almost overnight is difficult and stressful, but Jimmy and Marie had little choice. At least Marie had relatives, including her mother and a brother, in the Topeka area. After further discussions with Dr. Peck, they packed in November 1947 and moved to 2518 Kentucky Avenue in Highland Park, a suburb of Topeka. Marie and Hollis Miller kept in touch by letter, and clearly Marie was deeply depressed by the news about Larry as well as by having to start a new life raising her sons in an unfamiliar environment. It also seems that rather than bringing Jimmy and Marie together, Larry's diagnosis put further strains on their already fragile relationship.

Highland Park, although more populous than St. Francis, was similar in that it was a rather poor community partly surrounded by farmland. When

Ron started at the nearby Highland Park High School, he learned that many of his classmates were from farms in the vicinity. Jimmy Evans was able to use his connections with the St. Francis cooperative to secure similar work selling silos to the local farmers.

With all the problems surrounding his brother's illness, Ron's freshman year at St. Francis High School was disrupted after the first two months for the move to Topeka. He then had to hit the ground running at Highland Park in November. Emily Wolverton noticed the arrival of the tall, athletic new kid with the friendly grin. She and her friend Sue Benson found that their late-arriving classmate only lived a few blocks away from each of them, and although they soon became friends, Emily's first impressions weren't entirely favorable. An intelligent and competitive pupil, she had been scoring top marks in algebra class until Ron outshone her. His hard work and high marks earned the teacher's praise and made Emily seem almost lazy by comparison. She thought, "Hey, what's with this new kid, coming here and taking away our prestige?" But she soon got over it. Ron was very friendly, very easy to know, and always smiling. To Emily, he was an "all-round good guy."

He quickly piled into every aspect of school life at Highland Park. Perhaps to some extent this was to distract his attention from his younger brother's illness and the increasing coolness between his parents, but just as likely Ron did so because he took no half measures: if it's worth doing, do it well. In an era long before the mesmeric attractions of social media and when even television was a luxury, school sports and social activities were an important part of youthful lives.

Ron—or "Ronnie" as he is invariably called in the school yearbooks—missed out on the football team's selection in his freshman year, but he took up track sports with enthusiasm. His late start didn't hinder an impressive career in the Highland Park Scotties football team, and he also distinguished himself in basketball and athletics.

Jimmy and Marie came to realize that their eldest son was particularly interested and greatly gifted in mathematics and science. It was no surprise when Ron rose to be president of the Science Club in his senior year. Each club member worked on several projects of his or her particular interest, usually involving electric motors, model aircraft engines, and battery-powered radios. Pictures of the Science Club also show Larry Evans in his freshman year, working on a radio, with no obvious signs of the condition that was overwhelming him.

Ron seemed to stretch himself in his junior year, trying his hand at amateur dramatics in a melodrama titled *The Gay Senorita*. Wearing what looked like one of his father's suits and appearing in the role of Dr. Forsythe, he seemed somewhat ill at ease and didn't repeat the exercise in his senior year.

He appeared much more comfortable on the Student Council. Among the twenty council members were Emily Wolverton and Larry Evans, with the group's yearbook picture being one of several school photographs showing both brothers. In his junior year, Ron also served as class president.

During his nearly four years at Highland Park, Ron was a diligent member of the Hi-Y Club, the purpose of which was to "encourage Christian attitudes among the boys of the school and the people of the community." The club promoted "clean speech, clean sports, clean scholarship and clean living," and by his senior year, Ron was its president. Larry was also a member of the eighty-five-strong group. The boys were encouraged to develop a sense of personal integrity and self-discipline "by maintaining perspective in a world of shifting values" to "prepare in thought and action for the post-war world, to conserve and to strengthen democratic traditions, and to co-operate with other groups of young people to achieve these objectives." The school had a similar group for girls called the Y-Teens, and one of the more popular events was the annual get-together of the two clubs' members.

While modern schoolchildren might smirk if presented with such a program for living, even a casual examination of the life and career of engineer, naval aviator, and astronaut Ron Evans suggests that much of the Hi-Y ethos influenced his approach to life. By way of example, some of his fellow navy pilots noticed that Ron almost never swore, at least not within their hearing.

As noted, Jay would prove to be the stand-out athlete of the Evans family: after graduating from Highland Park, he received a full-ride scholarship to play football at Kansas State University and later played professionally for the Denver Broncos. But Ron was no slouch at football, playing in the school's A-team as a guard. In Ron's senior year, the A-team finished second in the Jayhawk League but only after a shaky start, which saw ten of the players, including Ron, promising not to shave until they secured their first win. Fortunately, that victory came quickly, and the yearbook picture of those rebels requires careful scrutiny to discern much evidence of beard growth. It is also worth pondering on the particular attributes of one of those smiling adolescents who in time would fly a quarter of a million miles away from Kansas.

A casual modern-day observer looking at the ten Highland Park footballers in the Class of '51 would have little or no basis for guessing which face would one day be gazing back at a distant Earth from a spacecraft orbiting the moon.

But Ron's prowess at athletics could have had an unexpected impact on his future astronaut career long before it started. The Highland Park track team frequently traveled to other towns in the area for track meets, which sometimes involved up to four schools. On one journey, the car in which Ron was a passenger crashed, and several boys, including Ron, were thrown around violently enough to need medical assessment. Ron suffered a back injury that, fortunately, seems to have been limited to soft tissue bruising. But in later years, he found himself worrying about that old injury, first when he was about to be examined by navy doctors and later when he faced grueling medical tests at the National Aeronautics and Space Administration (NASA). His lingering concern was that some new X-ray technique might identify a tiny imperfection arising from that road accident and give the doctors an excuse to exclude him in favor of someone else.

By 1951 neighbors and school friends alike clearly saw that Jimmy and Marie Evans were raising two strapping sons. Ron would eventually top out a shade under six feet; Jay, five inches taller. But the middle boy, Larry, was increasingly unwell despite the efforts of the Topeka doctors. Ron looked out for his brother in school and helped his mother take care of Larry at home. Jay was only eleven years old and did not understand exactly what was going on, but however much the rest of the family tried to shield him, he could tell that Larry was very sick. He and Ron didn't discuss Larry's health. Giving emotional support isn't something boys have ever been much good at doing.

Larry Joe Evans died in the spring of 1951 at just sixteen years of age. Even though he knew it was coming, his brother's death hit Ron very hard. The school closed on the day of Larry's funeral as a mark of respect and to allow pupils to attend. Emily and Sue joined many of their classmates at the service, although the event was something of a blur to Emily. Ron's friends and neighbors could see how affected he was. They all felt very sad for him, but apart from the ritual exchange of condolences and thanks, he didn't really want to talk about it.

In most lives there comes a moment, an event, or a tragedy that, in retrospect, clearly marks the end of childhood. For Ron Evans, the death of his brother was that event.

2. Jayhawk

It was great! First class! It was the finest
experience I have ever had!

—Theodore Roosevelt (1858–1919) after an airplane
flight on 11 October 1910

No records survive of any discussions between Ron Evans and his teachers as he progressed at Highland Park High School, but as such a high achiever, he most likely would have been advised to go to college. Certainly, Ron himself realized that expanding his horizons would mean going to a university. As much as he had enjoyed growing up in a small farming community, he didn't want to be a farmer, always at the mercy of the weather and other factors outside his control.

Neither Jay Evans nor Marv Miller could remember talking to Ron about his going to college, perhaps because it did seem such an obvious course for him. But if by his senior year at Highland Park Ron had mapped out a personal plan for the future, it still wasn't pointing even remotely toward a career in aviation, let alone "outer space." When Ron decided he wanted to go to college, he clearly had to consider a few key questions.

Where did he want to study? His family was in Topeka, and both Kansas State University in Manhattan and the University of Kansas (KU) in Lawrence were relatively close. If distance was important, KU was closer, and that factor probably *was* important given his family's situation. What did he want to study? Faced with the same question, there was no doubt in the mind of fellow Kansan and future astronaut Joe Engle. K-State was essentially an agricultural college and at that time did not offer an aeronautical engineering degree. Engle wanted to fly, and KU offered aeronautical engineering. His choice was clear, and he headed to Lawrence.

Why Ron Evans chose to study electrical engineering at KU is not entirely clear. His brother Jay didn't know, and if Ron ever told him, it could only have been some passing remark that didn't register. Ron didn't tell Marv Miller, either, but Marv wasn't particularly surprised by his friend's choice. After all, hadn't Ron spent all that time with his grandfather as an unpaid, unofficial apprentice electrician, winding wires onto all those electric motors?

That left the most important question of all: how would he pay for it? Whatever Ron's choice of college or major, it would all be academic without the key element of funding. The family was poor, and Jimmy seemed more and more remote, spending increasing periods away from home. Ron's parents were unable to pay for his pursuit of a university degree. Marie's sister Wilma was willing to provide some financial help, and Ron, like most students through the ages, was prepared to work to supplement his income. But that still wouldn't be enough. He needed a scholarship.

If, even at that point, Ron had developed a desire to fly, he could have applied to the U.S. Air Force for a scholarship through its Reserve Officers Training Corps (ROTC). The preferred option for many would-be pilots—including Joe Engle—ROTC allowed those who were youth rich but cash poor to learn to fly at Uncle Sam's expense.

Having looked at the various options, Ron concluded that his best chance would be to apply for a U.S. Navy ROTC scholarship to study electrical engineering at KU in Lawrence. It seems reasonable to assume his father's naval service during World War II was at least an influence. Remembering all the neighborhood children running around in those sailor hats from Jimmy Evans, Marv Miller was not at all surprised when Ron pursued a navy scholarship. But Jay Evans wasn't so sure his father's wartime service had much, or any, influence on his older brother's plans. "I don't think so. When he was at high school he wanted to put in for some scholarships. One of the better ones was with the navy. I think that's probably what enticed him more to go into the navy program—for a scholarship, which he got. It paid for his education. I don't think there was any influence there from Dad being in World War II."

As events would play out, a navy career would eventually prove to be an excellent gateway to the moon, but whatever motivated Ron Evans's application for a navy ROTC scholarship, he did not originally intend to fly, whether within or beyond the earth's atmosphere. The mascot of the University of

Kansas, the Jayhawk, is a mythical bird, and the Jayhawk from St. Francis expected to be flightless.

Meanwhile, as Ron considered his options, teenager Bob Settles from Wakefield, Kansas, was one of forty thousand young men who had applied in 1950 for the two thousand scholarships. Bob initially fell short with "alternate status" but then got a telegram from the navy confirming his acceptance. He was a year ahead of Ron and never became acquainted with the future astronaut during navy midshipman training. However, Bob later discovered that Dick Ross, his wife's cousin, had been a navy ROTC cadet at KU and the fraternity roommate of fellow cadet Ron Evans.

Ron applied to KU to study electrical engineering and was offered a place in the four-year course due to start in September 1951. That was the easy part. He then completed his application for naval officer's candidate training, typing the form himself with the skills he had acquired during a typing course at school. (His C grade in typing was the lowest grade he obtained in his senior year.) He submitted the application on 30 January; then he waited.

He had become a part-time truck driver with the Kaw Valley Produce Company of Topeka in the summer of 1950 and continued to make what money he could on weekends and during holidays, but he could hardly drive the trucks while studying at KU. Aunt Wilma was still willing to contribute to the cost of Ron's education, but it seems unlikely her generosity would have stretched to four years without a scholarship. It would be no exaggeration to say that the entire course of his future career rested on that application or that (all things being equal) he probably had a one-in-twenty chance of success. But all things aren't equal: the process was designed to seek out the best candidates. Ron left no account of his journey through the selection process, but Bob Settles has graphically described what happened to him.

Two months after submitting his application, he was summoned for psychological testing, which involved a very long day of written tests designed to establish exactly what made him tick. A month later, he had a rigorous medical examination and a career interview. Applicants had to fill in forms detailing their medical history in minute detail. Spare a thought for eighteen-year-old Ron Evans agonizing over what to say about the back injury he suffered at school. "Is it in my medical records? What if I don't tell them, and they find something? What if I *do* tell them, and they decide not to take a chance on me?"

Bob Settles noted that "a military group medical examination is a phenomenon impressed indelibly into memory for a lifetime." His group of fifty was instructed to remove all their clothes, including their shoes, and stand at attention in a straight line while a group of doctors examined them in humiliating detail.

Those who survived the application process in 1951 were notified over the summer, and among those accepted into the U.S. Navy Reserve Officers Training Corps, much to his delight and relief, was one Ronald Ellwin Evans. Aunt Wilma must have been pleased, too.

It was time to register for college. Hatched from their nests in high schools around the state and beyond, the fledgling Jayhawks arrived at the KU campus in the fall of 1951. Ron was relieved to see at least one familiar face. Emily Wolverton from Highland Park was a talented clarinetist and had won a scholarship to study music. As their lives moved in very different directions, they didn't see much of each other but always stayed in touch.

During the fraternities' rush week, Ron attended several rush parties, partly because freshmen students tend to be attracted by free food and drink but mostly because it was a great opportunity to meet people. Ron didn't have the advantage of a family history in any of the fraternities, but he liked what Sigma Nu fraternity was offering. He impressed the brothers with his enthusiasm and academic history enough to receive a bid, which he accepted. Emily was rather surprised to learn that Ron had joined a fraternity, as membership usually required money and influence. Sigma Nu offered a vision of "excelling with honor," but what probably mattered most initially to a young student trying to find his way in the world was a roof over his head and new friends, including some whose friendship and loyalty would be lifelong.

While two Kansans meeting at a university in Kansas is not an amazing coincidence, it is at least interesting that the future NASA astronauts Ron Evans and Joe Engle first met at KU. They crossed paths at a couple of functions and became distantly acquainted. Joe remembers Ron introducing Emily to him as a friend from high school, and Emily recalls that Joe was "outgoing and had a great smile."

Like much of the student body, Ron smoked at KU, but he wasn't just a keen consumer of tobacco but also a campus representative of Chesterfield's. Then one of the leading cigarette manufacturers, Chesterfield hired Ron to pass out free samples of its brand to encourage existing smokers to switch brands. For this he earned twenty-five dollars a month. As a promoter of Chesterfield's, Ron was in good company: Ronald Reagan, Humphrey Bogart, Bob Hope, Bing Crosby, Dean Martin, Jerry Lewis, and many other stars lent their names to boost sales of the brand.

Ron also arranged to have a Rowe cigarette vending machine installed in the Sigma Nu house and made further modest profits from the sales, all carefully recorded in a set of handwritten accounts. In his final year, Ron was one of several students profiled in the in-house KU magazine *Kansas Engineer.* The article included the advice: "If you're a smoker, see Ron—he's campus representative for Chesterfield cigarettes and probably has lots of free samples."

At the University of Kansas, Ron's life progressed along two distinct tracks: the teenage freshman student attended lectures and seminars on the intricacies of electrical engineering, while the U.S. Navy cadet was committed to naval science classes. And drill. The navy took this ritual very seriously. For one hour each week throughout his four years at KU, Ron had to dress in full uniform (which had to be clean and pressed) and master close-order drill, rifle drill, marching drill, and parade drill. This thirty hours of drill annually not only cut into the navy cadets' study time but—for better or for worse—also established an awareness of their presence on campus.

As his first academic year came to an end, Ron packed his bags and prepared for one of the most important features of ROTC life—his first midshipman cruise, commencing on 9 June 1952. Along with many other cadets with no alternative form of transport, he made the journey to Norfolk, Virginia, by train at the navy's expense. He wasn't too impressed with the standard of accommodations, referring to the crowded Pullman cars as "the oldest railroad cars I've ever seen in my life." Perhaps a couple of nights in a "sardine can" were somebody's idea of preparation for spending six weeks in the belly of a battleship.

The USS *Wisconsin,* an *Iowa*-class battleship and one of the most powerful ships ever to patrol the seas, was launched in December 1943. The ship displaces fifty-two thousand tons at full load and is 880 feet long and 108 feet at the broadest point. Heavy armor plating gives it a low and surprisingly elegant profile in the water, while its nine sixteen-inch guns and twenty five-inch guns delivered devastating firepower in wartime. *Wisconsin* had seen heavy action against Japan and took an artillery hit in the Korean War. The final chapter in its fighting history came in 1991 with the launching of Tomahawk cruise missiles and one-ton shells at Iraq during the First Gulf War.

In the early 1950s, *Wisconsin* interspersed periods of action in the Korean War with a series of training cruises for a new generation of naval officers. Ron could hardly have ended up on a more fascinating or impressive ship, and its sheer bulk made it more stable in heavy seas than a frigate or a destroyer. That was a godsend to Ron. As he would later admit to those who weren't actually witnesses, the rolling Atlantic swell sometimes made him seasick.

In his fraternity house back in Kansas, Ron only had to share a room with fellow ROTC cadet Dick Ross. On board *Wisconsin*, the crew quarters housed 750 young midshipmen in dormitory-style compartments with upper and lower bunks and a locker for each bunk. With a crew of almost four thousand men, waiting in line was the order of the day for just about everything, whether washing, eating, or visiting the "head." After breakfast, served from 6:30 a.m., everyone reported to the muster station to shout "Here!" when called. All officers had to be saluted. For the first time in most of their lives, the cadets had to wear a uniform all day, every day. While the enlisted men had a white hat, the midshipmen's hats were marked by a one-inch blue stripe around the upper edge. Some of the crueler enlisted men spread the rumor in ports that the blue stripe marked men with "social diseases."

On that 1952 cruise, *Wisconsin* crossed the Atlantic in three weeks before arriving at its first port of call, Greenock, near Glasgow, where it stayed for five days among what the ship's cruise book referred to as "the hospitable Scots." Ron was able to stretch his legs ashore and buy presents for friends, including a scarf for Emily that she still has. "With a touch of nostalgia," the ship and its company left Scotland to steam south. On 12 July *Wisconsin* docked at the port of Brest in northern France for a four-day stay, allowing most of the midshipmen to visit Paris, where they gawked at the sights by day and then sampled the city's nightlife. Ron's scrapbook of his early career contains one souvenir of Paris—a used ticket to the Folies-Bergère.

From Brest, *Wisconsin* sailed back across the Atlantic, conducting speed trials and firing the giant sixteen-inch guns as part of an endless round of training for possible future hostilities. The crew had one more brief stop at Guantánamo Bay, Cuba, in those pre–Fidel Castro days before making the run northward to Norfolk. Ron and his shipmates then had a few weeks' break before the beginning of sophomore year.

Ron had no desire to repeat the cattle truck experience in the antiquated railroad cars, so he and his fraternity housemate Gary Irish decided to make their way to Andrews Air Force Base (AFB) near Washington DC and try to bum a flight to Kansas. They were hoping someone would take pity on two young midshipmen trying to get home, and they got lucky. After hanging around for a while, they were offered seats in a B-25 bomber. It just happened to be the base commander's aircraft, which had been converted to carry senior staff around the country in comfort. The bomb bay doors had been welded shut, carpets had been fitted, and there were other little luxuries, such as a couch to sit on. The aircraft had two pilots but no other passengers, so the first time Ron Evans flew in an aircraft was in conditions bordering on luxury, certainly compared with his outward journey to Norfolk.

Ron relished that first flight. "Man, oh, man! Looking down, you could see the earth whipping by, you could see the fields and the cities, and we also had a little night-time and you could see the lights. Of course, in a B-25 you didn't fly that high. I don't know what altitude we were, somewhere below ten thousand feet because I know we didn't wear oxygen masks. It was really great to be able to get up there and fly!"

That trip would be the first step in a major change of direction for the careers of both young men. After Ron and Gary made it back to Kansas, they "talked about that flight for the next year!"

In preparation for sophomore year, Ron managed to buy an old car and returned to Highland Park to see friends and family. Occasionally he gave Emily a lift to see her own family. By 1953 Ron saw that his parents' marriage was obviously in trouble. When they broke the news to him on one of his trips home that they were getting divorced, he was understandably upset, but as he was taking care of himself at KU and was motivated to do well in his studies, he tried to rise above the family crisis.

As with the death of Larry, the two surviving Evans brothers didn't really know how to share their emotional burdens. According to Jay, "Not too much would keep Ron from accomplishing what he was going after. He had the scholarship program and he wanted to finish that up, so I never did notice any major change in him. Men just deal with things; you just do what you have to do."

Ron did talk to Emily about the divorce, and she could see that it bothered him a lot. She was the only person he knew at the university who also knew his family. She was a good listener and a good friend at a time when he needed someone to talk to.

By the end of his sophomore year, Ron was riding high in his electrical engineering classes, and his affability and resourcefulness made him popular with his Sigma Nu brothers. As the academic year ended, he looked forward to a summer with his fellow ROTC midshipmen, but it would be rather different from their transatlantic cruise of the previous year. Their second summer training program was designed as a recruiting opportunity for two of the navy's special forces. Both the Navy Air Corps and the U.S. Marines would be wheeling out their stalls and trying to persuade the budding sailors to join them.

First came three weeks at Corpus Christi Naval Air Station (NAS) in Texas. The heat and humidity of a Texas summer came as a shock to many of the trainees, but it was a free holiday and, for most of the young men, a first opportunity to fly. Although Ron has left few details of his experiences at Corpus Christi, his 1952 predecessor Bob Settles noted that each man had a two-hour flight in a Martin PBM Mariner flying boat, mostly over the sea. Each trainee was given a turn at the controls. Flying three thousand feet above the waves, Ron found it a thrilling experience that brought back memories of the previous summer.

Next the midshipmen were given individual instruction in the single-engine AT-6 advanced trainer (known in the navy as the SNJ-4). Flying eight thousand feet over the coast of the Gulf of Mexico, Ron was subjected to demonstrations of turning, stalling, and spinning. Despite the effects of the rolling seas on his stomach, he thoroughly enjoyed being thrown around in an aircraft, even the noisy SNJ-4.

After being treated to a spectacular display by the navy's Blue Angels aerobatic display team, the midshipmen packed their bags again and transferred to Little Creek, Virginia, and the U.S. Marines' headquarters for training in amphibious landings. The midshipmen were kitted out with the familiar khaki uniforms and helmets, and were assigned accommodations in Quonset huts. Few could resist having their pictures taken dressed as marines, and Ron was no exception. Smiling broadly in one portrait, he has his helmet strapped under his chin and a canteen slung on his hip like a holstered six-gun. In a group shot with nine of his fellow midshipmen, all wearing khaki baseball caps, the background shows several white-painted Quonset huts and a high wire fence with sturdy wooden posts. It could be a scene straight out of the film *The Great Escape* except that they all seem to be enjoying themselves.

The midshipmen were all issued M1 assault rifles, trained in their use, and taken to a firing range to test their marksmanship. The main purpose of the exercise was to mount a simulated beach invasion, involving a small fleet of navy ships and landing craft of the type made familiar in 1944 newsreels of D-Day. Ron found himself taking up temporary residence on the USS *Rockwall*, an attack-transport ship that had participated in the invasion of Okinawa and had later supported the atomic tests at Bikini Atoll. As on the much larger USS *Wisconsin*, Ron found that every human need was prefaced by waiting in line.

The amphibious exercise began well before dawn. Bob Settles wrote of his experience: "We were awakened at three o'clock in the morning and instructed to prepare for our assault. It all happened just like you see it in the documentaries of WW2. As soon as [the landing craft] were all loaded, they formed a straight line and moved toward the beach so as to arrive simultaneously."

Since the exercise was intended to be as realistic as possible, some level of genuine risk was unavoidable. One of the midshipmen in 1952 had died, having fallen into the water before reaching the beach. Fortunately, Ron's group all made it safely ashore.

That summer of 1953, the special forces were successful in persuading at least one midshipman to look beyond the basic navy life. Dick Ross volunteered to become what Jan Evans would affectionately call a "gung-ho marine." As for Jan's future husband, undoubtedly Ron was enthralled by his airborne experiences, and he began to think seriously about becoming a naval aviator. Certainly, from the beginning of his junior year in the fall of 1953, Ron's favorite topic of conversation was again flying. He talked to Emily Wolverton about flying. He talked to his fraternity brothers about flying. He talked to strangers about flying. At any opportunity, he would turn a conversation around to his new passion. Ron seemed to have fallen in love with flying and even expressed his concerns to Emily that too much studying at night might strain his eyesight and risk failure in any future naval aviator medical tests.

Ron's obsession didn't derail his studies or his life outside the classroom. He was excelling at electrical engineering and, by the end of his junior year, was listed in the top 10 percent of the class. He was elected not only to the Sigma Tau engineering honor society but also to the ranks of the oldest and most prestigious national engineering honor society, Tau Beta Pi, and to Eta Kappa Nu, the international honor society of the Institute of Electrical and Electronics Engineers.

The summer of 1954 saw another transatlantic voyage for the midshipmen. This time, Ron was detailed to participate in a nineteen-ship naval exercise in European waters. He was assigned to the USS *Gwin*, a destroyer-minelayer that had survived multiple kamikaze attacks in the final months of World War II. With a displacement of 2,200 tons and a length of 376 feet, *Gwin* was tiny in comparison with a battleship, a fact that became all too obvious to the midshipmen in the ocean swell of the mid-Atlantic.

Gwin visited Lisbon, Portugal; Le Havre, France; Valencia, Spain; and Torquay, England. The shore visits were fun, but at sea Ron found himself once again waiting in line for meals that did not always stay down. As Jan Evans later noted, "That was something you did in ROTC—you went out on the ships during the summer months, and I know that was something he thoroughly enjoyed. He got sick every time he was on a ship, but he enjoyed it."

Friends and family later joked that Ron must have taken up flying to avoid sea duty. Many a true word is spoken in jest! Looking back on his career, Ron explained, "I really didn't decide to be a pilot until they put me out on the third cruise ... after the junior-to-senior year when they put me on some crazy thing like a destroyer. And you float around on that thing for two months and you know damn good and well you want to do something else in the navy! And that's when I *really* decided I wanted to be a pilot!" No one who knew him doubted he had been well and truly bitten by the flying bug, but he obviously realized that every hour spent in the air was an hour less on a heaving deck.

Back at KU for his final year, Ron seemed to have both halves of his career under control. His ROTC classes were going smoothly, and in college he was selected to serve as the electrical engineering representative to the university's Engineering Council. He was even appointed an assistant instructor in engineering drawing, a position he held for seventeen months.

Socially, he was a mainstay of Sigma Nu and was elected president of the KU chapter in his final year. Contemporary photographs show a happy household with many formal social gatherings, but whether the revelers are wearing white tuxedos or black tuxedos, Ron Evans is always wearing a happy smile.

Meanwhile, the conditions back in Highland Park were difficult. Ron's father had moved to Bird City, where he took up work similar to his farming cooperative work in St. Francis. Ron made frequent visits to see his mother and brother. Larry's death and Jimmy and Marie's divorce hung like a black cloud over the family, and Marie, in particular, took the breakup badly.

The end of Ron's final year at the University of Kansas was in sight. Of course, in addition to earning his degree in electrical engineering, he had to be offered a commission in the U.S. Navy before becoming an ensign. But he was not interested in just *any* commission. Ron by now was completely wedded to the idea of becoming a naval aviator. He had already submitted another self-typed application form requesting to serve as a pilot, but it must have occurred to him that the navy couldn't assess his potential as a pilot until he took flying lessons. He could be the next Charles Lindbergh, but the navy wouldn't know it until it had incurred the expense of training him. All the navy had to go on at that point would be his university degree, his naval science class results, and an overall assessment of his proficiency as a well-drilled naval cadet who was a little prone to seasickness.

Available profiles of Ron Evans are surprisingly contradictory about the year of his graduation. His NASA and U.S. Navy biographies say he graduated in 1956, but his KU permanent record states that he was "assigned to the class of 1955." The program for the eighty-fourth annual commencement ceremony, held on 4 June 1956, at the University of Kansas lists Ronald Ellwin Evans of Topeka as one of thirty-five graduates awarded a bachelor of science in electrical engineering. Ron was one of twenty-two graduates whose degrees were "granted since June 1955."

Ron himself always said that he "graduated mid-semester," and the university records, although a little opaque, confirm that he completed his coursework in the fall of 1955. He might have deferred some of his coursework to allow time for his duties as an assistant instructor in his department. He also had taken a job working in a factory manufacturing plate glass in the summer of 1955 to gain general engineering experience.

Typically, students who finished their coursework in the fall came back for the graduation ceremony the following year, but the one thing we can be certain about is that Ron's degree was awarded in absentia. On 4 June 1956 he was flying over Pensacola, Florida.

After completing his university studies, Ron took a temporary job with General Electric while he waited to hear from the navy. Ironically, General Electric paid better than the navy, but the company didn't see much of Ron.

He received a letter from the chief of naval personnel dated 12 December 1955 offering him a commission to serve as an ensign in the U.S. Navy. In the full page of military jargon, the following words jumped out at Ron: "You are hereby designated as a student Naval Aviator." The training class would convene on 19 February 1956 at NAS Pensacola.

Ron lost no time advising the chief of naval personnel that, yes, he most definitely would accept the offer. That produced written confirmation of his appointment as an ensign and trainee naval aviator effective 3 February 1956. (The document still survives because lacking proper writing paper, Ensign Evans later wrote a letter on the back of it to his future wife. But that is another story.)

This Jayhawk would fly after all.

3. The Pilot and Miss Pollom

Flying might not be all plain sailing,
but the fun of it is worth the price.

—Amelia Earhart (1897–1937?)

When Ens. Ron Evans arrived at NAS Pensacola in February 1956 for flight training, he was entering the "Cradle of Naval Aviation," the primary training base for all U.S. Navy, Marine Corps, and Coast Guard officers. Pensacola lies by the warm waters of the Gulf of Mexico, just about as far west as you can travel along the Florida panhandle without entering Alabama.

Ron was one of thousands of young men passing through Pensacola on their various trajectories, although typically almost one in three of those trajectories would end back on the ground (or, more likely, on the heaving deck of a navy warship). For some the change would be due to undetected medical issues. Others would fail to keep up with the relentless training schedule. Still others would find that flying simply wasn't for them.

Many of the successful trainees would serve together and become lifelong friends and squadron mates, but their trajectories would not always intersect immediately. Ron never met Howard Brown at Pensacola, for example, but they would later fly in the same squadron.

The navy had more than one way for the men to earn their "wings." Unlike Ron, Howard did not have a university degree but did spend two years at college before joining the U.S. Naval Aviation Cadet Program as a "NavCad." He arrived at Pensacola in September 1955, but before he could learn to fly, he had to study to become a naval officer. That meant undergoing four months of concentrated classroom work at "ground school."

Both young trainees began their preflight training around the same time. While it is unsurprising that those two, out of many hundreds, couldn't remember meeting at Pensacola, Howard jokes that he and the university graduates were "probably traveling in different worlds." Be that as it may, they all now faced the same daunting hurdles of exacting academic study, physical training, and water-based exercises. Ron was in his element with the mathematics and engineering studies, but even those few students who had previously obtained civilian pilots' licenses found the aviation science classes daunting.

While the U.S. Navy generally preferred its pilots to keep their aircraft in the air or at least on a solid surface, a great deal of time was spent training would-be pilots to escape from aircraft that had ditched in the sea. Trainees had to be able to jump off a thirty-foot tower into the swimming pool and swim the equivalent of one mile in one hour without touching the sides of the pool. This feat often required cajoling, encouragement, and coaching by the instructors.

The trainees also had to escape the clutches of the Dilbert Dunker, essentially a modified cockpit from an SNJ trainer aircraft mounted on an inclined track above the pool. Wearing clothes, shoes, lifejacket, and parachute, the "victim" mounted what must have seemed like a scaffold before being strapped into the cockpit. When the Dunker was released, it entered the water and flipped upside down. While holding his breath, the trainee had to extricate himself and struggle to the surface without the help of the instructor. Failure meant repeating the exercise for as long as it took. Howard Brown never liked swimming underwater and wryly describes the experience as "interesting."

One part of preflight training that seemed to presage Ron's future career was altitude training. After learning how to use oxygen equipment, trainees entered a low-pressure chamber and were taken to a simulated altitude of thirty thousand feet. On the way "up," they were asked at twenty thousand feet to remove their masks for ten minutes and to perform a series of simple manual tasks. The ensuing mental confusion and reduced hand-eye coordination were the best teachers of healthy respect for properly maintained oxygen equipment.

Ron and Howard survived their preflight training and were delighted and relieved to advance to actual flying. Both were taken to Whiting Field, an outlying air base about thirty miles northeast of Pensacola. There were actually two airfields. At South Whiting Field, trainees flew in the SNJ, the aging and very noisy single-engine trainer that Ron had experienced at Corpus Christi three years earlier. At North Whiting Field, trainees would fly the T-34 Mentor and specifically the navy's new, improved T-34B model, which had only been delivered in 1955. Ron was pleased to be sent to North Whiting to fly the newer aircraft. Howard "ended up being lucky" and drew North Whiting as well, although his self-effacing description may be reminiscent of the old golfing adage that "the more I practice, the luckier I get."

Following aircraft familiarization classes, the trainees' first trip aloft was a demonstration flight during which the T-34B was firmly under the control of the instructor. Some might have considered it a joyride, but Howard ruefully recalls, "About half the students taken up on the first flight got airsick—me included!" While Ron Evans's records do not indicate whether he was among those who kept their breakfasts down, his aviation logbook shows that his seventy-eight-minute demonstration flight took place on 30 April with an instructor named Havlek at the controls. Ron's first instructional flight took place the next day, and in the first half of May, he took to the air another eight times, gradually learning to master the T-34B. For any trainee pilot who has found his vocation, this should be a period of great exhilaration as the world rotates gloriously beneath him like a living map. At the same time, he can become quite frustrated on realizing how much he still has to learn before flying becomes instinctive and natural.

On 18 May Ron had his last pre-solo check flight, this time with a different instructor named Sankey. Over the next two hours, Ron had to persuade this relative stranger that he could be trusted to fly on his own without breaking either the navy's expensive aircraft or his own neck. He had to demonstrate proper aircraft handling techniques and his ability to land and take off safely several times. Meanwhile, the instructor was undoubtedly weighing how he would feel if he sent this youngster solo too soon, and it all ended in a loud bang and a patch of melted runway.

The instructor seems to have been satisfied, although the traditional scene of the man in the back climbing out and saying, "Okay, son, take her up again. I'll wait this one out!" didn't actually happen. For whatever reason—possibly because of poor weather or possibly just because it was late on Friday, and the navy didn't do weekend flying instruction—Ron had to wait three days for his first solo flight. It came on Monday, 21 May 1956; lasted one hour and thirty-six minutes; and included four demonstration landings. On such a special occasion, civilian pilots have been known to mark the achievement with some appropriate comment in the logbook entry. But in his government-issue navy logbook, Ron confined himself to a small letter *s* in the "remarks" column. However, unless North Whiting Field existed in a different universe from the rest of the world of aviation, Ron and anyone else who soloed that day undoubtedly visited the bar afterward and toasted each other's well-deserved achievement. And in accordance with another tradition, each new solo pilot had his tie cut off at throat level by his fellow trainees.

Of course, he was only at the beginning of a long journey. Ron spent the next five weeks flying mostly with his original instructor. Typically, the instructor would demonstrate new procedures and flying techniques on a dual flight; then Ron would try to duplicate the effects on a solo flight. One by one, he had to master stall and spin recovery, inverted flight, aerobatics, recovery from emergencies, and general airmanship in another thirty-four flights, seventeen of which were solo.

Ron then progressed to a more high-performance aircraft at another of Pensacola's outlying airstrips, Saufley Field. The T-28B Trojan was a single-engine tandem trainer, but there the similarities with the T-34B ended. With around six times the horsepower and close to twice the maximum speed, the Trojan was the first navy training aircraft designed to transition pilots from propellers to jets. After a week of groundwork to familiarize himself with the vehicle, Ron took to the air again on 11 July with a new instructor.

Over the next ten days, he learned to appreciate the power and maneuverability of the muscular Trojan in an adrenaline rush of loops, barrel rolls, wingovers, stalls, and spins. He was then introduced to formation flying, which seemed downright counterintuitive after many previous warnings to keep well away from other aircraft. Marine aviator Gerald "Jerry" Carr, who would become an astronaut with Ron in 1966, had passed through Pensacola the previous year. He recalled going into formation flying with "a good deal of trepidation" and added, "I always returned from those flights wringing wet with sweat. . . . It was a real workout."

The navy seemed determined not to let grass grow under the feet of its trainees. After the whistle-stop visit to Saufley, the trainees were relocated north of Pensacola to Corry Field (which would soon be decommissioned as the city expanded). There, Ron was introduced to the 1-CA-1 flight simulator, built by Link Aviation and described decades later in a navy journal as a "clicking, hissing, electro-mechanical vacuumized monstrosity . . . prone to frequent mechanical failure and misalignments." But when it wasn't hissing or misaligned, it was actually a very useful training device. Ron spent sixteen hours in late July in the "blue box," becoming sufficiently proficient in instrument reading and navigation to progress to the practical application of those lessons.

In a series of thirty-six "hops" in August, he practiced flying in various weather conditions, including several night flights to make it even more interesting. He had to learn to make cross-country flights using a map to navigate his way to distant towns or airfields before finding his way home. A proper understanding of the aircraft's instruments was crucial, particularly when flying in darkness or above the clouds. Emerging unscathed from Corry, Ron had his logbook endorsed by the base commanding officer with the message: "No flight violations or accidents for the period."

During twenty-two instructional and fourteen solo flights, Ron had gained both increasing confidence as a pilot and a good grasp of the quirks and handling characteristics of the powerful T-28B. When Ron and his fellow trainees moved from Corry Field to Barin Field, Alabama, some twenty miles west of Pensacola, it must have been wrenching for them to abandon the Trojan for the older, slower, less powerful SNJ, which they had managed to avoid at Whiting Field. This apparently backward step was not a deliberate navy ploy to disconcert any young pilots who might have been getting a little cocky. It was simple practicality: the T-28 was not capable of landing on aircraft carriers, but the SNJ was. The whole point of the navy's having its own pilots was to extend America's military presence across the globe, using aircraft carriers as floating runways.

Before taking to the air at Barin, the trainees endured yet another classroom session to learn the more modest characteristics of the older aircraft. Ron had his first demonstration flight in the SNJ on 1 October. Following were seven instructional flights and a ninety-minute solo to demonstrate his ability to absorb all the classroom and instructional lessons. Then came night formation flying, air-to-air gunnery practice, and simulated aerial attacks in which the trainees dropped bombs and fired rockets. That exercise graphically demonstrated that they were learning not just how to fly but also how to deliver devastating firepower from the air, as and when required by decisions taken many steps above their pay grade.

Ron and his fellow trainees must have realized that even the air force taught its pilots to do everything they had done up to that point. What really mattered at Barin Field—what really distinguished these would-be naval aviators from *ordinary* pilots—was their field carrier landing practice (FCLP) and what came after it.

On arriving at Barin, the trainees couldn't help noticing a worryingly short section of the runway marked in white paint in the shape of an aircraft carrier's flight deck. They were now at the crucial stage of their training where landing a little short or a little long wouldn't be good enough. The man in charge of the FCLP exercise was the landing signal officer (LSO), an experienced navy carrier pilot who stood at the far end of the white box and used a set of orange paddles to signal approaching pilots their deficiencies in height, speed, or direction. He could, and frequently did, bellow additional instructions over the radio.

First, however, came more classroom briefings on every aspect of landing an aircraft on the deck of a moving ship. The key to the trainees' success would be their ability to fly low and slow. In this potentially lethal combination, the aircraft is close to stalling speed and too low to recover in the event of a stall or incipient spin.

In the second half of November, Ron made thirteen flights. On the first, his instructor demonstrated the techniques of the low-and-slow approach, lining up with the white-painted "carrier deck" and touching down at the point where the arresting cables would be strung across a real deck. Then he gunned the engine, flew back into the circuit, and repeated the exercise three more times.

All but one of Ron's next twelve flights were solo, if you could call it solo after merely swapping the back-seat controller for the shouting man on the ground. A year earlier, Jerry Carr had found FCLP "really demanding" and added that "it took several days of practice and a number of tongue-lashings for us to get enough FCLP landings to satisfy our LSO's requirements." In Ron's case, his solo flights produced a total of sixty-nine touch-and-go simulated carrier landings, after which he was ready for the final crucial challenge.

Ultimately, the only true test was to land on the real thing. The USS *Saipan*, an old straight-deck carrier stationed in the Gulf of Mexico, served as the target. On 1 December the trainees took off in groups and flew east in formation, setting up their approach to the ship as it steamed into the wind. Although maintaining the right height and right speed on final approach still wasn't easy, it wasn't new either. But as the deck expanded to fill Ron's view, nothing could have fully prepared him for that first sudden jerk as his tailhook caught the arresting cable and brought the aircraft to an abrupt halt. He barely had time for a moment of elation before his SNJ was being repositioned for launch, which had to be at full throttle to clear the flight deck. Then he had to circle and repeat the exercise five more times. His success was evidenced by another low-key logbook stamp: "Qualified this date in carrier landing. Made six landings aboard the USS *Saipan*." Ron didn't feel the need or the entitlement to add anything to the page to mark the occasion. It was enough that he knew the day had ended well.

His basic training was now complete, but that left the crucial question of where he would be going next. Various options were available to successful trainees. Based on their flight grades, they would proceed to careers in propeller aircraft, helicopters, or jets. Ron was delighted to receive orders to report to Naval Air Station Memphis in January for jet training.

First, he had two weeks' leave over Christmas. He made preparations to return home to share his good news with the family and to look up old friends and neighbors.

Janet Merle Pollom was born into a happy household in Topeka, Kansas, on 21 August 1933. Her father, Harry, and her mother, the former Marjorie Mohler, had been high school sweethearts, and the Oakland High School building where they had met and fallen in love was where Jan later went to grade school. Marjorie was a town girl, but Harry, the youngest of seven children, was raised on a farm beyond the Topeka suburbs. During World War I, Harry had to drop out of school for a year to help run the farm because his brothers were all at college or off serving in the army.

When he was still in junior high, Harry helped out at the Gibbs Clothing Company during weekends, running errands for Frank Gibbs. Harry was a hard worker, and after finishing his education, he quickly rose to become the manager of one of Frank Gibbs's growing chain of clothing stores. Harry worked for Gibbs for over fifty years, always managing one store or another, and Jan remembers Mr. Gibbs as a wonderful man and fine employer.

Jan and her elder sister, Marian, enjoyed their childhood in Topeka. Harry's job allowed the family to live in relative comfort during the Depression, and Jan fondly recalls the obvious mutual love and respect that her parents shared throughout their lives. They were a genuinely sweet and besotted couple.

Harry Pollom was a sports enthusiast who loved following all kinds of sport and ignited a love of sport in his daughters. Marjorie had also been active in school sports, which was one of the reasons Harry had noticed her. She had an adventurous spirit, most notably demonstrated just after she left high school around 1923. In that barnstorming era, hundreds of former wartime pilots traveled around the country offering plane rides for money. Marjorie couldn't afford the price, but her less-adventurous friends clubbed together and paid the twelve dollars so she could take to the air in a rickety biplane high over Topeka. She had a ball.

One might assume that except for the ubiquitous rationing that emptied shelves in stores across much of the nation, Kansas, lying right in the middle of the continental United States, would have escaped most of the effects of World War II. Not so. After the attack on Pearl Harbor, the West Coast had to address the threat from the expanding Japanese empire, while the East Coast faced the rather less obvious threat from a distant Nazi Germany. Just eight years old when her country went to war, Jan never heard of Operation Drumbeat, but starting in January 1942, German U-boats eventually sank hundreds of Allied ships along the Eastern Seaboard and in the Gulf of Mexico. At night the enemy found it easy to spot target ships silhouetted against the lights of coastal cities, so the U.S. government imposed blackout restrictions along the coasts. This order was extended to, and rigidly enforced in, midwestern states such as Kansas, if only to instill a sense of national solidarity with the endangered coasts. All forty-eight states were "in it together."

The young Jan Pollom could see the effects for herself. She vividly recalls feeling her way through a darkened home during the war, peering through the blackout curtains as the wardens did their rounds while on the lookout for transgressors letting chinks of light escape. It was a magical time for stargazers as Topeka slept in darkness. Jan was mesmerized by the beauty of the winter moon illuminating icicles hanging from the trees and casting its radiance on the snowy ground. For a priceless moment, she could forget about the reasons for the blackout and enjoy the spectacle of the city bathed only in moonlight.

After the war ended, the world had changed forever, but education carried on. Jan was a bright student, and by the time she had completed high school, she had the grades and exam results to study at nearby Washburn University. She initially chose a four-year course in psychology, which she liked, but after she took a summer job at Forbes Air Force Base on the outskirts of Topeka, she decided to switch to a business and secretarial course at Washburn. She went back to the base the following year and eventually had a permanent job working for a full (bird) colonel.

Topeka's status as the capital of Kansas is probably best known (at least outside the United States) to ardent quiz addicts. Most foreigners have never heard of Topeka and would probably guess that Kansas City is the state capital. Nevertheless, on Monday, 21 May 1956, for one night only, the Topeka Municipal Auditorium became the center of the rock-and-roll universe as it hosted a concert by the rising young star Elvis Presley (billed by the tour organizers as "the Nation's Only Atomic-Powered Singer"). Jan was one of the 2,500 (mostly female) audience members who got to hear Elvis sing "Heartbreak Hotel," "Blue Suede Shoes," and other songs that he would make immortal. She was there with several girlfriends, clutching tickets that their parents had bought for them. A press report of the audience's reaction to Elvis mentioned "continuous squeals and clapping." Looking back, Jan is amazed it was as late as 1956 because "we were all so crazy and excited you would have thought we were 14 or 15 years old!" Little did she know she would get to meet Elvis in person seventeen years later.

During her years at Washburn, Jan met a young woman named Sue Benson from Highland Park in Topeka. They became firm friends. Shortly before Christmas 1956, Sue contacted Jan and said that her parents were going out of town for a while. Would Jan come and stay for a few days to keep her company? Jan readily agreed, packed a small suitcase, and made the thirty-minute journey to Highland Park, looking forward to a few days of girl talk.

That first evening, there was a knock on the door. The caller was a tall young man with a military-style haircut whom Sue introduced as Ron Evans, a former neighbor and school friend who was now in training to be a naval aviator. Jan was quite taken by this visitor with his friendly eyes and big broad grin that lit up the room. Ron explained that he was on leave and catching up with old friends. He wanted to hear how life had been treating Sue and the rest of the gang from Highland Park High School. It seemed to Jan that he very thoughtfully included her in the discussion, and far from feeling the outsider, she was quickly engaged in a lively three-way conversation. Ron asked all about her family, and they identified several mutual friends in addition to Sue. Bearing in mind what Ron had been doing during the previous ten months, he could easily have dominated the conversation with exciting stories from Pensacola, but Jan was impressed that he seemed keen to hear what she and Sue had been doing. He came across as a modest young man with very little to be modest about.

A cynical fly on the wall might have concluded that this meeting was not a coincidence and that Sue Benson was engaging in a little matchmaking, but the suggestion makes Jan laugh. "I don't think so. I don't think so at all! It was the way of life, particularly in that day and age when we all grew up. People would come home from college, and you always got together with your friends. They had gone through high school together and been friends all those years. I think it was just natural [that] one of the first things he would do would be to come over to see Sue."

But Jan never asked Sue the question, and Sue never claimed any credit for the match. In this *Happy Days* era, groups of young men and women would all hang out as friends without necessarily dating. By another coincidence, among Jan's closest friends all through high school was a boy named Al "Buss" Brodecker, and they hadn't dated. Buss later married Sue.

One thing for sure was the obvious chemistry between Ron and the slim, attractive young woman with the engaging smile whom he had just met. In Jan's own words, "That was it." They met again the next day and on every day of Ron's leave either at Sue's house or at Jan's family home. Such an early introduction to Jan's parents clearly didn't deter Ron.

Does that seem like love at first sight? Jan laughs, but it is a soft, surprisingly girlish laugh. "Well, it does, doesn't it? I guess I never thought about it that way, but . . . yeah." Her voice seems to be arriving from somewhere far away in time.

It seems natural, if a little impertinent, to wonder when Jan first realized Ron was the man she would marry. She explains that after Ron's leave ended, he headed to Memphis for advanced flight training, but they exchanged letters "just about every single day." Then in June 1957, before taking his first squadron assignment, he had another two weeks' leave, "and that pretty well did it."

4. Wings

Up, up the long, delirious, burning blue
I've topped the wind-swept heights with easy grace,
Where never lark, or ever eagle flew—
                    —John Gillespie Magee Jr. (1922–41)

Ron arrived at NAS Memphis late on Friday, 4 January 1957, along with Joe Delmus, another young Kansan pilot transferring from Pensacola for jet training. They soon discovered that classes didn't actually start until 16 January, which left Ron with time to reflect on his changed circumstances.

Until his leave in Topeka, Ron had been totally and absolutely focused on learning to fly—ideally the hottest, highest, and fastest aircraft the navy could provide. Now, only a few short weeks later, his priorities had been . . . *adjusted*. Throughout 1957 he would write a series of letters to Jan Pollom that provide a candid and fascinating insight into the future Apollo astronaut's hopes and fears, his love of flying, and the new love that had ignited so unexpectedly and with such potential to distract his attention and his focus on flying.

That weekend Ron tried to set out his thoughts in his first letter to Jan: "At first I just thought I would try to date you for something to do while I was home; however, after a short time I would have done almost anything so that I could have been with you. It didn't matter what we did, just as long as I was near you. I know that I want to see you again, all I have thought about is you."

He certainly wasn't the first young man to experience the exquisite agony of separation from the source of his turmoil. A few days later, he wrote again with almost a hint of resentment at his new circumstances and in terms that might have scared off a less worthy recipient: "Why did you have to do it to me? Here before my leave I was a happy, carefree bachelor without the slightest thought of getting married for a long time. And then what happens? I meet you and now all I can think of is you and how I wish you could be with me."

But there was no "Dear Ron" reply. The attraction was definitely mutual, and the two continued to exchange regular letters and occasional phone calls. Of course, phoning into or out of a military base was more involved than phoning a friend down the street. At the time, it was also expensive. Ron's monthly income as an ensign was $370, out of which he sent regular sums to his mother and his aunt Wilma. After the monthly payment on his car, he had very little money left.

Although his ultimate goal at Memphis was to transition to jet aircraft and earn his naval aviator's wings of gold, he would first have to reacquaint himself with the T-28B piston-engine trainer and undergo a grueling course in all-weather flying and flying "blind," or by instruments alone. He found to his frustration that the skies over Memphis seemed permanently leaden; eleven inches of rain were recorded that January. All too frequently a low cloud ceiling grounded all training flights.

When the weather did relent, Ron's flights were all instructional, and this time the instructor always sat in the front seat. The pupil was in the back seat "under the bag." A canvas hood blotted out his view of the sky and the ground, leaving only the readings on his instruments to guide him. Before starting these flights, Ron expressed his concerns to Jan: "Hope I will be able to get all the way through here without getting a 'down.' A 'down' is given by the instructor if the student doesn't fly the 'hop' satisfactorily. Sure am going to try to do my best. This flying is what I like to do, and I don't see any sense in just doing it half-way."

He observed that while he was sitting under the bag, his instructor would act as the safety pilot and "pull me out of any graveyard spirals if I get into them." But after the first flight, he lamented that his limbs seemed to have lost the ability to control an aircraft. The instructional work assailed him from all directions, and the only consolation was that he had been assured it "would all tie together in about two weeks." He also recognized that the training was all aiming toward the day when he would fly a single-seater jet through the clouds and back to his aircraft carrier without getting lost or disorientated.

On 30 January Ron wrote: "Flew a lousy hop today, but guess it was just my off-day. Will try to do better tomorrow."

The sessions under the bag were leading up to an important ground-controlled approach (GCA) flight during which he would navigate by instruments and make an approach to the runway while being talked down by a radar controller on the ground. He would descend to two hundred feet before being waved off and would then go around for another approach.

Four days before the GCA test, Ron went some way toward baring his soul in his letter to Jan: "Honestly, Jan, the only thing I think about any more of any importance is you and flying. I must confess that most of my time is taken up on the flying end of it right now. However, this does not mean that I do not think of you and wish you could be with me.... It is too late to turn back in this flying game and it means a great deal to me that I get my wings of gold.... I don't want you to get the idea that flying will always come first in my life and you second, because that will not be the case. In other words I believe that when the time comes, a happy and enjoyable home life is the first thing to strive for."

Monday, 11 February, was the day of reckoning. Ron reported the outcome to Jan: "Flew two 'hops' today and got a 'down' on the GCA hop. I flew real lousy and deserved it, I guess. Will just have to settle down and keep things under control. Will get another chance for GCA later.... I will try to do better next time. I know that I can do it, but just couldn't relax and get it done today."

If it seemed from his letters that Ron was at a low ebb halfway through the first part of his course, a telephone call to Jan apparently worked wonders. They discussed two visits to Memphis. Not only would Jan, her mother, and Ron's mother drive from Topeka to Memphis later that month but also Ron was determined that Jan would come back to pin his wings on him—not *if* he won them but *when* he won them. When Jan wondered how long the drive to Memphis would take, Ron said that he and Joe had managed it in twelve hours "driving 70–80 mph."

That drive to Memphis took longer than twelve hours, but it was a golden opportunity for Jan and her mother to get to know Marie Evans. Jan found Marie to be "a very sweet lady" who hadn't had an easy life but didn't bemoan the fact. Although not the most outgoing of individuals, she was perfectly chatty in a one-on-one or two-on-one setting. The only subjects she didn't want to talk about were the breakup of her marriage and Jimmy's whereabouts.

Ron and Jan were delighted to see each other again, even with the presence of two chaperones. They toured Memphis, which was something of an eye-opener for Marie, but the visit was over all too soon. All three women in their different ways had really enjoyed seeing Ron again, and he later wrote: "Don't know how to tell you how much I enjoyed having you down here this weekend. I miss you already."

Three days later, he opened up even more about his feelings: "I was sorry to see you leave Sunday. I guess I am not real emotional. I used to be when I was a kid, but have hardened myself, I guess, because I always used to think it was sissy for a boy to be emotional. Of course when my brother died I sort of broke down. Actually I guess I am more emotional than I like to admit sometimes. In fact, I miss you more than ever now that I have just been with you."

If Ron had suffered any loss of focus since December, his letters covering the final two weeks of instrument and all-weather training show a more upbeat assessment of his prospects. In spite of appalling weather, he made five more flights, including a two-hour formation night flight. But the day of reckoning was upon him: on Monday, 4 March, he had two hops scheduled, including the repeat of the crucial GCA test. As soon as he had an opportunity, he wrote to Jan with his news: "Well, what a relief! Got an 'up' on my GCA this morning and got an 'up' on my final check this afternoon. Sure feels good to be through."

Now at last he would fly jets at Memphis before returning to Pensacola to complete his training. Looking even further ahead to his future in the navy, he told Jan of his intention to apply for a posting on the West Coast. He concluded the letter: "Well my darling, about out of page and need some sleep."

The navy's TV-2 Shooting Star was a two-seater jet aircraft that was developed from the first operational jet fighter flown by the U.S. Air Force. With a single Allison turbojet engine delivering 5,400 pounds of thrust, it would give Ron and his fellow pilots a far bigger kick in the pants than anything they had previously flown. But first came the inevitable training sessions.

After lamenting to Jan about "more damned gauges, levers, switches, warning lights and dials" and about fighting a flu virus, he finally set off with his instructor on the first of twenty-three flights. That evening he wrote to Jan: "Jets are great. Faster than hell and have to be alert to keep ahead of them. But a lot of fun. Had two hops today and am tired. Get two more tomorrow, A-3 and A-4 hops. A-4 is a 'safe for solo' hop. Hope will get an 'up' on it. Two 'down' and you are out of program and floating on destroyer in 'black shoe' Navy. Not for me, I hope!"

He obviously felt confident enough to plan Jan's second visit to Memphis, this time—if all went well—to pin his gold wings on his chest. His logbook records two more flights with his instructor, followed by his first solo in a jet aircraft on 4 April that lasted 102 minutes and included multiple touch-and-go landings. Typically, the log entry is unembellished by any personal comment, not even a little *s*.

Jan, her mother, and Marie Evans repeated their long journey from Topeka to attend Ron's wings ceremony on Friday, 12 April. Rear Adm. Frank Akers, the chief of the Naval Air Technical Training Command, handed Ron and his fellow pilots the certificates designating them naval aviators, but Jan pinned the gold metal wings onto Ron's uniform. In the photographs of the ceremony, it is difficult to tell who looks more delighted. Even a casual observer would have had no trouble guessing where their relationship was headed.

A few days later, Ron wrote: "I hope your weekend was as wonderful as mine. I enjoyed every minute we were together. Be sure to tell your mother that I enjoyed her being there."

He had seen a vision of a settled, happy family life back in Topeka the previous December at the Pollom home. He had probably heard the story of Marjorie's barnstorming flight three decades before he ever entered an aircraft. One evening Harry and Marjorie had invited Ron to join the family for dinner. Marjorie had made a tuna casserole, and Ron wolfed down his serving. It took a year before he confessed that he loathed tuna.

For the remaining twelve flights in the TV-2, five of which were solo, Ron honed his instrument-flying skills and practiced flying in formation wingtip to wingtip with three other pilots. They even made an enjoyable cross-country visit to Miami with an overnight stay. He now assessed himself as in the top three of his class. His final TV-2 flight was another GCA test, which he passed with flying colors, bringing his association with Memphis to a happy conclusion.

Ron checked into Advanced Training Unit 206 at NAS Pensacola on Thursday, 2 May 1957, to complete his training. He was looking forward to flying the Grumman F9F-2 Panther, a carrier-qualified jet flown extensively during the Korean War. Its Pratt & Whitney turbojet engine could push the aircraft to 575 miles per hour, but what the Panther *lacked* excited Ron: it had no second seat. The Panther was a true fighter jet.

First, inevitably, came two weeks of ground school and a new set of gauges, levers, switches, and dials to master. The actual flying would follow later in May, but Ron was getting twitchy about his ultimate destination in the navy. He knew he wouldn't find out until mid-June and confided his concerns to Jan: "Guess I told you I may be stuck here in Pensacola as an instructor for the next 15–16 months. . . . They only have four squadron billets for the ten navy people in my class. This means that six of us will either get plowed back as instructors or [go] to some air station on some island as ship's company and not be in a squadron."

On 20 May Ron fired up a turbojet and made his first single-seater solo flight. His next letter was much more upbeat: "You are getting your first letter from a 'Panther' jet pilot. Got two hops today. The first one had an instructor chasing me in another plane. Then the second one was completely solo. Really had a ball up there!"

He told Jan that the Panther wasn't flown by the fleet any more, but it was still an agile performer. Stiff and sore from pulling g's during aerobatics, he remarked: "Sure is a great plane. Somewhat of a dog for a jet, but still pretty good."

He was particularly exhilarated by tight formation flying: "Really having a ball flying the old Panther. It is really hard to explain to some 'ground pounder' the sensations you get flying around with other airplanes. In the parade position there is two feet between wing-tips. . . . The most fun is doing all kinds of aerobatics in a tail-chase when the planes follow each other about 100 feet behind."

Ron soon received some good news about assignments: he and his friends Joe Delmus and Nelson Gillette learned that they would be sent to the West Coast and have something to do with the Pacific fleet. Ron realized he might still get stuck in a fleet air supply squadron, but that post, he observed, would "still be better than spending my whole navy career in Pensacola."

In contrast to the mostly wet and cloudy conditions at NAS Memphis, Ron had been enjoying glorious weather since his return to Pensacola. He even found time to take up golf, but whether he did so because of peer pressure or because he knew Jan's father was an avid golfer is not explained in the letters. Neither is it clear if Jan had the heart to report on Ron's "progress" to her father. Ron wrote: "Joe and I played another 18 [holes] in the afternoon. Got down to a 67 on the back nine and 70 on the front. Am improving somewhat but still not very good."

By now he was counting down his remaining hops. He was anticipating the completion of his training, the crucial news of his deployment, and then the long drive of over a thousand miles from Pensacola to Topeka, with a pit stop in Little Rock. Jan remained very much in his thoughts, and one candid remark left little doubt about his future intentions: "Sure wish you could be waiting for me in our home when I finish flying, that is how much I miss being with you."

In the last week of his training, Ron finally received his orders from the navy and shared the details with Jan: "The good news came on Friday. I am going to a squadron—VF-142 at Miramar in San Diego, Ca. They are flying FJ-3 Fury jets now and supposed to get new F11F Tiger jets soon. Am really happy about the orders."

This was no exaggeration. Originally established at Memphis in November 1949 as VF-791, the squadron was redesignated VF-142 in February 1953 and selected the nickname Fighting Falcons. The squadron went on to distinguish itself both in peace and war.

After making his final Pensacola hops on 18 and 19 June and bringing his training to a successful conclusion, Ron spent two days completing any unfinished business; then he set off on the long journey to Topeka. Taking no chances, he had already rotated the car's tires and changed the oil.

For obvious reasons, Ron's letters temporarily stop during his leave. Jan recalls they spent "two wonderful weeks together in Topeka," although there were no dramatic developments. Harry and Marjorie easily saw that their younger daughter had fallen for a genuinely nice guy. Ron was very polite and very natural, and they never had any doubt that he had not just feelings but also respect for Jan. They could all tell he wasn't simply acting as the nice guy for the parents' benefit: he was the real deal. Jan herself saw Ron as a very honest, happy, and caring individual who was a good listener. He didn't always have to be the one doing the talking. She also valued his sense of humor, which was fairly dry. In a roomful of people all trying to deliver the funniest wisecracks, Ron had the knack of delivering the best one-liner at just the right moment.

He hit the road again on 10 July, heading west toward San Diego, but had personal matters to attend to on the way. He stopped in St. Francis to see family and friends; then he made for Denver, where his father was temporarily based. As Ron later informed Jan, the visit was rather awkward, as by then his father was living with a woman who had once been a good friend of his mother's.

Ron's new life as a naval aviator at NAS Miramar began on Sunday, 14 July, when he checked into the bachelor officer quarters, saddle sore after the long journey. His start was slightly anticlimactic, for the squadron was at sea on the USS *Hornet*. Ron was temporarily assigned to Fleet Air Supply Squadron 12, and a familiar face was there to greet him, as it was the home of his old friend Gary Irish.

Having expected to remain a ground pounder until his own squadron returned, Ron was able to fly with Gary in a Douglas AD-5 Skyraider. As a jet pilot, he was slumming it in the propeller-driven twin-seater, but he had an enjoyable experience and earned his flight pay for the month.

To add to his satisfaction, he heard on the grapevine that VF-142 would be acquiring the brand-new Vought F8U-1 Crusader supersonic jet fighter, which had only been available to the navy since March of that year. He lost no time informing Jan. Then, in one of his more startling literary segues, he continued: "Did I tell you that you are the only girl that means anything to me in my life? In other words I love you."

Their time together in Topeka had certainly been romantic, but Ron had not actually used the "L word." If his letter sent Jan's pulse racing, that was nothing compared to the impact of his letter eight days later that ended with the words: "Well, my one and only wife to be, must get some sleep." While we don't know exactly how the Jan Pollom of 1957 reacted to this not-quite-a-proposal, six decades later she is sure that it was "the first time anything like that had been mentioned."

Far from frightening himself into any backtracking, five days later Ron raised the possibility they had previously discussed of Jan's visiting Miramar: "About your coming out—it would be to get married and stay with me." Well, it still didn't meet the technical definition of a marriage proposal, but Ron didn't leave too much room for doubt. Jan remembers his letter gave her "butterflies." Seemingly, Ron had filed his flight plan and was preparing for his copilot to come on board.

On returning to VF-142, he was still just about the only pilot there and killed time learning about the squadron and what each department head actually did. Finally, on 25 July, VF-142 flew in from the *Hornet* cruise. As the squadron was still a few pilots short, the first thing the top brass did was find the new boy and put him on watch duty. This didn't actually require Ron to stand at a guard post with a rifle, asking, "Who goes there?" But it did mean he had to be available to answer the telephone and to respond if something happened—for instance, if any of the men got into a fight in town. In fact, there *was* some trouble involving a young sailor with a colorful record who was court-martialed. To his surprise, Ron was appointed assistant defense counsel at the hearing, supporting a lieutenant (junior grade) whose lack of legal experience rivaled Ron's. They lost, and their "client" got four months' hard labor, a pay cut, and a demotion.

The squadron had a long wait not only for new pilots but also for the F8U-1. Half promised for September, it was now looking as though the planes would arrive in November. To fill the gap, the pilots trained to fly the North American FJ-3 Fury, which represented another step up in performance for Ron. The aircraft had some odd quirks and had suffered several accidents during early testing, but it had a particular capability achievable in a full-power dive that Ron gleefully reported to Jan: "Finally got my hops in Fury today. . . . Sure is great to be flying again. The old FJ-3 is really a good airplane. Went through the speed of sound this afternoon. The plane has a tendency to roll as you pass through the sound barrier, and then it will roll again as you slow down and pass back through it. Flew most of the time up around 40,000 feet. Sure is nice up there."

On the ground as the new boy, Ron was still getting more than his fair share of standing watch. He was also appointed the squadron's flight officer, education and personnel officer, operations officer, and training officer. He wrote to Jan: "Sure wish more junior officers would get here."

Howard Brown had done well in Pensacola's basic training and had been sent for advanced flight training to NAS Beeville, Texas, where he received his wings of gold in March 1957. He had qualified at the top end of his class and, just as Ron had, was wondering where the navy would send him. After a five-month assignment at the Pensacola Photographic Reconnaissance School, he received orders to join Fighter Squadron 142 at Miramar, where he arrived in September 1957.

Meanwhile, Ron was sounding off to Jan: "4th September: We were supposed to get our first F8U today." He wrote again: "7th September: Squadron still isn't all together yet. Still have three more guys to check in. That is why I had so many jobs the last week."

Howard settled in as one of VF-142's latest ensign acquisitions. Among the pilots he quickly got to know was Ens. Ron Evans from Kansas. "I thought he was a real gentleman, very friendly, kind of a quiet gentleman. He was well-educated, he had a great technical background and he was assigned, I believe, as the electrical department officer. I know Ron was a very smart man—you never wanted to argue with him!"

Ron, Howard, and all their squadron mates had a shared frustration that the promised F8U Crusaders had not yet arrived. At least they were able to keep up their proficiency and flying hours by carrying out combat training flights in the TV-2 and the FJ-3.

On 7 October Ron stretched the definition of "marriage proposal" to its elastic limit. He wrote: "I know that I am going to marry you some day and I will want to live with you for the rest of our lives. In other words I hope you realize that I love you very much and want you to be my wife."

Any trace of lingering doubt about his intentions was swept away a few days later: "Don't know if I will get more than four days leave at Xmas for sure yet. Do you think we could get married and back here in four days?"

This proposal was what Jan had been hoping for, although she hadn't really expected it via letter. Her parents were equally excited by the news, and her mother suggested that Jan reply by telegram. It was a very simple reply: "Yes sir. Till then. Love Jan."

Then the problems arose. The Polloms really should have realized that the base communications officer would be the first one to see the telegram. After that, it went to the executive officer via his secretary and then to just about everybody else on the base. Ron was probably the last person on the base to see Jan's reply. The telegram wasn't exactly in code, and everybody seemed to know Ron was engaged to be married before he did.

In his next letter, he pretended not to understand the telegram, joking that he could never remember what he wrote in his letters. Then he ran up the white flag a few days later: "I don't know why but I had the longest time getting up the nerve to ask you to marry me. I guess I just couldn't get used to the idea of my being married. I feel much better now that I have taken the step of asking you to be my wife."

There followed a great deal of discussion about the when, the where, and the guests, not to mention several increasingly urgent requests about the size of the third finger on Jan's left hand. Ron also went house hunting and eventually rented an apartment in La Jolla for eighty-five dollars a month. Most important, they fixed the date of the wedding—Sunday, 22 December 1957—in Topeka.

Finally, VF-142's new aircraft turned up. Manufactured by Chance Vought Aviation, the F8U Crusader would become known as the "Last of the Gunfighters," being America's last fighter aircraft to use guns (four 20mm cannon) as its primary weapon. Those guns caused an early problem for marine corps test pilot and future Mercury astronaut John Glenn. On a gunnery test in 1956, he fired all four cannon, resulting in nineteen square feet of starboard wing surface detaching and requiring him to nurse the crippled aircraft back to base. It turned out the vibrations from the simultaneous firing of all four guns had sent a shock wave along the wing that was similar to the cracking of a whip. This experience, in a nutshell, was the purpose of test flying, and the fix was relatively simple.

In July 1957, during Project Bullet, Glenn briefly set a transcontinental speed record by flying an F8U-1 at supersonic speeds with the afterburner lit all the way (except during aerial refueling). The Crusader's average speed on the flight was 723 miles per hour. Glenn noted that when the F8U-1 was allocated to squadrons, the pilots welcomed it as "the best fighter they had flown."

That opinion was certainly shared by the pilots of VF-142, but first they had to attend ground school to learn the intricacies, performance characteristics, and quirks of the navy's latest superjet. The Crusader was a very expensive single seater, so the navy needed to be sure would-be Crusader pilots knew the vehicle inside out and (literally) blindfolded. In April 1958 a special training squadron, VF-124, would be established at NAS Moffett Field to train Crusader pilots, but until then the navy employed a more rudimentary technique. VF-142's three most experienced pilots went to Moffett Field, where Crusaders were already in service. Once checked out in the aircraft, they returned to Miramar and proceeded to train the rest of the pilots, including Ron and Howard. New pilots would be accompanied on their first few flights, like fledgling birds; then they would be allowed to fly off on their own.

VF-142 was only the third squadron on the West Coast to acquire the Crusader. But not all of the pilots were able to master the sleek new machine. Several had to leave the squadron.

Ron made his first flights in the Crusader on 16 November 1957. That evening he made an expensive phone call to Jan. She could tell from his first words that he was elated: "I did it! I did it!" Caught up in the excitement, she remembers gushing something about his getting to fly his "favorite bird!" Jan realized that her only rival for Ron's affections had silver wings and a tail.

Ron's love affair with the F-8 Crusader started that day with two escorted flights of around an hour each. Many years later, when his old friend Marv Miller asked what it was like to fly the Crusader, Ron just smiled and said, "Marv, you don't fly the Crusader. You *wear* it."

Howard Brown leaves no doubt how *he* felt about the Crusader. He almost uses proverbial hushed tones to praise it: "It was a *beautiful* airplane. I just *loved* that airplane! It was so advanced: I mean, as a new aviator, it was hard for me to describe the advanced nature of it, because everything in the fleet would have been advanced to me. . . . It had such a powerful engine, such a clean aerodynamic fuselage, that you had to be very conscious all the time, especially on landings, of your airspeed versus the weight of the airplane. . . . It was very sensitive when you got into combat maneuvering. If you got to pull high g's, putting a lot of pressure on the wings, if you weren't careful, next thing that would happen to you is you would find yourself in a spin."

He wouldn't volunteer it, but he accepts the proposition that the Crusader was a high-performance aircraft that needed an equally high-performance pilot to fly it properly. It was also the best aircraft he ever flew.

Ron was granted longer wedding leave than he had expected—15 to 30 December 1957. Shortly before his departure, he made an expensive visit to a jeweler. He still planned on the long drive back to Topeka but now joked in his last bachelor letter that the cost of the ring meant he could no longer afford the gas. Jan would just have to carry on without him.

On Sunday, 15 December, Jan's bridal shower was under way at her friend Jody's house when the doorbell rang. Jody answered, and to everyone's surprise, in walked Ron, wearing his flying gear. He had managed to catch a service flight to Forbes AFB and had called Harry Pollom, who said nothing to his daughter and drove out to collect his future son-in-law. Ron reached down to one of the zippered pockets of his flight suit and with a flourish produced Jan's beautiful diamond engagement ring. The groom had made it after all.

5. Fighting Falcon

Nothing great was ever achieved without enthusiasm.

—Ralph Waldo Emerson (1803–82)

Ron and Jan were married at Westminster Presbyterian Church, Topeka, three days before Christmas 1957. Despite the short notice, they had a large gathering of friends and family. Several of Ron's old buddies from ROTC and Sigma Nu were there, including Dick Ross and Gary Irish, who served as Ron's best man. Jan's friend Sue was her bridesmaid, and Jan's brother-in-law, Jim, was an usher. It was a happy day for Harry and Marjorie Pollom and for Marie Evans, accompanied by her son Jay, who was now towering above everyone else. No doubt to Marie's relief, Jimmy had decided to stay away, avoiding any awkwardness.

The wedding wasn't lavish, but everything was arranged just the way Jan wanted it. Ron wouldn't have had it any other way. Jan wore a white satin dress, which a seamstress she had known in college made for her, and had a matching white fur muff that proved useful on the chilly winter's day. Ron had wondered whether to wear a suit, but he finally opted for his navy uniform, now bearing the insignia of a lieutenant junior grade (JG) as well as his prized gold wings. Several of Jan's friends designed and made the wedding cake and guest book.

Ron was due back on flying duty on 31 December, so the newlyweds had limited opportunities for a honeymoon. After the ceremony and the celebrations, they spent two days in Kansas City before returning to Topeka to share Christmas Eve with Harry, Marjorie, Marie, and Jay. Early on Christmas Day, they set off on the long drive to California, calling on relatives who hadn't been able to attend the wedding. Farther along the road, they resumed their honeymoon for two days in Las Vegas, where Jan remembers getting dizzy watching the roulette wheel.

Mr. and Mrs. Evans arrived at their first home, the tiny rented apartment in La Jolla, in time for Ron to complete his 1957 tally with two flights. Jan had no doubt that in marrying a naval aviator she was also marrying his job. They had discussed his intentions, and she knew flying was Ron's career. He loved being a naval aviator, and he was in it for the long haul. Jan was proud to support him through thick and thin.

Although America was at peace, the never-ending Cold War rumbled on. Jan realized that if international relations soured further, a future "hot" war (maybe even hotter than Korea) was in the cards. And that would pitch her husband and his squadron into combat. It was, of course, what the men trained for every day (if not "for" then certainly "in case of"). Theodore Roosevelt had once said, "Speak softly and carry a big stick." Ron was now part of President Dwight Eisenhower's big stick.

Jan settled happily into her new life, aware that Ron had really landed on his feet with VF-142. The team spirit and camaraderie made both of them feel at home. The glue that binds a squadron together has many ingredients, but a good commanding officer, or "skipper," is certainly important. To Jan, Cdr. George Whisler was "a heck of a neat skipper," and Betty Whisler was the perfect "skipper's wife." Wives had no formal role in the U.S. Navy, but Jan quickly learned that a good skipper's wife provided friendship, guidance, advice, and calming reassurance for the young wives of junior aviators who were no longer wet behind the ears but perhaps still a little damp.

Five days a week and sometimes at night, the squadron's pilots honed their flying skills, but on Friday afternoons, all of the pilots and senior nonflying staff gathered for happy hour. In fact, attendance was compulsory. The men all gathered round a table and began rolling dice until someone lost. That unfortunate individual had to buy drinks for everyone else. Then the dice were rolled again, with the men passing them around the table. Even at ten cents a drink, a run of bad luck could be expensive. Depending on how the dice fell, victory could be intoxicating and cheap. It wasn't unusual to see four or five martinis lined up in front of some bleary-eyed pilot. The happy hour tradition was intended to cement camaraderie, and to that end, one more rule rigidly applied: no wives or girlfriends could join the fun until six o'clock. Woe betide any young woman entering the hallowed ground even ten seconds early; her man would have to buy another round for everybody.

The women usually gave it an extra ten minutes, just to be sure. Then couples could get together and head off to dinner or parties. Jan fondly remembers this early part of Ron's career as "really fun times."

Howard and Bev Brown lived several miles farther from the base, and Howard quite often gave Ron a lift in the mornings. The two young officers took this opportunity to get to know each other, and Howard always saw Ron as "one of the good guys." But here was the paradox of Ron Evans: while he would pitch in at all the squadron social events and was certainly one of the guys, Howard knew him to be "a very quiet fellow, never super-exuberant, very professional." Even in the midst of the squadron's merriment while others might have been shouting and bawling to be heard, Howard couldn't ever remember Ron shouting. About anything.

Back in September 1957 in one of his letters to Jan, Ron had bemoaned the fact that he was still a "lowly ensign." By 1958 he and Howard had progressed a rung up the promotional ladder, but it wasn't a difficult step. Promotion to lieutenant (JG) had only two requirements: keep your nose clean, and don't get killed. Eventually, the navy's Bureau of Personnel (BUPERS) would write a letter saying, "Congratulations. As from such-and-such date you are a Lt. (JG)."

In another letter, written shortly after he began flying the Crusader, Ron told Jan that he had to go to NAS North Island, San Diego, to be fitted for a high-altitude pressure suit. Peering unwittingly into his own future, he joked that he would look like a "space man" and would often later refer to his "space suit." He wasn't exaggerating: the pressure suits that the Crusader pilots of VF-142 wore were essentially the same suits that the Mercury astronauts would wear into space. The only obvious difference was that the navy's suits were green, not silver.

Evaluating the Goodrich Mark IV suits was one of the squadron's main tasks in 1958. On one air defense intercept exercise, the Crusader pilots climbed to fifty-five thousand feet, very close to the aircraft's operational ceiling. Although the cockpits were pressurized, the new suits provided an extra measure of safety. The aircraft was difficult to handle in the thin air, and only by firing afterburners were the pilots able to maintain altitude. The U.S. Air Force mounted a simulated attack with F-104 Starfighter jets, but when the formation of Crusaders banked to the left, the Starfighters, which were also close to their ceiling, were unable to match the rate of turn and lost their intended prey.

Howard Brown vividly recalls flying his Crusader at what seemed to be the edge of space. The sky above was a deep blue and the bright band of the horizon almost three hundred miles distant. Even while admiring the awesome vista, he was acutely aware that in the thin air excessive stick movements could induce an accelerated stall, a mishap that befell a Crusader pilot from another squadron who was flying a fraction too slowly when turning at fifty-five thousand feet. His aircraft dropped into a spin, and he lost twenty-five thousand feet before recovering.

The Crusader was built to operate off aircraft carriers, and inevitably the moment of truth came for Ron and the other junior pilots who had not yet landed a jet on a carrier deck. Throughout February and into March, the pilots carried out almost daily field carrier landing practice with the ever-present landing signal officer bawling into his microphone when the approach was even slightly off target. Then came the real test, a series of landings on the USS *Midway* as it steamed through the Pacific swell.

In mid-March Ron launched and made four landings followed by steam-catapult launches. The Crusader was not an easy aircraft to land safely on a carrier, and doing it for the first time was downright difficult. The captain did his best to help by steering the ship into the wind, but all too frequently the ocean swell was rolling in from a different direction, forcing the approaching pilot to make constant adjustments to his lineup.

Lt. Gene Geronime, having flown the FJ-3 Fury on the USS *Hornet* cruise the previous year, was one of the more experienced aviators in VF-142. He had been among the first pilots to convert to the Crusader at Moffett Field, but even he had concerns about his first Crusader landing on *Midway*. The pilot's goal was to snag one of the four steel arresting cables stretched across the landing area; usually he tried to put the hook down between the second and third cables, catching the third. Controlling the aircraft's descent was absolutely crucial, but flying propeller planes or jets with straight wings really wasn't a good way to prepare for that first Crusader carrier landing. In those less sophisticated, more conventional aircraft, he could control his final approach with the throttle and by adjusting the position of the nose—too high, drop the nose a little; too low, pull the nose up a little—but not with the Crusader. The aerodynamic properties of the F-8 were such that if a pilot was landing short and pulled the nose up, he got very little extra lift but a lot more drag. The net result was a plunging aircraft, and if it struck the approach ramp, the pilot would have a very bad day. That is what happened to Lieutenant (JG) Lindsay, who had a ramp strike on 12 March and was forced to eject into the ocean. Aircraft No. 143744, which Ron had successfully landed the previous day, became the squadron's first Crusader loss. In fact, so many Crusader accidents, usually involving inexperienced pilots, occurred on U.S. Navy carriers that the Last of the Gunfighters acquired another altogether more sinister nickname, "the Ensign Eliminator."

For the pilot making his final approach to the carrier deck, the way to control the rate of descent was with the throttle, but the powerful J-57 turbojet engine, which could accelerate the aircraft through the sound barrier in level flight, had an important drawback: it was slow to wind up and unleash the thrust to flatten an over-steep approach angle. Anticipating the need to adjust the throttle was crucial. On top of all that, the pilot had to land on the main gear and not on the rather flimsy nosewheel. It was perhaps not surprising that a few of Ron's early landing attempts produced "bolters," meaning he missed the arresting cables and had no option but to accelerate off the deck on maximum thrust, go around, and try again.

When not sharpening their carrier-landing skills, the squadron concentrated on gunnery practice, attacking targets on the ground and drones trailing behind towplanes. At that time, the Fleet Air Gunnery Unit (Pacific) was based at Naval Auxiliary Air Station El Centro in Southern California, a desert base with hot days, cold nights, and scorpions in your boots if you left them on the floor.

In October 1957 Ron learned that the squadron was going to deploy on the navy's new super ship, the *Forrestal*-class aircraft carrier USS *Ranger*. Commissioned on 10 August 1957, it was 1,063 feet long and, when launched, was the largest and most powerful warship ever built. For such an enormous brute, *Ranger* could outpace many smaller ships, with a top speed exceeding thirty knots. It was also the first of the modern carriers in the U.S. Navy, having been designed and built with an angled flight deck for greater efficiency. But such enormity had some disadvantages. Constructed and launched in Newport News, Virginia, *Ranger* unfortunately could not fit through the Panama Canal and therefore had to transfer to its new base at NAS Alameda, California, by going the old-fashioned way.

The men of VF-142 were proud and delighted they would be in the first Crusader squadron to operate off a *Forrestal*-class carrier, but the path to Ron's first ship deployment was not a simple one. *Ranger* first had to conduct a shakedown cruise south to the Caribbean. Squadron VF-142 was then supposed to fly to the East Coast and, beginning in June 1958, join the carrier for a two-month passage around Cape Horn at the tip of South America. This would have been a great honor except the squadron had no carrier-qualified Crusaders. As Howard Brown recalled: "Chance Vought was building F8Us, and as soon as they had built a bunch of them, they would send them out to the navy, but they would come out without some mandatory things that had to be on the airplane to make it carrier-qualified. The first group of fourteen airplanes we had were not qualified to land on a carrier. And then we got another group of fourteen airplanes, and they still weren't qualified for carrier work. We went through three groups of fourteen airplanes, brand new from the factory, before we got carrier-qualified airplanes."

By then it was too late for the Cape Horn adventure, so VF-142 had to be replaced. The men were frustrated and disappointed to miss the experience, but at least they would deploy on *Ranger* for its first cruise of the western Pacific (known as a WestPac cruise). While the ship was steaming around the Horn, the pilots resumed their field carrier landing practice training. This passed without incident except on one occasion Ron missed his footing exiting his aircraft and fell onto his gradually balding head, fortunately injuring only his pride.

On 20 August *Ranger* took up station at Alameda in San Francisco Bay. In late September, finally in possession of correctly configured Crusaders, the pilots took turns making practice landings on the deck of the carrier. Ron made nine launches and landings without any bolters.

The next few months were hectic for Ron and Jan. In preparation for the cruise, squadron families had to give up their rental accommodations, so Jan said goodbye to the apartment in La Jolla. She and Ron traveled with the other families to Alameda, where they were allocated temporary housing in navy-owned Quonset huts. Then the cruise was delayed, and the families had to move back to San Diego. As they now lacked accommodations, the navy arranged to reopen part of the Cyane Naval Housing Project at Pacific Beach, San Diego. Condemned and previously scheduled for demolition, the apartments became home to some of America's finest pilots and their families.

Shortly before the cruise, Ron and Jan joined two other squadron couples on a trip to San Francisco. Jan had intended to stay with her parents while Ron was away, but her friends Norma Bellinger and Dusty Bailey were going to fly to some of the Pacific ports of call to see their husbands. Not to be outdone, Ron surprised Jan by presenting her with an airline ticket to Hawaii, the first port of call. But she had already sent all her possessions, including her clothes, to her parents' home in Kansas. When the USS *Ranger* weighed anchor on 3 January 1959 with Lieutenant (JG) Evans on board, Mrs. Ron Evans was "headed to Hawaii with a wool jumper, a wool sweater and a wool suit."

At sea during the early weeks of the cruise, the six squadrons of Carrier Air Group 14 participated in *Ranger's* operational readiness inspection, testing the ability of the ship's complement to enter combat if called upon. The ship passed with flying colors, and VF-142, under its new skipper Lt. Cdr. L. R. "Tom" Mix, received the highest grade. When the ship's 1959 WestPac cruise book was produced, the entry for VF-142 noted: "This squadron is for hotshots."

Meanwhile, the wives were accommodated at Fort DeRussy, an army base near Waikiki beach, Honolulu. Jan and four others bought an ancient, ramshackle Packard automobile with retread tires to go sightseeing and to visit their husbands at Pearl Harbor when *Ranger* was in port. The dealer saw them coming and charged them twenty-seven dollars.

The holiday ought to have been carefree and relaxing for Jan, but something wasn't quite right. She felt sick most mornings. Eventually the others drove her to Tripler Army Hospital for a checkup. A formidable but sweet marine nurse finally announced, "Honey, you're pregnant." Jan was also anemic, so the others drove her back and forth to Tripler for five daily iron shots.

Lieutenant (JG) Evans left Hawaii for the Far East knowing he was a member of the ship's top squadron and, even more important, that he was to become a father. Meanwhile, Jan gathered her few possessions and headed to Kansas to stay with her parents for the duration of the cruise. While the ship was at sea, most of the officers and men communicated weekly by letter, but Ron had always been fascinated by electrical gadgets and had brought a reel-to-reel tape recorder with him. His weekly letter to Jan contained a taped message, and she had a similar machine at home to record her replies.

The pilots of VF-142 continued to test their "space suits" during the cruise and were the first to make a carrier deployment equipped with full pressure suits.

The pilots had an air-conditioned ready room for suiting up where the temperature and humidity were maintained at comfortable levels, and they were featured in a navy magazine article that included a picture of Ron, Howard Brown, and others preparing for a mission. With his helmet on, Howard was described as resembling a "man from Mars."

The *Ranger* cruise was a peacetime operation, but if you are going to carry a big stick, then you have to practice swinging it. Just in case. As retired captain Gene Geronime later put it: "Aviators have to fly, or they lose their capabilities quickly. It's like racing drivers: you don't just drive in the race, you practice!"

During the course of the WestPac cruise, as part of a Seventh Fleet operation, *Ranger* launched seven thousand sorties in all. Bearing in mind the primary task of the pilots was to protect the fleet and their own ship, Ron and his colleagues practiced air intercepts of "enemy" aircraft, escorted ground-attack aircraft, and fired their 20mm guns at towed targets. Every hop was fully briefed and had a specific purpose. The pilots also practiced their instrument flying. This time the need to be able to find their way back to the ship was not a theoretical exercise. As big as it looked from the quayside, *Ranger* seemed to shrink to a very small refuge when viewed from high above and surrounded by a vast ocean.

Pilots on their first carrier deployment were known as "nuggets" to the more experienced men, but Ron was building a good reputation. According to Gene Geronime, "He was a very honest, bright individual, not a 'party guy' at all, but very sociable, very professional, and always had a smile on his face. He was just a great guy to have in the squadron. I don't think he ever changed. I knew quite a few of the astronauts, a lot of them my year group, people I knew from either the Naval Academy or aviation associations. A few of them were 'hot shots,' but that was never Ron. He was a very dedicated guy who went out, did his job extremely well, and was a team player, the sort of guy you like to have in the squadron. He was the kind of guy you could rely on."

This may explain why Ron was selected to take part in a "fly the flag" good-will exercise called Operation Flying Brothers held at Clark Air Base in the Philippines. Pilots from South Korea, Taiwan, the Philippines, and Thailand joined the Americans in demonstrating their flying, bombing, and gunnery skills. Ron was *Ranger*'s junior man, flying with Lt. Cdr. Charles Deasey and Lt. Richard Bellinger. The exercise was intended to send a message to other countries in the region that did not harbor brotherly thoughts. Relations between communist China and Taiwan had reached boiling point the previous September with Chinese MiG fighters and Taiwan's U.S.-supplied F-86 fighters clashing over the Taiwan Straits. Taiwanese pilots firing American Sidewinder missiles downed a number of MiGs.

President Carlos Garcia of the Philippines visited the ship and was treated to his own personal air display, which included several of VF-142's Crusaders. Unfortunately, Lt. R. J. "Dick" Peterson, suffered engine failure and had to eject in full view of the president. The pilot's relief on being fished safely out of the water may have been tempered by what happened next. When brought on board and still dripping wet, he was led to the bridge, where he was introduced to the president, who could have been forgiven for thinking that the U.S. Navy certainly went the extra mile to put on an exciting show.

The pilots of VF-142 were flying the F8U-1, the first production version of the Crusader. Superb performer though it was, the F8U-1 was essentially a day fighter, neither designed nor properly equipped to fly at night. Even for a second-cruise pilot like Gene Geronime, flying the F8U-1 at night was challenging. The radar was not suitable for night flying. The front canopy, made of bullet-proof glass, was long and sloping, and it easily fogged over, reducing visibility. The cockpit's warning lights could not be dimmed at night. The last thing a pilot needed on a night flight was the instrument panel lighting up like a Christmas tree. On later deployments, Geronime learned to stick tape over some of the lights. No one wanted to be dazzled by a low-fuel warning on a final approach.

Nevertheless, the plan was for all pilots of the squadron to make two practice night landings during the cruise. Not every pilot got the opportunity, but Ron completed both of his night launches and landings without incident, all within an hour.

Meanwhile, 1959 was a big year for NASA and the burgeoning space race. On 9 April the world was introduced to America's seven Mercury astronauts. They made headlines all over the United States, and the news even reached U.S. Navy ships patrolling the Pacific. Whether the supersonic fighter pilots thought of the astronauts as "spam in a can" or as lucky dogs, they most certainly knew about them.

Even before Alan Shepard, John Glenn, and the rest were ready to fly into space, NASA planners were thinking ahead and considering the need for more astronauts in the future. At some point during the cruise, Gene Geronime received a letter—not from NASA but from the navy—asking if he was interested in having his name forwarded to NASA as a possible future astronaut. Gene discussed it with Ron and showed him the letter but told Ron he wasn't interested. Anyway, Gene thought he would be too old if or when more astronauts were needed. But Ron was definitely interested. He even asked Gene if he could keep the letter. This appears to be the first time Ron is known to have expressed any interest in extending his flying beyond Earth's atmosphere.

But he doesn't seem to have mentioned the idea to anyone else, not even to his wife. Jan cannot remember Ron's showing any interest that early in becoming an astronaut. The explanation isn't too hard to fathom. If Ron didn't already know from Gene Geronime's letter, he could easily have found out that in those days astronaut candidates needed to be test pilots. As it was too early for Ron to pursue the idea, there was no need to discuss it with anyone, least of all Jan, who was shortly due to give birth.

Ranger's cruise ended on 27 July 1959, and the pilots of VF-142 flew their Crusaders back to Miramar. Meanwhile, in Kansas, Jan was being pampered by her parents. She had lived in Topeka all her single life, but around the time of her wedding, Frank Gibbs had confided in Harry Pollom that one of his clothing stores was having problems. Would Harry be willing to manage that store in Salina? It would mean moving house, but Harry and Marjorie talked it over and agreed to the move. Harry went on ahead, while Jan and her mother stayed in Topeka until the old house was sold. Then they followed Harry to Salina. In years to come, newspapers and magazines profiling the crew of *Apollo 17* would describe Ron Evans as "a modish dresser" who embraced the fashions of the early 1970s, but the reporters probably didn't realize Ron's father-in-law still managed a men's clothing store.

As his squadron settled into life ashore, Ron took a period of leave and made it to Salina on 18 August. Three days later, Jan gave birth to their daughter, Jaime Dayle Evans. Her name was a nod to her uncles James and Dale. Fighter pilot Ron Evans had been promoted to the rank of family man.

6. Top Guns of *Oriskany*

And let's get one thing straight. There's a big difference
between a pilot and an aviator. One is a technician;
the other is an artist in love with flight.

—Elrey Jeppesen (1907–96)

Big changes occurred in Ron's life in 1959 other than the birth of his daughter. Almost as soon as the squadron returned to Miramar, most of the officers received orders to proceed to new assignments. Howard Brown would fly Crusaders in photo reconnaissance. Gene Geronime was off to the Naval Postgraduate School in Monterey, a posting that Ron would regularly request. Even the skipper was moving on.

Only four pilots from the *Ranger* cruise were staying with VF-142: Lt. Max Bailey; Lt. Lowell R. "Moose" Myers; Lt. Cdr. Dick Peterson, fully recovered from his unplanned swim in the Pacific; and Ron, now with one cruise under his belt and no longer a nugget.

A whole batch of new pilots arrived including a dozen fresh from Crusader training. The outgoing and incoming pilots passed like the proverbial ships in the night. Among the new men were Lt. (JG) John Holm and Ens. Jimmie Taylor. The new skipper was Cdr. Ted Dankworth.

Even the aircraft changed. Most records show that VF-142 took delivery of the improved F8U-2 Crusader in September 1959, but Ron's first flight was on 31 August. The uprated Crusader had an improved engine delivering higher thrust (16,900 pounds on afterburner) that allowed a maximum speed of just over 1,100 miles per hour. The plane had a few more bells and whistles, but for all its upgrades, the new Crusader remained essentially a day fighter.

Domestically Ron had time to settle into the lifestyle of a husband and father, although Jaime's first home was the less-than-palatial Cyane housing project. Jan made friends with several other navy wives living nearby, although one conversation about future assignments could have gone better. Her new friends' husbands had graduated from the U.S. Naval Academy. When Jan mentioned that Ron hoped to go to the Naval Postgraduate School to obtain a master's degree, the other women hinted, ever so politely, that because Ron wasn't an academy graduate, he would never get assigned to Monterey. Jan kept her cool but thought, "We'll see . . ."

Like a broom with a new head and a new handle, Squadron VF-142 swept on. Any disruption from so many personnel changes was minimal, and the old sense of camaraderie continued. For newcomer John Holm, the squadron was a very easygoing, sociable outfit. Everybody got to know each other, and the Friday afternoon happy hours and squadron parties continued under the guidance of the "very active and gregarious" new skipper Ted Dankworth and his wife, June.

But it wasn't all fun and games. September was one long training exercise for the Falcons. Then the squadron was assigned to a WestPac cruise that would be conducted in 1960.

The USS *Oriskany* was a far cry from the shiny new super carrier on which Ron had cut his naval aviator's teeth. More than fifty yards shorter than *Ranger* and not much more than half of its fully laden displacement, the "O-Boat," as the ship was affectionately known to Ron's squadron, was one of a class of carriers built to fight a Pacific war that was won before the ship was even launched. But a new era of jet fighters was dawning, and the U.S. Navy either needed many expensive, new purpose-built carriers to accommodate jets and nuclear bombers or would have to make do with what it had. Between 1947 and 1955, fifteen older carriers were extensively modified under the Ship Characteristics Board (SCB) Program 27, with the unfinished *Oriskany* as the prototype. It was first commissioned in September 1950, in time for action in the Korean War.

The rest of the mothballed carriers were modified under Program SCB-27A or SCB-27C. The latter provided more powerful, British-designed steam catapults. While still in the shipyards, three of the SCB-27Cs were fitted with a further British innovation, the angled flight deck, which dramatically improved operational safety. These changes in due course were retrofitted to all but one of the fifteen carriers. *Oriskany's* turn came in 1957, and it was finally recommissioned in March 1959.

Those carriers modified under SCB-27C became known as "27 Charlies." Strictly speaking, *Oriskany* was not a 27 Charlie, but the designation stuck and is still used by some of the naval aviators who served aboard. Getting the terminology correct was the last thing on the minds of Crusader pilots who were trying to land on those smaller carriers. If they caught the fourth arresting cable rather than the third, they would still come to an abrupt halt but with a dramatic view of the ocean as the aircraft's nose rocked over the edge of the flight deck.

Ron had his first introduction to *Oriskany* on 7 October 1959. Over the next few weeks, he completed eleven landings without a mishap or even a bolter.

In early December Ron was included in VF-142's team for the fourth annual Naval Air Weapons Meet at Marine Corps Auxiliary Air Station Yuma, Arizona, hosted by the Fleet Air Gunnery Unit. The competition drew the top squadrons from the Atlantic and Pacific coasts to test their combat readiness against friendly foes. The competition was known as Top Gun but should not be confused with the U.S. Navy Strike Fighter Weapons School established in March 1969 at NAS Miramar and later popularized by the movie of the same name.

The squadron's team was led from the front by Ted Dankworth and included not only a future Apollo astronaut but also two future admirals, Moose Myers and Jimmie Taylor. Jimmie vividly remembers representing the "world-famous Fighting Falcons," although some of the other squadrons got a kick out of calling them the "Fluttering Flamingos." The friendly rivals dropped bombs, launched Sidewinders, and fired streams of 20mm bullets at a large banner at the end of a cable that was long enough to reassure the towplane's pilot. The Falcons didn't win Top Gun, but they acquitted themselves well against tough competition.

In early 1960 the squadron participated in work-ups, routine training exercises between one cruise and the next. April saw more flying visits to *Oriskany* in preparation for the cruise. Ron then said his goodbyes to Jan and Jaime for seven months before the carrier departed on 14 May.

Lt. John Allen, one of the more senior pilots on the cruise, was not actually attached to a squadron. He was on the staff of Carrier Air Group 14, serving as a landing signal officer. He had been on the *Ranger* cruise, learning the ropes as a trainee LSO, but on board *Oriskany*, he was not only controlling landings but also flying aircraft with VF-142. He notes, "The Crusader was the most challenging airplane we had to operate on board the carrier, so I wanted to be flying the airplane to be sure I was up-to-date on how it operated."

John got to know Ron Evans very well on the cruise. He recalls that "Ron was a very, very competent aviator. He was very calm, and very gentlemanly. For instance, I never once heard Ron use profanity, which was a little unusual for a naval aviator perhaps!"

The days of an LSO controlling landings just with handheld paddles were in the past. By the late 1950s, every U.S. aircraft carrier had an optical landing system, yet another British invention, that was known to pilots as the "meatball."

In essence, the system involved four bright yellow lights mounted on the carrier deck that were aimed at an optically precise concave glass mirror; there, they converged and were visible to an approaching pilot as a single, bright orange light, or the meatball. On *Oriskany*, the mirror was mounted on a gyro-stabilized platform to allow for the ship's movement. A horizontal line of green lights showed the central position. The approaching pilot would descend to five hundred feet and, at five miles out, would lower his landing gear, flaps, and tailhook. During his final approach, while constantly adjusting for lateral motion and wishing for X-ray vision to see through the black smoke from the ship's funnel, the pilot had all of twenty seconds to line up on the meatball. If the orange light was above the green line, he was too high. Below the line, he was too low and risked a ramp strike. If the light was in the middle, he should in theory catch the third arresting cable and feel the brutal deceleration of the perfect landing, going from 145 miles per hour to zero in half a second.

As the LSO, John Allen still had ultimate authority, even with the meatball in operation. Via his telephone-like radio handset, he gave any necessary additional information or guidance to the pilot. With another handset—called the "pickle"—he could illuminate a set of red wave-off lights that communicated a mandatory order to the pilot to abort his approach and go around again, perhaps because of an obstruction on the flight deck not visible to the pilot.

The Crusader had an innovative design that allowed the entire wing structure to pivot seven degrees out of the fuselage on takeoff and landing to allow increased lift without reducing forward visibility. This was a further pre-landing adjustment the Crusader pilot had to make on his approach to the carrier. If the wing failed to rise, perhaps through combat damage, a safe landing would have been problematic at best. Fortunately, this was very rare. But not unknown.

In choosing to fly the Crusader, John had no doubt that he was testing himself with the best. "It was a very high-performance airplane, and as a result of that, it took a great deal of practice and skill to operate it safely on the carrier. So I guess that's one reason we did as much practicing as we did! It was a challenge, it was really a challenge! The accident rate for the Crusader throughout the fleet was kind of high, but it was really a treat to fly, because it was the first airplane we had that could go a thousand miles per hour in level flight. It was quite an experience to have that much power at your fingertips. We had a club called the 'Thousand-Miles-per-Hour Club,' and we got pins and all that sort of thing. It didn't make a heck of a lot of difference, but we kind of pumped ourselves up by doing that!"

John chose his adopted squadron wisely. On 30 June VF-142 became the first F-8 squadron to complete a fiscal year without an accident or the loss of an aircraft. It was still accident free by the end of the cruise, and the squadron was awarded the U.S. Naval Air Forces, Pacific Fleet (AIRPAC) Battle Efficiency Award as the outstanding day fighter squadron.

As one of the nuggets on the cruise, Jimmie Taylor paid close attention to the flying skills of the more seasoned officers. He had learned to fly and had even become a flying instructor in light aircraft before joining the navy and receiving his wings in September 1958. Although he was already a gifted and confident aviator on *Oriskany*, he knew he was surrounded by those who had skills in abundance. But Jimmie observed that if there was such a thing as a "typical fighter pilot," Ron Evans seemed to depart from the template. "Well, Ron was one of those less-vocal F-8 pilots. Most of us had a little swagger in the way we talked, the way we acted, and the way we flew! Ron was not as demonstrative as a lot of us. He was an outstanding pilot, obviously, but he didn't go through a lot of the gyrations, a lot of the show-off-type things that most of us did. He was quieter than most, let's put it that way. There's a movie called *The Quiet Man* with John Wayne. Well, I'd call Ron 'The Quiet Man.' It didn't mean he couldn't have a lot of fun when he got wound up, but he was on the quiet side compared with the loud, boisterous braggarts that existed in the F-8 community!"

Jimmie became attuned to the squadron's regular, almost metronomic flying schedule. As a day fighter squadron, VF-142 participated in a sequence of sorties on a twelve-hour cycle. Beginning at 6:00 a.m., several aircraft would launch, and an hour and a half later, those aircraft would return, and another batch would launch. This cycle continued until 6:00 p.m. Shorter, half-hour test hops could be fitted in during those ninety-minute cycles.

Jimmie soon realized that his squadron's day-only status did not go unnoticed. "Well, since the operating room aboard a small carrier is so critical, we used to get razzed about being 'day only' fighter pilots. We weren't trained to fly at night, and we were the only squadron on the ship that didn't fly at night. So we were called the 'gentlemen fighter pilots' because when everybody else was up flying at night, we were down in the wardroom watching movies, or we had launched and gone ashore for the night to give them operating room aboard the carrier."

Even on a ship as large as an aircraft carrier, a junior officer (JO) did not have the luxury of private accommodations. Lt. (JG) Ron Evans shared the JO bunkroom that lay under one of the ship's steam catapults. It was noisy, hot, and smoky. Jimmie Taylor recalls that Ron was "an incessant smoker." Jimmie endured it for a while before finding less noisy but even hotter accommodations.

Lt. (JG) John Holm had flown with Ron at NAS Miramar shortly after joining the squadron. He, Ron, and Moose Myers were formed into one of the several three-man training groups on *Oriskany*. They flew against each other, challenged each other, and tested each other in all of the maneuvers they had to master to become the accomplished naval aviators they all wanted to be. John accepted the heat, the noise, and the smoke, and joined Ron in the JO bunk room.

John says, "I got to know Ron very well. We flew a fair amount together in the squadron, and everybody got to know each other. He was a great guy. He wasn't a really outgoing person in the sense that he was a leader of a crowd or anything, but he was a good buddy, a good friend, and very smart."

As on the *Ranger* cruise, Ron communicated with Jan by exchanging taped messages, but this time he also had a baby daughter whose first birthday was in August, halfway through the cruise. Jaime was a fast learner, and Ron would include little messages to her. Jan had a framed picture of Ron in his flying gear and pointed to the picture as she played Ron's messages to the child. Eventually Jaime added her own little contributions to Jan's taped replies. In spite of the noise from the steam catapult, John Holm remembers hearing "a *lot* of baby talk" on Ron's tapes.

The two young officers were both in line for promotion. Such was the reputation of VF-142 that simply being in the squadron was a good starting point for a young officer eager for advancement. In those days, all commissioned officers were expected to continue their education, and BUPERS had a correspondence course program. Officers had to sign up, and they then received a long list of subjects, some mandatory and some optional. They had no computers, no internet, no search engines. Everything was handwritten and laborious. When Ron and John ordered a correspondence course, four weeks later each received a fifteen-pound bundle in the mail containing books, test papers, answer sheets, and even return envelopes for submitting the fruits of their labors to BUPERS. The target was to complete successfully around twenty courses in general knowledge and navy matters, and after that (and always assuming you were otherwise deemed worthy), you got notification of promotion to full lieutenant.

Jimmie Taylor may have had time to watch movies at night during the cruise, but Ron and John spent a great deal of off-duty time at their desks in the noise and the heat, completing their correspondence courses. Ron was several months ahead in the pecking order and had completed a lot of his tests ashore. He duly received his promotion in June 1960. John persevered and received his in October.

In the future, promotions would be based both on annual fitness reports submitted by their commanding officers and on some compulsory tests required of everybody. According to Lt. John Allen, though, the most important considerations would always be to "keep our noses clean, don't screw up too badly, and be competent in doing our job."

Like the *Ranger* cruise before it, *Oriskany*'s journey around the western Pacific was conducted in peacetime and was essentially an exercise in flying the flag and showing support for friendly nations. One odd difference is a significant lack of information available about the 1960 cruise. Apart from the bare dates of departure and return, a casual researcher will struggle to find any details about what the crew of the ship and its aviators actually did while at sea. Even the cruise book, usually a good source of information, reveals little beyond multiple photographs of the ports visited. It is more of a travelogue than a history.

One possible reason was the ongoing situation between China and Taiwan. Although hostilities had died down, the two sides were still regularly lobbing shells containing propaganda leaflets at each other. Another reason might have been the unpublicized nuclear weapons aboard *Oriskany*, although declassified U.S. Navy documents show that this deployment was basically the norm. Precise details for 1960 are not available, but a realistic estimate is that U.S. ships and submarines around the world carried 1,500 warheads. *Oriskany* certainly had its share. The issue of nuclear attack training was a particularly topical and sensitive one. Only two weeks before the start of the cruise, the Soviet Union had shot down a high-flying American U-2 spy plane over the city of Sverdlovsk (now Yekaterinburg) and had captured the pilot, Francis Gary Powers.

As a reminder that this peacetime cruise was being conducted in the shadow of the Cold War, one of the ship's attack squadrons, flying AD-6 Skyraiders, regularly launched aircraft at 6:00 a.m. that remained airborne for ten hours, practicing low-level nuclear deliveries. Everyone on board, from the captain to the lowliest seaman, was aware of the ship's nuclear capability. In years to come, U.S. warships arriving at Japanese ports would be met by strident antinuclear protests, but in 1960 as it toured the Pacific, *Oriskany* was warmly welcomed, never more so than in the bars of Japan.

Although Ron was not noted for being a "party animal," he was happy to honor an old navy custom, the "wetting-down" party. Traditionally, the newly promoted officer arranged and paid for a big party and invited his fellow officers and other guests to toast his move up the chain of command. As John Holm was also being promoted, they agreed to host a joint wetting-down party. *Oriskany* had paid a visit to Yokosuka, Japan, on the outward part of the cruise and would be calling there again for a week on the return journey. The two officers arranged to hold the party in the city's swanky New Kinko Hotel and ordered hot food and drinks for forty to fifty guests.

The ship's 1960 cruise book advised that Japan was "still essentially a man's country" with a seemingly endless supply of young women "assigned" to each bar visitor to pour his beer for him and to entertain him with conversation and "drinking games." The cruise book reassuringly added that "the girls are not necessarily 'loose women' (some of them are secretaries or students needing a part-time job) but are usually just friendly to men in general, without reserve or inhibitions."

Whether you thought of all the girls in all the gin joints in all the ports as occupational hazards of navy life or as perks of the job might depend on whether you were a young navy wife back home or a young fancy-free bachelor officer let loose far from home. Ron Evans was neither. He was now a very happily married man with his own two girls waiting for him in California. That is not to say he was an aloof sobersides who turned his back on the pretty Japanese women and refused to talk to them. His then-bachelor roommate recalls that Ron happily joined in the chatting, flirting, drinking, and general merriment, but John adds that Ron knew where to draw the line. He didn't chase the women. He "played it straight."

At the wetting-down party, around thirty Japanese girls turned up from the many Yokosuka bars to help pour the naval officers' beers. Many of the junior officers were single. People talked about that party for a long time.

When the uss *Oriskany* returned to San Diego on 15 December, VF-142's accident-free record had stretched to an almost unheard-of year and a half. But let the record show that even a peacetime naval cruise was a hazardous business: seven naval aviators in other squadrons lost their lives in off-ship flying accidents.

On their return to NAS Miramar, Ron and three of his fellow pilots—John Allen, Moose Myers, and Ken Baldry—received new orders from BUPERS. All four would transfer to Moffett Field in Northern California to join VF-124 as instructors. Whatever Ron's private thoughts were about the assignment, it was shore based and would allow him to be with his wife and daughter every day.

In an interesting postscript to the *Oriskany* chapter in Ron's life, one Harry Jensen, since deceased, posted a message in 2010 on a website dedicated to the Crusader. Jensen had been an F-8 plane captain on the carrier during that 1960 cruise. He wrote: "The pilot who flew the plane I was responsible for was later to become astronaut Ron Evans, who was the pilot of the command module on the last moonshot with Cernan and Schmitt. It was plain to see why he was picked for the program. A finer man would be hard to find."

7. Teachers and Pupils

You cannot teach a man anything, you can only
help him find it within himself.
—Galileo Galilei (1564–1642)

Like all young naval officers, Ron underwent a quarterly review of his work and progress. He regularly got the chance to discuss his future intentions and to list his preferred assignments. That didn't mean the navy actually listened. Some officers talked about filling in the "dream sheet," and like most dreams, their cherished plans often evaporated with the cold light of a new day. Jan knew that Ron routinely requested assignment to the Naval Postgraduate School in Monterey to further his education and to boost his promotion prospects. She also remembered what he had written in his letters from Memphis when he thought he might end up instructing novice pilots at Pensacola. If he couldn't go to Monterey, he certainly wanted to remain a fighter pilot. For a two-cruise Crusader pilot to fly a lesser aircraft would have seemed a waste of talent.

The assignment to VF-124 was not what Ron had requested, but he would eventually be training other pilots to fly the Crusader. Whatever his innermost thoughts, this was undoubtedly a feather in his cap. John Allen considered it an honor that the navy had such confidence in his flying abilities that the service entrusted him to pass his skill and knowledge to other pilots.

When the four new instructors arrived at NAS Moffett Field, south of San Francisco, John settled into the bachelor officer quarters, and Ron began a hunt for accommodations near the base. He knew it would be temporary as the squadron was about to make a long-planned move to NAS Miramar; only then would the Evans family be able to put down roots. Meanwhile, they moved into a duplex on Bay Shore Drive in Sunnyvale. The freeway heading to the base had so many accidents the locals called it "Bloody Bay Shore." Their neighbor was an older man with a pink Lincoln convertible and a driving disqualification. He cherished his garish vehicle and persuaded Ron to drive it to the base a couple of times a week to keep the engine running smoothly. Jan always smiled at the thought of one of America's top fighter pilots turning up at the squadron in a big pink automobile. Nothing fazed Ron.

Before he could start any instruction, Ron had to fulfill a regular, mandatory requirement for all naval aviators and renew his instrument flying qualifications. This involved classroom time followed by eleven flights in the back seat of an F9F-8T Cougar advanced jet trainer. On each flight, the familiar black fabric hood completely blocked Ron's view out of the windows. The last flight was the crucial test, which he passed, and the whole syllabus proved to be a useful introduction to the next phase of his career.

It almost certainly came as a disappointment to him that instead of progressing immediately to instructing in the Crusader, he had to be content with the twin roles of instrument flight instructor and ground school instructor in the front seat of the Cougar. Likely there just weren't enough openings at Moffett Field for Crusader instructors, and of the four pilots who transferred from VF-142, Ron was the junior man.

The Cougar hardly compared with the Crusader, although it did come with 20mm cannon and could (just about) go supersonic in a dive. The Blue Angels aerial display team had selected it between 1954 and 1957. Of course, by 1961 the Angels had traded up.

By the end of June, Ron had made 118 flights in the Cougar, mostly providing fleet-level instrument training to novice pilots and upgrading veteran pilots in new instrument procedures. Other flights—made with fellow instructors including John Allen and Ken Baldry—were either squadron utility flights or purely recreational, probably to keep each other sharp.

As a fleet readiness squadron, VF-124 was responsible for maintaining its instructors in a combat-ready condition in case of a national emergency. At least this requirement provided Ron with access to Crusaders, and he was able to saddle up his supersonic steed about once a week.

Ron enjoyed watching his baby daughter grow from a helpless infant into a real little person with an impish grin and boundless energy. After he and Jan married and moved into their cramped apartment in La Jolla, she had found a senior secretarial job with a security clearance at a military base. She sometimes missed sharing part of her day with other workers, but she was never bored looking after Jaime and the home. Although the arrangement seemed the classic 1960s family setup, with the husband going out to work and the wife staying home to look after the children, Jan never really saw it that way, particularly when Ron was away on training exercises or cruises. At those times, she had to assume the roles of both parents. To say she was an equal partner in Team Evans would be to attach glib twenty-first-century terminology to twentieth-century family dynamics. Jan had no qualms about telling people how happy she was to let Ron get on with the flying while she looked after their home. She never felt like "just a housewife," and the arrangement worked fine for both of them, even after Jan realized she was expecting their second child. They were both delighted but knew they would have to think about finding a more permanent home.

One morning in April 1961, Ron and Jan woke to the news that Maj. Yuri Gagarin, a Russian air force officer, had become the first man to fly in space. As America slept, the world had changed. At twenty-six years old, Gagarin was a few months younger than Ron and Jan. He had married the month before they did, and he and his wife had two young daughters and, no doubt, family dynamics of their own. Three weeks after Gargarin's feat, the United States responded with Cdr. Alan Shepard's dramatic suborbital flight shown live on TV. Ron could not have failed to note with pride that his country's first astronaut was a naval aviator.

At the end of June, VF-124 packed its bags and moved south. Navy families have always had to be very adaptable, and Ron and Jan certainly shared John Allen's preference for the familiar surroundings of Miramar. Ron scouted around for a new home, and they settled on 3262 Mount Tami Drive, just north of San Diego and convenient for work. As the house wasn't yet ready for occupation, they initially had to move into a condemned but temporarily reprieved Quonset hut on the base. They brought one particular reminder of their time in Sunnyvale: from the local pound Ron had picked up a dog that they called Moffett. She would be part of the family for thirteen years. People kept asking if she was a miniature border collie. Ron seriously doubted it but eventually found that he saved time by saying, "I think so."

By early July, Ron was back in the familiar embrace of the F-8 Crusader, beginning a new and more satisfying phase of his career as an instructor. On the ground, he also served as the squadron's avionics division officer and the aviation equipment division officer. But every evening when he returned to his family's temporary home, he could see the growing evidence of another major development in his life.

Even in the latter stages of pregnancy, Jan didn't spend much time taking the weight off her feet. The new house was theirs, but Jan wasn't going to move in the furniture until she had scrubbed the rooms to provide a germ-free home for the new baby. She spent a whole day with brushes and mops, and it may not have been a coincidence that she went into labor that night. Baby Jon was born at Scripps Memorial Hospital on 9 October 1961 and didn't have to live in a Quonset hut. As soon as Jan and Jon left hospital, the family moved to Mount Tami Drive. Their furnishings were sparse, but the house was spotlessly clean.

Craig Kintzel earned his naval aviator's wings of gold at NAS Kingsville, Texas, in August 1961 and joined VF-124 the following February, itching to fly the Crusader. His route to Miramar had started with a degree in aeronautical engineering from Purdue University. When he was a freshman there, one of his older fellow students was a former combat pilot by the name of Neil Armstrong, who had returned to the university to complete the degree that hostilities in Korea had interrupted.

When Craig was at Kingsville, the assignments officer who was handing out squadron and other postings was Tom Tucker, who went on to have an illustrious naval career. Every eager young pilot was hoping to draw a crack squadron, but Tucker had to find someone willing to go on a six-month maintenance course at Memphis. With his degree in aeronautical engineering and a year at North American Aviation before joining the navy, Craig had the perfect background. They worked out an arrangement by which Craig would go to Memphis on the maintenance course and then get his dream assignment of learning to fly the Crusader with VF-124.

When Craig finally made it to Miramar in February 1962, he found himself in Crusader ground school with a very mixed bunch of pilots. Most of the class of ten men were fresh from flight training, but one older face was well known to Craig. By a quirk of fate, Tom Tucker was reassigned and had to undergo Crusader qualification. Another older face belonged to Cdr. Forrest Petersen, aged thirty-nine, a navy test pilot and more recently one of the pilots of the X-15 rocket plane. Having flown at 3,600 miles per hour and having reached 102,000 feet, Petersen was now qualifying in the Crusader before returning to aircraft carriers.

VF-124 was also known as the F-8 Replacement Air Group Squadron and was a great leveler. Sitting cheek by jowl with Tom Tucker and Forrest Petersen, Craig started off with two weeks of basic ground school. They had no simulators, only blackboards and lectures. When the day arrived for Craig's first flight in the Crusader, he was introduced to his instructor, a smiling young Kansan named Ron Evans.

"Ron was very pleasant, soft-spoken, [with] an easy-going personality," Craig recalls. "The next drill was, he put me in the cockpit, he stood alongside, and I was blindfolded and then he gave me a blindfold cockpit check. I had to point to every instrument and then recite off all the limitations and this-and-that of the airplane. And once he was satisfied, now we're ready to go. And since it's a single-engine, single-seat supersonic fighter, your first flight is solo! You're going to figure everything out 'on the fly' so to speak!"

It should again be emphasized that anyone who was given the opportunity to fly the Crusader was already known to be a pilot of superior skills.

They took off with Craig in the lead and Ron on his wing as a chase pilot who would keep a close eye out for any problems. They crossed the Pacific coast and went through a set of basic maneuvers, with Ron giving instructions by radio and Craig following through. A trouble-free hour and a half flashed past, and they prepared to return to Miramar for a set of touch-and-goes to familiarize Craig with the landing characteristics of the Ensign Eliminator. At that point, despite a weather briefing that had promised clear skies, one of the region's notorious fogbanks materialized, blanketing San Diego and Miramar, and reducing visibility to below the safe minimum.

Ron as the instructor now had three alternatives, none of them appealing. The first was to let Craig make his first Crusader landing on instruments, with Ron observing as wingman and offering guidance. The second was for Ron to lead the way down through the pea-souper with Craig on his wing. He quickly rejected those options, as both placed too much responsibility on his pupil. The third option was to divert to Miramar's alternate landing site at Yuma, but they were now low on fuel. While gaining altitude burns fuel, a higher altitude reduces air resistance, so they climbed to twenty thousand feet and set the engines for maximum efficiency. As they completed the transit to Yuma, Craig only had enough fuel for one approach. His first Crusader landing had to be a good one. And it was.

Miramar stayed socked in all night, so the two pilots phoned their wives to break the news that they wouldn't be home. Then they booked rooms at the bachelor officer quarters and retired to the bar for a meal and maybe a few drinks.

Craig recalls, "We wound up closing the bar, sharing lots of stories, but mainly Ron was very impressed by my first landing and 'making him look good,' because if the student crashed, it was the instructor who was held responsible! So he reminded me two or three times that I saved *his* neck by making that landing!"

They flew back to a fog-free Miramar the next morning, and Craig made another textbook landing. He assumes their conversation the night before must have included the hot topic of naval aviator John Glenn's dramatic spaceflight on 20 February. Although the latter part of the evening got a little hazy, Craig is certain Ron didn't say anything about wanting to become an astronaut.

Ron again was assigned to Craig for his third or fourth Crusader flight. This time they would take off on afterburners and climb to forty thousand feet for supersonic flight. When Ron briefed Craig on the plan, he bet him a martini that he couldn't stay subsonic all the way up to forty thousand feet. Craig was used to gaining altitude with his aircraft's nose twenty or thirty degrees above the horizon. This time he knew he would need a much steeper climb to hold the acceleration down, but he underestimated the awesome power of the Crusader on afterburner. Even pulling back sixty degrees wasn't enough, in spite of the perception that he was going straight up. He broke through Mach 1 and lost the bet, as Ron had expected. Craig would have needed an angle of seventy degrees to stay subsonic, and his brain simply couldn't process such an extreme requirement so early in his training. But he never had so much pleasure losing a bet.

Craig admired and respected all of the Crusader instructors. The trainees recognized that their instructors were not just excellent pilots. They were the best of the best. Most, like Ron, would later fly combat missions over Vietnam. One of the most noteworthy was Lt. Foster "Tooter" Teague, a future skipper of Fighter Squadron 51 who would shoot down an enemy MIG, survive ejection into the sea when his own aircraft was hit by ground fire, and find room amid a chestful of decorations for two Silver Stars.

Years later when Ron was selected to fly to the moon, Craig Kintzel warmly remembered his instructor from Miramar: "He had an infectious smile and was just a very pleasant guy to be around. I was delighted when he was selected for

Apollo 17. I really wished him well, and I had nothing but the highest respect and admiration for him."

The Evans family spent a year at Miramar while Ron trained pilots to fly the Crusader to combat standard. Dick Cavicke, one of the senior instructors, describes the work as "exceptionally good duty: lots of flying, good people, and not much time away from home." John Allen's assignment at VF-124 lasted three years, but the navy had apparently paid attention to Ron's regular requests. He wasn't a Naval Academy grad, but the new orders he received from BUPERS were clear: "You will proceed to the Naval Postgraduate School, Monterey, to study aeronautical engineering."

8. Deke and Al

It is not what happens to you, but how
you react to it that matters.

—Epictetus (c. 55 AD–c. 135 AD)

Monterey enjoys a cool Mediterranean climate that is rarely very cold and rarely very hot, but after the heat of San Diego, it was still a bit of a shock to the system for the Evans family. Ron and Jan found a tenant for their home on Mount Tami Drive, intending to return in due course, and rented a house at 1805 Ord Grove Avenue in the Seaside district of Monterey. A contemporary family portrait, probably taken in early 1963 in the front yard, shows Jan enveloped in a heavy hooded coat, Jaime wearing a fur-lined hood, and Ron braving the elements with only a jacket over his shirt. Any natural protection for his head was rapidly disappearing. The family also acquired a second dog, a German shepherd they named Flieger (Aviator), which would prove to be a faithful companion and devoted guardian for Jaime and Jon.

The Evans family soon made friends with an older couple who lived nearby. "Nana and Poppa" became firm favorites of the children and would often babysit. They mentioned their daughter Jane and shared that she was married to Cdr. Robert Ferguson, who would later serve with Ron on the USS *Ticonderoga*.

The return to full-time education was, to put it mildly, a change of pace for Lieutenant Evans after the Pacific cruises and supersonic flight instruction. At least he would not incur a financial penalty. Ron's navy salary had risen from $720 to $770 a month in July 1962. He would be paid to study at the Naval Postgraduate School (NPS), but the navy expected value for its money.

This period in Ron's life introduced him to several other NPS students whose lives and careers would intertwine with his own in years to come.

Already studying since July 1961 were fighter pilots Eugene Cernan and Richard Gordon Jr. Like Ron, Gene had a degree in electrical engineering, had passed through ROTC at university, and had two WestPac cruises under his belt. He had even made his first carrier landings in 1958 on the USS *Ranger*. Dick, who was four years older, had a degree in chemistry and was a genuine twenty-four-carat test pilot, having been the first project test pilot for the F4H Phantom II before trading in his pressure suit for a pile of books.

Arriving in the same month as Ron, Paul Weitz had a degree in aeronautical engineering and had also received his navy commission via the ROTC. He had been assigned to Air Development Squadron 5 at China Lake, California, and worked on weapons delivery tactics. Flying five different aircraft types and making four or five flights a day, he was in seventh heaven until he received new orders from BUPERS.

Jack Lousma, another graduate in aeronautical engineering, had joined the marines and completed flight training in 1960. He was assigned to Marine Attack Squadron 224 (VMA-224), Second Marine Air Wing, based at Marine Corps Air Station (MCAS) Cherry Point, as a jet attack pilot flying the Douglas A-4C Skyhawk. He later deployed with VMA-224 to MCAS Iwakuni, Japan, during 1962–63.

The orders to attend the postgraduate school were met with different reactions. For Ron, his delight that the navy had actually listened to his requests swept aside any momentary sense of dislocation. The posting was equally satisfying for Jack Lousma. But wasn't it difficult for a hotshot pilot like Lousma to go back to school?

"No, it wasn't difficult at all," he counters. "I requested that assignment, I wasn't assigned involuntarily. I worked to get that assignment because I wanted to get a postgraduate degree, so to me it was a privilege."

But at China Lake in 1962, the letter Paul Weitz received from BUPERS was entirely unexpected: "Then I got unsolicited orders to the Navy's Postgraduate School in Monterey. I did not apply for it. I didn't want to go, because I had what I thought was a good job. I had no desire to go back to school."

Whatever misgivings he harbored, Weitz knuckled down and, in time, came to acknowledge that Monterey was a turning point in his career. Lacking test pilot credentials, he was convinced that his Monterey master's degree later tipped the balance in favor of his astronaut selection in 1966. The same may have applied to others as well.

Those five future astronauts' names seem to fit together like peas in a pod but only with the knowledge of hindsight. Lousma arrived as Cernan and Gordon were preparing to leave for Houston. He didn't know either of them, but he recognized Cernan around the campus because he had just been selected as an astronaut. Lousma did meet Evans and Weitz at a social event at the school, but they were a year ahead and shared no classes with him. The only two of those future astronauts who actually studied together were Evans and Weitz, although by coincidence their class also included Robert Overmyer, a marine air corps pilot who would join NASA in 1969. He would work with Ron as a support crew member on *Apollo 17* and the Apollo-Soyuz Test Project before making two space shuttle flights, the second as commander.

The postgraduate school had been established at Annapolis in 1909 but then moved to Monterey at the end of 1951, occupying the site of the former Hotel Del Monte. Once known as "the most elegant seaside resort in the world," the hotel had hosted heads of state, including American presidents, Hollywood film stars, famous artists, and other dignitaries who were wealthy enough not to worry about the price of their accommodations. By 1962 the echoing laughter of the rich and famous had been replaced by the footsteps and voices of earnest young naval officers hurrying from class to class.

It might seem a transgression of some natural law that seasoned fliers like Dick Gordon, Gene Cernan, and Ron Evans were not found regularly in the wee small hours sampling the beer in Monterey's hostelries, but it just wasn't like that. As Gene Cernan noted in his autobiography, "The school turned out to be damned difficult, and I hit the books harder than I had ever studied in my life. I would go to class at 8.00 am; stay there until 5.00 pm; come home and have dinner with Barbara, then study until midnight or one o'clock in the morning. Aeronautical engineering, the Navy way, was a bitch."

Looking back nearly six decades, Gene Cernan's then-wife Barbara confirms his account. "He was very studious, he didn't take anything for granted, he knew he had to learn, and he did! He was really smart, and he worked every day at it, I can tell you for sure. And I'm sure Ron did the same thing. They had to!"

In the spring of 1963, informed rumors began circulating that NASA wanted another batch of astronauts to work on the moon program that President John F. Kennedy had announced two years earlier. What had been an open secret in May became an official announcement on 5 June, and if Ron took his nose out of his books long enough to read it, he may have noticed that NASA no longer required its astronauts to be test pilots. A university degree in engineering or the physical sciences together with at least a thousand hours of flying time in jets were the key criteria, but—crucially—any would-be spaceman from the services had to be recommended by his military organization. It was very much a case of "don't call us, we *might* call you."

Ron's flying time in the Crusader alone comfortably exceeded a thousand hours. The navy was certainly aware of this. Interviewed in 1986, Ron explained what happened next. Basically, NASA approached the navy and said it needed new astronauts. Could the navy recommend any of its men? The navy went through its records and identified thirty-four officers, including Dick Gordon, Ron Evans, Gene Cernan, and two other classmates of Gene's: Bob Shumaker and George "Skip" Furlong. The navy then contacted those individuals, either by telegram or telephone, informing them that they met NASA's basic qualifications for astronaut selection and asked if they would like to volunteer.

If Ron had been an undergraduate at a university, he could have put his feet up and enjoyed the long weeks of Monterey summer sunshine, but the navy frowned on long periods of paid inactivity for its serving officers. That meant he was taking a summer class. Ron could have chosen an engineering option; instead, he signed up for a short foreign language course in Russian. It helped that the course was taught at the Military Defense Language Institute in Monterey. A head start in Russian would prove useful to Ron in the next decade. In retrospect, Jack Lousma thought it was "a smart thing to do," although at the time they had no inkling that American astronauts would one day be speaking Russian on board an orbiting Soyuz spacecraft. Ron's motivations were most likely Cold War–related, in the "know your enemy" category.

Some furlough time was allowed, and even as the navy was trying to contact its potential astronaut nominees, Ron and Jan took a break in Topeka. Fortunately, he had left a forwarding address and phone number. He candidly admitted years later that becoming an astronaut was not actually his career goal in 1963. It wasn't that he had thought about it and decided it wasn't for him; he just thought he would not meet the criteria. When he received the navy's telegram, everything changed, more or less instantly. He didn't actually reply, "Hell, yes!" But that was what he was thinking. Ron never felt the need to trumpet his own achievements, but neither would he try to hide them.

He knew he was "a pretty damn good fighter pilot," so if it was no longer necessary for a candidate to be a test pilot, then—hell, yes!—why *shouldn't* the navy put his name forward?

As Ron later recalled, "So, pleasant surprise, I guess, is the way to put it. Absolutely! And of course, once you find out, 'Hey! I'm in this process! Somehow I got selected in this process,' you think—'Let's charge!' You change your career goals, you know!"

This was a crucial facet of Ron's life and career: he would set a goal for himself and strive to achieve it, but if unexpected changes in circumstances arose, he was not so rigid in pursuing his goal that he refused to change his trajectory. Meeting Jan had led to a major change. Now a telegram from the navy suggested another.

When Ron discussed the possibility of becoming an astronaut with Jan, she immediately saw he was ready and willing for this new challenge. Some wives might have been apprehensive or even fearful at the thought of their husbands being rocketed into outer space, but astronauts were big news and were capturing people's imaginations all over the world. Jan was excited and happy to give Ron her total support. She also felt it was a great honor that the navy had expressed such faith in him.

Having been volunteered by the navy, Ron had to complete and submit a formal written application for federal employment for the position of astronaut. It requested details of his educational achievements, his employment record, and all the places in the world he had visited. At the height of the Cold War and only months after the world had teetered on the brink over the Cuban missile crisis, the application form also inevitably asked, "Are you now, or have you ever been, a member of the Communist Party?" For the record, he answered no.

Ron submitted his application form to NASA on 6 July 1963, comfortably ahead of the mid-July deadline for military applicants. What followed was "a hurricane of NASA paperwork," as Gene Cernan described it, designed as "a preliminary culling" of the final total of 720 applicants. That number included 228 mostly starry-eyed civilians (including three women) lured by the perceived glamor of becoming "spacemen." The military applications totaled 492, and it speaks well of the navy's preliminary selection process that the number included the five NPS students at Monterey and that all five were among thirty-four applicants who made it to the final stages of the selection process.

Brooks Air Force Base in San Antonio, Texas, was the venue for the grueling physical and psychological tests to which the applicants were subjected in early August 1963. Several astronauts who successfully negotiated the Brooks torture chamber have written graphic accounts of the medical tests. Walt Cunningham tells how the slightly too-tall Capt. C. C. Williams stayed up all night, jumping up and down to compress his spine enough to pass the height qualification. Most of the candidates seem to remember with particular displeasure the baffling test in which ice water was poured in one ear and warm water in the other. The conflicting messages to the brain apparently cause the eyeballs to swivel wildly.

During their passage through purgatory, the candidates had their hearts and brains monitored, took treadmill tests to measure their endurance, and suffered oxygen starvation, spinning chairs, and isolation chambers. They faced such standard psychological scrutiny as the infamous inkblot test and the blank sheet of paper test. Stories of the latter being described as "polar bears fornicating in the snow" or provoking the comment "it's upside down" may or may not be apocryphal, but no account of the selection process seems complete without them.

Six men failed the testing. Because of a rare medical anomaly, NPS student Bob Shumaker not only lost the chance of going to the moon but also was grounded altogether, demonstrating a classic example of why pilots are wary of doctors. He eventually won back his wings, only to be shot down over Vietnam in 1965. He spent eight years as a prisoner of war (POW) before resuming a stellar navy career in which he made admiral.

The surviving twenty-eight candidates were invited to Houston, where they were softened up at cocktail parties, their every word and gesture scrutinized. It probably unnerved some of the men to watch Dick Gordon being greeted warmly by Mercury astronauts Walter "Wally" Schirra and Al Shepard (who had been Dick's instructor at test pilot school). No candidate begrudged the best of the rest their future seats on Gemini or Apollo, but each man knew those seats were limited and that others had to fail for him to succeed. Donald K. "Deke" Slayton, one of the "Original Seven" Mercury astronauts, had been grounded by a minor heart irregularity and was now the coordinator of astronaut activities. Slayton had determined that he needed somewhere between ten and twenty new astronauts; so come what may, at least eight of these gifted and eager young men would not make the cut.

The final bunker was a series of interviews, evaluations, and written tests at the Rice Hotel, Houston. Ron, Gene, and the other NPS students probably felt as if they were back at Monterey as they wrote answers to exam questions about spaceflight and orbital mechanics. A NASA panel that included Slayton, Schirra, and Shepard then interviewed each candidate. They asked a series of probing questions, particularly about flying careers, and to avoid any misunderstandings, each man's flying record was lying there on the desk right in front of him. Cernan "felt like a prisoner before the parole board" but noted that the men behind the desk were all quite pleasant "except for Shepard whose cold eyes seemed to look right through me."

With all the prodding and probing completed, the would-be astronauts departed Houston, knowing that the matter was now out of their hands. They had all been ordered to make their way to Houston on separate flights, apparently to throw the press and others off the scent. Oddly, the candidates had no such restrictions on the way home, and Ron shared his flight with Dick Gordon and Walt Cunningham. They had a layover in Los Angeles, and Walt invited his friendly rivals to his home. They all began the process of converting their stressful and often uncomfortable experiences into war stories that would be told and retold over drinks in the years ahead. Human nature being what it is, the three men also spent time looking at each other and wondering. The only one who didn't think Dick was a shoo-in was probably Dick.

The twenty-eight applicants were told that they would learn their fate by telephone. Deke Slayton would deliver the good news; any other voice would signify failure. There were exceptions. Air force pilots Dave Scott and Ted Freeman, while touring test-pilot schools in France and Germany, actually received their good news in the form of letters from Deke Slayton delivered to their German hotel.

In Monterey on 14 October, the first day of the fall term, Ron Evans and Gene Cernan were both called out of class to take long-distance telephone calls. They arrived at the aero-engineering office at the same time, and each was offered a seat and a separate phone extension. They must have realized Deke couldn't be calling both simultaneously. Sure enough, Gene heard "the gruff voice of Deke Slayton" offering him a job at Houston if he was "still interested."

Ron saw that Gene was "all smiles" after the call.

On the other line, Ron heard the voice of Al Shepard, the bearer of bad news, and was "totally disappointed, obviously!" Shepard was saying Ron had done a great job and should keep up his interest in the space program. Maybe NASA would call him later.

Gene recalled the bittersweet moment when, in the middle of his own jubilation, he "saw the face of a very dejected Ron Evans." The two men shook hands and exchanged congratulations and commiserations. Dick Gordon had also received good news from Deke. In all, fourteen out of the twenty-eight finalists were selected for NASA's third astronaut group.

What happened next is open to debate but certainly involved some consumption of alcohol. According to Cernan, before he could even phone Barbara, he was "swept out the door, down to the bar at the old Mark Thomas Inn and got falling-down drunk without buying a drop of booze." Barbara certainly remembers waiting nervously to hear Gene's news and getting anxious and annoyed when he didn't phone. While Gene recalled that he did eventually phone and that an irate Barbara hurried to the bar to join the celebrations, she is pretty sure she didn't go to the bar. She *does* remember Gene's state when he returned home. "He was in a happy condition. He wasn't drunk or anything like that. You know, they had had a couple of drinks—he was just in a happy, happy mood!"

Ron's emotions were entirely different. To have had the prospect of an even more exciting career as an astronaut dangled in front of him, only to see it yanked out of reach, was painful. Nine years later, *Time* magazine quoted him as stating, "That night, Gene and I went out and got totally sloshed." In 1986 he explained that he returned to class to finish up and went out later—not to cry in his beer but certainly to share a few drinks. And, yes, he ended up drunk.

Jan's memory is that Ron called her right away. He wanted her to hear the news from him, not from somebody else. She remembers him saying, "Oh, I didn't get selected for the program." Then he went back to class. She assumes he must have had "a few drinks" with Gene and the others, but when Ron came home in time for a late dinner, he certainly didn't seem drunk to Jan. Like the fisherman's account of "the one that got away," the levels of intoxication that evening seem to have been a little exaggerated over the years in the retelling.

Jan could see her husband was devastated, and she felt so sorry for him. But Ron didn't wallow in self-pity. He simply said, "Second best is not good enough." They talked some more before going to bed, and from then on, it didn't seem to worry him anymore. There was nothing to be gained by dwelling on what might have been, and Ron had a master's degree to earn.

Until the mid-1960s, a navy master's degree required three years of study, with the third year usually taken at a civilian engineering school such as Princeton University or the California Institute of Technology (Caltech). Unfortunately, Ron did not have the luxury of three years' study. While ground assignments in the navy did usually last three years, he had already spent half that time as an instructor at v f-124. Completing only two years' study at the postgraduate school would not earn him a master's degree. Paul Weitz, who hadn't wanted to go to Monterey in the first place, was in the same position and summed up the dilemma succinctly: "I didn't have three years. I didn't want to spend two years in Monterey and come out of there with nothing."

Already into their second year, Ron, Paul Weitz, and several others "found a couple of sympathetic professors" who told them that if they completed all of their second- *and* third-year requirements in that second year *and* taught a course as well, the school would confer a master of science degree on them. It would not actually be a master's in aeronautical engineering, but it was the best deal available.

Ron himself said he "worked his tail off." Jan recalls his efforts clearly: "We never saw Ron! He left when it was still black in the morning, and he didn't get home until after midnight. But he did it! And they had to write a thesis also. He had one engineering professor [who] was an absolute demon, and you could not have a misspelled word in your thirty-five- to forty-page thesis. Your margins had to be exact. You could have no erasures. And guys paid like five dollars a page to get their theses typed. I typed four different theses for fellows, and I didn't charge five dollars a page! Anyway, I was always very, very proud of those guys. There were five of them allowed to do this. Anyway, they worked their hearts out!"

In March 1964, as if the task wasn't hard enough already, Adm. Hyman Rickover dropped a bombshell. Rickover, known as "the father of the nuclear navy," was testifying before a House appropriations subcommittee in Washington. Often a controversial figure, he referred to the Naval Postgraduate School at Monterey as "largely a correctional institution for the academically deficient." He then rubbed salt in the wound by adding, "Some graduates of this institution, after three years, have received the equivalent of about one year of what would be considered true graduate work at a good engineering college." When the admiral's comments hit the headlines, Jan Evans vividly recalls Ron and his fellow students being very upset at his implication that their hard work was just a walk in the park. But they persevered.

Graduation day was on Monday, 1 June 1964, and among the recipients of the hard-won master of science degrees were Lt. Ronald Evans, Capt. Robert Overmyer, and Lt. Cdr. Paul Weitz. Ron was relieved to have survived the academic pressure cooker. Despite his characteristic tendency not to dwell on the past, he later allowed himself the comment: "I don't have so many fond memories of Monterey."

9. Screaming Eagle

'Tis all a checkerboard of nights and days
Where Destiny with men for pieces plays.

—Edward Fitzgerald (1809–83)

Marlo Holland first met a future *Apollo 17* astronaut at NAS Miramar in 1958. Assigned to Attack Squadron 126 (VA-126) to check out in the A-4D Skyhawk low-level nuclear bomber, he soon encountered another young lieutenant (JG) by the name of Eugene Cernan. Marlo wasn't sure if Cernan was actually running the flight department, but he certainly gave that impression. Charismatic and impressive in the way he handed out duty assignments, Cernan was already acting like a full lieutenant.

As it turned out, Marlo never got to fly the Skyhawk. That was a relief to him because trying to outrun his own mushroom cloud didn't appeal to him, and anyway he was too tall to fit comfortably into the aircraft. Gene Cernan notified him of his change of assignment to a fighter protection squadron on the carrier USS *Hornet*. Five years and a couple of assignments later, Marlo received orders to report to Fighter Squadron 124 (VF-124), to transition to the navy's top fighter, the F-8 Crusader, before joining Fighter Squadron 53 (VF-53). Known at the time as "Crusader College," or the F-8 Replacement Air Group (RAG), VF-124 was part of the West Coast Replacement Air Group and was responsible for providing fleet-level training for new and transitioning pilots and refresher courses for veterans. Reporting to Miramar, Marlo found himself in Crusader ground school with an interesting mix of pilots. Some, like himself, were experienced aviators. Others, such as Ens. Phil Vampatella, had only recently earned their wings, but Marlo would remember that name two years later when Vampatella shot down a North Vietnamese MIG-17 with a

Sidewinder missile. Also sitting in class with Marlo was a slightly older, balding pilot with a friendly smile and an air of calm authority—Lt. Ron Evans.

After more than three years on ground assignments, Ron's next carrier deployment would be with Fighter Squadron 51 (VF-51), the Screaming Eagles, but first he had to report to VF-124 to requalify for carrier operations. Marlo Holland remembers Ron as seasoned, easygoing, and confident. They shared a few wry smiles when he learned that Ron had actually been a Crusader instructor, but now they were all pupils together as Ron went through the process of having two years of Monterey rust scraped off his wings. Former instructors had no privileges. They all went through the whole syllabus, starting with the basics and working through instrument training in the Cougar before Ron could finally ease himself back into the cockpit of the familiar Crusader. Familiar but improved—the upgraded F-8E could operate at night.

As Ron was restoring the cutting edge to his flying skills, events unfolding almost eight thousand miles away in a Pacific inlet called the Gulf of Tonkin were about to cast a shadow over the region that would profoundly affect a whole generation, leaving wounds that continue to bleed nearly six decades later. Just as a single assassination in Sarajevo did not cause the First World War, the conflict in Vietnam was perhaps inevitable given the Cold War climate of the mid-1960s. The spark that lit the powder keg for the United States just happened to flare in the patch of sea where VF-51 was already deployed with the USS *Ticonderoga*.

The world learned of an attack on the American destroyer USS *Maddox* on 2 August in international waters. The crew of *Maddox*, assisted by aircraft from *Ticonderoga*, warded off the attack. Two days later, the world heard that *Maddox* and a second destroyer, the USS *Turner Joy*, had both come under attack from North Vietnamese PT boats firing machine guns and torpedoes, again in international waters. No Americans had died, and once again the U.S. Navy had struck back, sinking enemy boats and giving the communists a bloody nose.

A typical American report on the events of 4 August appeared in *Newsweek* magazine and was later quoted in the memoir of VF-51's skipper Cdr. James Bond Stockdale: "By 9:52 pm both destroyers were under continuous torpedo attack. In the mountainous sea and swirling rain no one knew how many PT boats were involved as they rose and fell in the wave troughs. The

U.S. ships blazed out salvo after salvo of shells. Torpedoes whipped by, some only 100 feet from the destroyers' beams. A PT boat burst into flames and sank. More U.S. jets swooped in, diving, strafing, flattening out at 500 feet, climbing, turning . . . and diving again. . . . The battle was won. Now it was time for American might to strike back."

The retaliatory strike took place the next day. The attacking force, including Crusaders from *Ticonderoga* led by Commander Stockdale, bombed a number of PT boat bases and their oil storage depot at Vinh, North Vietnam.

On 7 August the U.S. Congress passed what became known as the Gulf of Tonkin Resolution, which President Lyndon B. Johnson signed into law three days later. The resolution was not a formal declaration of war, but it authorized the president to take "all necessary measures" to repel any armed attack against U.S. forces, to prevent further aggression, and to assist South Vietnam and other signatories of the Southeast Asia Collective Defense Treaty.

Those who seek a detailed history of the origins, causes, and consequences of the Vietnam War are invited to look elsewhere, but it would be almost impossible to write about a naval aviator's preparing to join VF-51 on the USS *Ticonderoga* in late 1964 without acknowledging the experiences of some of that ship's pilots. Commander Stockdale would later endure 2,714 days of hell on Earth as a POW in North Vietnam but would return to receive the Medal of Honor in 1976 and make admiral. Unlike *Newsweek*, Stockdale was actually there. He saw what happened. And what didn't happen.

On Sunday, 2 August, Stockdale was leading a group of four Crusaders on a training exercise when they intercepted a call for assistance from the USS *Maddox*, which reported it was under attack by North Vietnamese PT boats. The Crusaders arrived just after the PT boats had broken off the attack and were speeding back to their base. Attacking with rockets and 20mm cannon, the pilots damaged two PT boats and sank a third. President Johnson and his advisers decided to take no further action.

Two days later, *Turner Joy* joined *Maddox* on patrol in the Gulf of Tonkin. Both ships were in sight of North Vietnam but in international waters. Understandably, both crews were jittery and reported that they suspected an imminent attack. *Ticonderoga* was ordered to launch aircraft as a precaution. As Stockdale approached the location of the destroyers in very poor weather, he heard reports of multiple radar contacts. *Turner Joy* was firing at targets that *Maddox* couldn't track, while *Maddox* was dodging torpedoes *Joy* couldn't detect on sonar. Flying no higher than a thousand feet above the waves and always within sight of the ships or their wakes, Stockdale spent ninety minutes searching in vain for any attackers. Both ships were firing into empty darkness, intermittently lit up by flares, but even with his superior aerial view, Stockdale saw no torpedoes, no PT boats, and no evidence of enemy activity.

Returning deeply frustrated to the ship, he reported the absence of any attacking vessels. He was satisfied that the "attack" had been the result of spooked radar and sonar operators. Expecting an attack, they then started seeing torpedoes instead of waves, whales, currents, or thermal layers in the water. Stockdale was assured that his "no attackers" report would be transmitted to Washington, and it was. The captain of *Maddox* submitted a long report that expressed doubts that there had been any attackers, and he urged a complete evaluation of the mix-up before taking any further action. Eighty-one minutes later, he contradicted his earlier report, concluding that an attack *had* occurred, but when Adm. Ulysses Sharp, commander in chief of the U.S. Pacific Command, later received a telephone call from U.S. secretary of defense Robert McNamara, the clear advice of the admiral to the politician was to hold off any retaliation until there was a definite indication of what had happened.

Interviewed for this book in March 2018, another VF-51 pilot, then-lieutenant (JG) Tom Klein, confirmed that when he arrived over the two destroyers, "the weather conditions were just horrible, with very low overcast." He found it was tough just avoiding hitting the water. He looked for at least the wakes left by attacking boats, "but the only wakes that we could find were the destroyers' wakes, even though they said they were being attacked by torpedoes and also weapons of some sort, machine-guns or whatever."

Lt. (JG) Ralph James of VF-53 vividly remembers being scrambled on the evening of 4 August and circling the two destroyers. He is emphatic: "We couldn't find anything. There were no boats to be seen."

In Washington, President Johnson accepted the advice of his secretary of defense that based on *two* unprovoked attacks on U.S. ships in international waters, a retaliatory strike should be launched. Despite his conversation with Admiral Sharp, McNamara failed to advise the president before the strike that the senior naval officer in the region was recommending no action without a more thorough investigation. Instead, McNamara confirmed the strike order. Had LBJ been misled? *Can* the leader be "misled," or does Harry S. Truman's dictum apply? Where did the buck stop?

Stockdale was astounded to receive orders to mount a retaliation based on an attack that he knew had not happened. He was well aware of the wider geopolitical realities of the Cold War and understood the concerns about the so-called dominoes of South Vietnam, Laos, Cambodia, and even Thailand falling under the onslaught of communism. But based on his unique personal insights, he concluded: "The fact that a war was being conceived out here in the humid muck of the Tonkin Gulf didn't bother me so much; it seemed obvious that a tinderbox situation prevailed here and that there would be war in due course anyway. But for the long pull it seemed to me important that the grounds for entering war be legitimate."

On the morning of 5 August, Ralph James and other Crusader pilots on *Ticonderoga* were awakened at five o'clock and ordered to report to the ready room, where they were briefed on a strike against the bases from which the attacking PT boats had originated. Ralph was certainly not alone in thinking the whole thing was a mistake. But Washington was telling them it wasn't, and it was way above Ralph's pay grade to start questioning the orders not just from his senior officers but also from his commander in chief. They had to believe LBJ knew what he was doing.

After the successful raids, no further attacks occurred on either side while *Ticonderoga* remained on station, but tension was high. Every night, three Crusaders mounted a patrol above the Gulf of Tonkin to protect the ship from any sneak attack by MIGs. On one of those nights, Ralph James was tasked to intercept a bogey crossing the gulf from the direction of China and heading in the general direction of the ship. It was traveling at high subsonic speed and was not identifying as any civilian airliner. Approaching on afterburner to catch up with the fast-moving target, Ralph was ordered to arm his Sidewinders, and he thought he was about to engage with an enemy MIG. The missiles locked onto the bogey, but as Ralph broke through a thin cloud layer, he spotted a rotating beacon and heard an urgent voice on his radio barking: "Switches safe! He's friendly!"

Moments later, he drew alongside a TWA Boeing 707 en route from Hong Kong to Bangkok. Some of the airline's pilots had decided it would be really smart to switch off their transponders while hightailing across the Gulf of Tonkin "so as not to attract MIGs." The passengers in the Boeing never knew they had been targeted by heat-seeking missiles. Fortunately for them, nervous pilots with itchy trigger fingers never got to fly the F-8 Crusader.

Back at NAS Miramar, any private forebodings about developments in the Far East were drowned out by patriotic support for the actions of *Ticonderoga*'s pilots. Gene Cernan noted an eagerness for action among former squadron mates who were still-serving naval aviators. Some of the pilots even maneuvered to be assigned to WestPac squadrons in case the shooting ended before they got into combat at all. With the painful knowledge of hindsight, this sounds naive at best, but some young Royal Air Force pilots on the airfields of southern England in 1940, yearning to test their Spitfires against the Luftwaffe, had voiced similar thoughts.

In this febrile atmosphere, Lt. Ron Evans completed his retraining with two days and nights of carrier-qualification flights on the USS *Midway* based off the coast of California. He knew that *Midway* had once been the largest ship in the world and was therefore not the best comparison for the smaller *Ticonderoga*. As a veteran of the 1960 *Oriskany* cruise, Ron already had experience on a smaller carrier, although the days of the gentlemen pilots who didn't fly by night had ended with the arrival of the F-8E version of the Crusader. Demonstrated by VF-51 in the Tonkin Gulf, sorties were flown day and night, depending on operational requirements.

On receipt of Ron's new orders, the Evans family gave notice to their tenant in San Diego and moved back into their home at 3262 Mount Tami Drive. They would live there for another two very eventful years.

Jan Evans had always been aware that her navy pilot husband might one day take part in a shooting war. It went with the territory. Now it looked as if the shooting would be in the skies over North Vietnam. Among the pilots she knew, she recognized a certain gung-ho determined attitude, a readiness to test their training in the superb flying machine that the navy had provided for them. They all *loved* the F-8. For most of them, nothing else they had ever flown came close.

Jan didn't consider Ron and his fellow pilots to be warriors as such, and she knew they were smart enough to harbor varying degrees of apprehension about the unknowns that lay ahead. But they also possessed, in varying degrees, that youthful arrogance that imbued them with a sense of personal invulnerability. And they were all fit and healthy specimens who were as well qualified, as well trained, and as ready to enter a war zone as anyone could be.

Having graduated from the VF-124 Crusader College, the pilots, both shiny new and refurbished, signed out of the squadron and dispersed to their new assignments. Marlo Holland and Ron Evans had the same destination, the USS *Ticonderoga* (*Tico*), which was already well into the second half of its 1964 cruise. Although they had trained together at Miramar, they would now be flying in different squadrons, so they made their way by separate military transport flights to NAS Cubi Point in the Philippines. On 12 October Marlo flew his Crusader to *Tico*, which had briefly broken off its patrol in the Tonkin Gulf to visit Japan and the Philippines. The next day Ron climbed aboard a Crusader that his new squadron had set aside for him and made the hundred-minute transfer flight to the ship. Thanks to its high landing speed and "difficult" reputation, any Crusader deck landing tended to draw an audience, but when a new pilot was joining the ship, an audience was virtually guaranteed, with most perched on part of the superstructure known as Vulture's Row. The onlookers were rewarded by a typically spectacular but otherwise uneventful Crusader landing, after which Ron reported for duty to his new skipper Cdr. James Stockdale.

However surprising, Marlo Holland does not recall ever bumping into Ron on board *Ticonderoga* during that cruise. They were in different squadrons operating on different schedules, and it was certainly no training cruise. As he points out, "Operations were so hectic—we were actually 'doing war.' At war, really."

The day before Ron departed Cubi Point, Marine Fighter Squadron 122 (VMF-122), including Crusader pilot Lt. Jerry Carr, flew in for a period of weapons training and, weeks later, several days of carrier requalification flights on *Ticonderoga*. Those two F-8 pilots also passed like ships in the night and never knowingly met until both later arrived in Houston as newly selected astronauts.

"Ron and I never even ran across each other on the *Ticonderoga*," Jerry noted. "I'm kind of surprised because the F-8 pilots, we were all pretty 'thick.' But I never met him. We never had a chance to meet each other on the ship at all. It's quite a coincidence."

Ralph James certainly remembers Lt. Ron Evans after he came on board. Ralph was both a pilot and a landing signal officer. On days when he served as the LSO for the whole ship, he guided all the returning pilots down onto the sometimes-heaving deck. He made notes on each landing and debriefed all the pilots on their landing techniques. Nobody, not even a squadron skipper, was so perfect he couldn't use a little friendly advice from a detached observer. Even though Ralph wasn't in Ron's squadron, the two pilots got to know each other well as a result of the frequent landing debriefs. Ralph remembers Ron as "a heck of a nice guy, one of those very nice people you meet in life."

Jan Evans was again settling into the life of a squadron wife. As when Ron went on his first two WestPac cruises, the camaraderie was not limited to the pilots themselves, and Jan appreciated the feeling of being a part of a big extended family. The wives had a real sense of belonging. Most navy pilots valued a shore assignment when they had young children to watch growing up, but when the whole family was together at Monterey, Jan had missed the comforting embrace of squadron life. She had quickly discovered that the relentless requirements of the NPS courses meant that for most of the week, Ron might as well have been away on a ship for all they saw of him.

As the wife of a Screaming Eagle pilot, Jan soon met Sybil Stockdale, the quintessential skipper's wife. Jan recalls an elegant lunch at the Stockdales' home with Sybil as the perfect hostess, offering the same reassuring friendship, guidance, and advice that had been such a feature of squadron life in VF-142. Jan knew that Commander Stockdale was very well respected by all who worked with or for him, but she never actually met him. Only a couple of weeks after Ron reported for duty, Stockdale returned to San Diego and became the carrier air group commander on the USS *Oriskany*. On 9 September 1965 he was shot down and captured by the North Vietnamese and became the most senior American pilot POW.

While the ship was away, Ron and Jan continued to exchange taped messages, although the mail flights weren't as regular as on training cruises. Unlike the old-fashioned letters from Ron's bachelor days, the more "high-tech" recorded messages have not survived, mostly because each message erased the last. Whatever else Ron told Jan in his recordings, he didn't say anything about operational matters, and he particularly didn't say anything about what had happened in the Gulf of Tonkin in August. Many of the men on board *Ticonderoga* were able to draw their own conclusions. Commander Stockdale's frustration on returning that night of 4 August had been witnessed, but the old World War II admonition that "loose lips sink ships" still made sense. As far as Jan was concerned, she had no reason to doubt the accuracy of the reports she read in the newspapers and saw on TV. When Ron came home from his first combat cruise, he limited his account of the military operations to generalities, and if he had indeed pieced together what happened ten weeks before his arrival on board the ship, he didn't discuss it with Jan or anyone else, to her knowledge. Further, she knew not to press Ron about military goings-on.

Even the ship's official cruise book, perhaps a victim of the "fog of war," is of very limited value for historians. Unlike other cruise books, the 1964 edition does not contain pilot names, photographs, or details of the squadrons that took part in the cruise. The official cruise chronology also bears little resemblance to the ship's final itinerary.

The increased tensions in the region meant that the U.S. Seventh Fleet maintained a permanent presence off the coast of Vietnam, initially at the coordinates 16 degrees north and 110 degrees east. Officially called Point Yankee, it was known to all as Yankee Station. Even more colloquially, a carrier at Yankee Station was "on the line."

Whatever they knew of the events of August, the fighter pilots of *Ticonderoga* who were on the line during October knew or had good reason to believe that each new day could pitch them into conflict with an enemy whose shores were usually visible from a patrolling Crusader. Marlo Holland graphically recalls how it felt: "Well, it was like we were at war! We weren't officially at war, but we were on combat duty. And every day we would sit to have a briefing, and we had to wait for the president of the United States and Defense Secretary McNamara. They were micro-managing every day, and we were taking off on flights all day long, going out—and I think we were trying to pick a fight."

Ron and the other pilots now received extra combat pay, which indicated to them what might be expected of them. Every day, and sometimes twice a day, they launched in their Crusaders with a full load of 20mm ammunition, four Sidewinder missiles, and a batch of five-inch Zuni air-to-ground rockets. That weaponry meant they could engage with enemy aircraft, or attack a target on the ground, or both. What actually happened was that they flew up the Gulf of Tonkin, with North Vietnam to the left and Red China's Hainan Island to the right, then returned to the ship. They were "all dressed up with no place to go." Sometimes they orbited *Tico*, providing aerial protection from an enemy that did not appear. At other times, they carried out air-refueling exercises. While not an unusual thing for naval aviators to do, for all Ron and Marlo knew, it might have been intended as a signal to distant watchers that they were ready to mount long-distance raids.

No MIGs attacked, although aircraft movements were visible over Vietnam and Hainan. The American pilots wondered whether Washington's plan was to flush out the MIGs. They weren't actually sure that North Vietnam *had* any MIGs at that early stage. Even if they did appear, would they be flown by North Vietnamese, Russian, or even Chinese pilots? Marlo was convinced they were preparing for the opening round of a proxy war with the Russians.

Phil Morrison was never a pilot. He didn't even finish high school, but he joined the navy just as his older brother Tim had. At the age of eighteen, Phil found himself alongside Tim, working as an aviation electrician in Fighter Squadron 51. On the night of 4 August 1964, he and a plane captain were assisting Commander Stockdale after his return from circling *Maddox* and *Turner Joy*. He vividly recalls an obviously frustrated Stockdale summing up the mission: "No boats, no torpedoes. I didn't see a damned thing!"

Phil revered Stockdale, not least because the skipper interceded with *Ticonderoga*'s captain so Phil could travel to the States and be with his family following the death of his father. Away on compassionate leave, Phil missed the arrival of Lt. Ron Evans, but he soon encountered the new officer upon his return to duty. Ron had been doing the rounds of the VF-51 electrical shops, introducing himself as the squadron's new avionics officer. He answered to the maintenance officer but had responsibility for the electrical shops, aircraft electronics, radio, and the like. Phil hadn't gotten along with the previous avionics officer but found the new guy to be friendly, personable, and willing to treat everyone with respect, even an aviation electrician third class like Phil. It didn't hurt that Ron also quickly recognized Phil's accent. The young noncommissioned officer (NCO) hailed from Arkansas City, Kansas.

Phil worked the night shift. It was a busy and tense time aboard the carrier, particularly on Yankee Station with a regular flow of Crusaders needing to be checked out for morning launches. Sometimes the early hours had lulls, and when Ron had checked the status of the aircraft, he would chat with his fellow Kansan about baseball or catfishing. (Apparently, you're not a genuine Kansan if you don't go catfishing.) Phil found Lieutenant Evans to be a very down-to-earth individual who did not "lord his officership" over the NCOs. He was clearly very astute about anything electrical but was always interested in *how* the engineers fixed particular problems. He was very much a hands-on officer who was happy to roll up his sleeves to help get a tricky job done. To Phil, he was "one of the finest men I ever had the pleasure to serve with."

On 29 October the USS *Constellation* had taken over at Yankee Station from *Ticonderoga*, which then steamed to Subic Bay in the Philippines, arriving on 1 November with a shipload of sailors and airmen eager for shore leave. When news broke of a Viet Cong mortar attack that same day at the Republic of Vietnam Air Force's Bien Hoa Air Base, south of Saigon, *Tico* was briefly ordered back to Yankee Station. Then the alert was canceled, and the ship made short visits to Hong Kong and Subic Bay before yet again returning to Yankee Station on 22 November. During its final six-day stint, Ron made seven more flights. Four are recorded in his logbook as attack missions, but on each occasion, Washington vetoed the potential target. Ron spent many hours keyed for action that never materialized, flying a Crusader bristling with firepower up and down the Gulf of Tonkin.

The ship came off the line on 28 November and was replaced by Ron's first WestPac ship, *Ranger*. After a one-day stop at Subic Bay, *Tico* finally began its journey home, arriving at San Diego on 15 December, which was twenty days later than reported in the cruise book.

Even as Ron and Jan settled back into squadron life under a new skipper, Bill Doak (whose wife, Charlotte, stepped deftly into Sybil Stockdale's shoes), events were moving quickly. Whether it would be considered a war or a conflict, full-scale and very real hostilities were looming in Vietnam. Following a number of Viet Cong attacks on U.S. assets, Gen. William Westmoreland, the senior U.S. military commander, requested the deployment of troops to protect Da Nang Air Base. On 8 March 1965, with the consent of Saigon, 3,500 U.S. Marines came ashore, marking the arrival of American fighting troops in Vietnam. It was all supposed to be kept low key, but the marines were met by sightseers and Vietnamese schoolgirls offering them flowers. It would not be like that for very long.

10. Rolling Thunder

War is regarded as nothing but the continuation
of state policy by other means.
—Karl von Clausewitz (1780–1831)

By early 1965, President Johnson's Vietnam policy reached the point at which he had to make a major decision: turn back or escalate. The first option meant abandoning South Vietnam and the other dominoes to the assumed communist takeover. He chose escalation in the form of Operation Rolling Thunder, a sustained campaign of aerial bombardment with a gradually increasing tempo.

Once the military campaign was escalated, how far would it go? The Soviet Union had tested its first nuclear weapon in 1949, provoking anxiety around the world that the Korean War could turn nuclear. By 1965 the Soviet Union was one of Hanoi's key allies and had become a nuclear superpower with an arsenal of hydrogen bombs. North Vietnam's other powerful ally and immediate neighbor, China, was within sight of U.S. jets over the Gulf of Tonkin and had successfully tested its first nuclear weapon in October 1964. General Westmoreland referred to "an almost paranoid fear of nuclear confrontation with the Soviet Union," but he acknowledged that a more likely threat was a direct intervention by China to prop up Hanoi in much the way the United States was propping up Saigon.

These concerns manifested themselves, particularly during the early years of Rolling Thunder, in a policy of self-denial: no strikes could be conducted within thirty nautical miles of Hanoi or ten nautical miles of the port of Haiphong. Many other apparently obvious targets also remained off-limits. The belief was that a carefully controlled policy of bombing selected targets, while simultaneously putting out diplomatic feelers, would keep China at bay and bring Hanoi to the negotiating table. As Johnson argued, "By keeping a lid on all the designated targets, I knew I could keep the control of the war in my own hands. If China reacted to our slow escalation by threatening to retaliate, we'd have plenty of time to ease off the bombing."

These early plans would establish the guidelines for the conflict into which the men of VF-51 and so many others would be pitched. But Rolling Thunder had an inauspicious start, with six American aircraft shot down during raids on 2 March 1965. That five of the pilots were rescued failed to disguise the concern.

After returning to San Diego in December 1964, *Ticonderoga* needed an overhaul. Ron and his VF-51 colleagues would be land based until September, carrying out workup training exercises to keep them combat ready for whatever lay ahead.

If Ron returned to Miramar thinking he was still a fighter pilot, he soon discovered that Washington, or the navy, had decided otherwise. "When we came home," Tom Klein recalls, "they had brand-new airplanes waiting for us that were fighter-bombers. They had modified the wings, with a bomb-rack system on the wings, and it absolutely ruined the fighter. Still the F-8E but modified so it could do air-to-ground. The air force had been famous for this. They'd build a beautiful fighter—very fast, very sleek, very capable—and then they'd hang as many bombs on it as they could, and then it becomes a big, slow barge. And that primarily is what happened to the F-8."

Amazingly, a single-seat, supersonic F-8E Crusader could carry the same weight of bombs and other armaments as a B-17 Flying Fortress of World War II on a typical long-range mission—and the B-17 had four engines and a crew of ten.

The squadron gained several new pilots between cruises. Lt. (JG) Wayne Skaggs was a nugget who had just completed flight training and conversion to the Crusader. He joined VF-51 around year's end and well remembers meeting and getting to know Ron Evans. "I would call Ron kind of 'laid back.' He wasn't excitable, just a very pleasant personality, a nice guy to get along with and work with. You couldn't ask for a nicer guy. He was very friendly, the sort of guy you could talk to about anything."

Lt. Jack Allen had been flying Crusaders in a shore-based squadron and had never landed on a carrier. Then in 1964 he was sent to VF-124 at Miramar, went through all of the necessary carrier qualifications, and was delighted to receive orders to join VF-51, where he reported in July 1965. He received a warm welcome, none more so than from Lt. Ron Evans. "I remember Ron introducing me to everyone. He was that kind of guy. He was a leader. He was one of the more senior guys in the squadron, and we all knew Ron as a leader. He was serious, but he always had a smile. And he seemed older, more mature, mainly because most of us had more hair than he did! But he seemed a little older, and we treated him with respect."

It's always nice to be liked, but in the navy, as in most other jobs, success and hard work are what set you apart from others. Ron knew he was a good pilot and had been receiving good annual fitness reports. He had worked hard as the avionics officer, and his efforts did not go unnoticed. During that shore-based summer, his skipper Bill Doak called him into his office for a chat. Ron said nothing to Jan at the time, but then one day in August, she answered the phone, and it was Ron calling from the base. The normally laid-back and unexcitable Lieutenant Evans was clearly excited about something. "I've got an early promotion! I'm going to be a lieutenant commander!"

This "spot promotion," on the recommendation of Bill Doak, was based on Ron's merit and achievement. The skipper had Ron earmarked for the position of maintenance officer, a squadron position usually held by a lieutenant commander. That wasn't a strict requirement, but Doak had clearly decided that Ron had earned an early promotion and a commensurate senior squadron job. A delighted Jan felt the promotion was a great honor and fine compliment to her husband.

Another new officer joining the squadron, apparently as the executive officer-in-waiting, was Cdr. Bob Ferguson. The name immediately rang bells with Ron and Jan, and the children were excited that the son-in-law of their former babysitters from Monterey—Nana and Poppa—would be flying in their father's squadron. Bob and Jane Ferguson bought a house just down the road from the Evans family. It wasn't quite ready, so as was the way in military families, Ron and Jan offered to share their home with them for a couple of weeks.

By July the fully refurbished *Ticonderoga* was back in action. VF-51 received orders to conduct pre-cruise carrier qualifications on the familiar deck, by day and by night, and then to embark "for temporary duty in connection with WestPac deployment for a period of about nine months starting on or about 25 September 1965."

As preparations were being finalized, several of the younger squadron wives approached Jan Evans with a question. They had all become close friends, and the younger women recognized that Ron was now one of the four most-senior officers in VF-51. The wives wondered whether Jan had heard anything about the forthcoming cruise that might affect their husbands. Had something happened, or was something planned that they ought to know about? They were undoubtedly concerned about the consequences of the escalating Operation Rolling Thunder. Jan found the question easy to answer: she had heard nothing, from Ron or anyone else, that gave her any better insight into what lay ahead for their men. The squadron was going back to war. That was enough to worry about without cranking up concerns with empty speculation.

The ship actually departed on 28 September, but first VF-51 underwent a change of command. After a year as skipper, Bill Doak was heading on an upward trajectory to the aircraft carrier USS *Bonhomme Richard*. He had been something of a mentor to Ron, who was sorry to see the departure of the quiet, congenial Doak. The new skipper, moving up from executive officer (XO), was Cdr. Charles McDaniel, a man with even less hair than Ron and whose tough, unsmiling demeanor had earned him the nickname "Nails." Wayne Skaggs thought the name was well deserved. Jack Allen, who hadn't been in the squadron long enough to know Bill Doak well, found McDaniel to be demanding but fair. He considered him a good combat leader and liked him for that. Bob Ferguson succeeded McDaniel as the XO.

Back home, Charlotte McDaniel became the new skipper's wife. Unfortunately, she was profoundly deaf. Although she could lip-read, the squadron wives recognized that Charlotte simply couldn't be the guiding light that Sybil Stockdale and Charlotte Doak had been. Sitting with a group of women all trying to express their concerns, Charlotte could never hope to gauge the mood of the room when she could only follow one speaker, assuming that speaker was facing her.

Further, an early rift developed with Charlotte and Jane Ferguson on the one side and the rest of the wives on the other. As the ship steamed toward Vietnam, a meeting was called to discuss porch lights. Whether it was Charlotte's idea or a suggestion from the navy was never clear, but she announced to the wives that in the future they had to leave their porch lights lit. Night and day. When (not if) word came that a pilot had been killed in action,

Charlotte or Jane would then be able to locate his family's home more easily. Furthermore, Charlotte and Jane would be mounting a series of trial runs. "Click those switches, ladies!"

This instruction went down very badly with Jan and the other wives. They unanimously decided that none of them would leave their porch lights on. No one wanted a permanent reminder of what they all knew *could* happen. Their husbands were superbly trained naval aviators, and that was their best protection. The wives weren't going to hang around, waiting to be told they had been widowed. Their relations with Charlotte and Jane never really recovered from that.

In early November *Ticonderoga* arrived in the South China Sea and took up position off South Vietnam. Since May 1965, Dixie Station—located at 11 degrees north, 110 degrees east, about eighty miles from the nearest air base at Cam Ranh Bay—had operated as a counterpart to Yankee Station. While missions from Yankee struck targets in North Vietnam, strikes from Dixie generally were targeted at Viet Cong insurgents and provided aerial support for American and South Vietnamese ground troops.

On 5 November Ron's first operational mission at Dixie Station was supposed to be a four-plane, high-altitude training exercise. It turned out to be a low point in his navy career. As the sun dipped in the west, Ron's wingman, Lt. Roy Miller, made the first launch. Ron was the flight leader and made the last launch. The plan was to rendezvous at twenty thousand feet, but as Ron arrived at the meeting point, Roy was temporarily in the lead. Ron indicated by radio and hand gesture that he was taking the lead, but for some reason, this gave rise to confusion. Ron described what happened in a 1986 interview: "I assumed I had the lead, and he still thought he had the lead. So I'm tootling along and here he comes—wham!—and hits my wing. Knocked my wing sideways. And the end of my wing hit his nose-wheel, or in the wheel-well area. The end result was he couldn't get his nose-wheel down. And he also hit my tail and knocked a piece off."

As the squadron's maintenance officer, Ron often sat down for dinner with Lt. Cdr. Gail Bailey, his opposite number in VF-53. Gail heard a blow-by-blow account of the collision and learned that the two Crusaders briefly locked together, causing both to stall and spin. By the time the pilots had separated and brought their aircraft under control, they had lost ten thousand feet. Each pilot checked out his own aircraft; then he conducted a visual inspection of the other.

The only significant damage to Roy Miller's plane seemed to be a jammed nosewheel, so he went back to the carrier, where the crash barriers were raised to prevent the F-8 from sliding too far on its nose. He was uninjured, and the aircraft suffered relatively minor damage.

For Ron the consequences of the collision were much more severe. His whole wing was slightly skewed, requiring him to apply more rudder to fly straight. The loss of part of the tail didn't seem too significant, but he had one major problem: the damage meant that he was unable to raise the Crusader's wing to make a normal landing. He talked it over by radio with the ship, and they decided that a landing on the carrier was too risky. Ron was ordered to fly to Tan Son Nhut Air Base outside Saigon, where he made "a beautiful landing considering the condition of the airplane." This was an important qualification. Without the wing raised, he had to make a much faster approach than normal and with a very nose-high attitude. The tail of the aircraft then dragged along the runway in a shower of sparks.

Slowing to taxi speed, Ron maneuvered the aircraft toward the hangars. He was met by the base duty officer, who stared incredulously at the bent and buckled Crusader.

"My God! What hit you?"

"My wingman!"

To add insult to injury, Crusader No. 104 actually had his name, Lt. Cdr. Ron Evans, stenciled on it for all to see. It was eventually shipped to the States on the USS *Bonhomme Richard* for repairs. Bill Doak had joined the ship by the time the plane arrived in San Diego, and he invited Jan Evans on board to have dinner with him and his wife, Charlotte. Jan even got to see Ron's battered aircraft for herself.

There were no hard feelings between Ron and Roy Miller, and Jan and Diane Miller obviously saw a funny side to what could have been a serious incident. The wives recorded a duet of the popular Bing Crosby song "Side by Side" and posted the tape to their husbands, adding to the ribbing they both received from the other pilots.

The collision certainly shouldn't have happened, but because no one was hurt and neither pilot had to eject, the incident faded into the background. The surviving pilots of that cruise barely remember it. None correctly recalls that it was Roy Miller who landed on the ship and Ron in Vietnam. Worse things happen at sea.

According to Wayne Skaggs, "I think the sort of thing that happened there could have happened to any of us. I thought I got hit one day by one of my wingmen. And he swore he didn't, but I tell you it pushed my airplane, and I felt the bump. It's just the case that at times, when you're flying in an area like that, you're all eyeballs all over the place all the time, even when flying in formation, and something gets your attention for just a second, causing you to be somewhere you shouldn't be."

It takes two to tango, but as U.S. Navy property had been damaged, reports had to be written. The bottom line was that Ron was the senior man and, more significantly, the flight leader. He therefore accepted responsibility for the collision.

Ron Evans was never a man to dwell on the past or waste time agonizing about things he couldn't change, but it must have irked him that his logbook now contained a single blemish after nine accident-free years. And it all happened on his first day on the line.

Ron was back in action four days later on the first of fourteen attack missions over South Vietnam. There were often two missions per day. On some of the flights, the targets were briefed in advance, but it was not uncommon to be told of the target only after launch. Sometimes V F-51 would attack Viet Cong supply trucks or convoys bringing men and weapons into South Vietnam, but the most satisfying attack missions from Dixie Station were in support of U.S. Army or Marine ground forces. The enemy often fought an ambush style of warfare, cutting off and trapping U.S. forces. The navy pilots would then take on the role of the cavalry, riding to the rescue. A forward air controller would identify the location of the enemy troops, and the Crusaders would attack with a combination of bombs, 20mm cannon, and Zuni rockets. The Zuni unguided rockets had to be lined up on the target while the Crusader pilot was in a steep dive, using his gunsight as an aiming guide.

The satisfaction from those ground-support missions came when word filtered back that such-and-such army or marine unit had sent a big thank you to the navy guys who had saved the day for them. Ron had a personal stake in those missions; he knew that his younger brother Jay was somewhere down there on the ground with the U.S. Marines. Interviewed only a few months before his death in December 2017, the much-decorated Jay Evans couldn't resist pointing out with a chuckle that when Ron and his fellow pilots had completed their missions, they got to fly back to their ship and spend the night in their nice, comfortable bunks. Meanwhile, the leathernecks had to make do with the muddy ground of a jungle clearing and the constant risk of an enemy counterattack.

For their actions during November 1965, Ron and thirteen other VF-51 pilots received Air Medals in a ceremony conducted on board *Ticonderoga* by the ship's commanding officer Capt. Robert N. Miller. The award citation, bearing the name of Vice Adm. Paul Blackburn, Jr., outgoing commander of the U.S. Seventh Fleet, referred to "meritorious achievement in aerial flight . . . during missions in support of combat operations in Southeast Asia against the insurgent communist guerrilla forces." It concluded: "Your courage and devotion to duty in the face of hazardous flying conditions were in keeping with the highest traditions of the United States Naval Service."

The pilots didn't fly combat missions to generate awards. If they got one, good, but nobody openly kept a tally of who got what. An award wouldn't add much to an obituary in the *Saint Francis Herald* or some other local newspaper. Most men of the squadron came home with similar awards because they all flew the same missions, and it just wasn't in their DNA to do it badly.

In their early months over North and South Vietnam, *Ticonderoga* pilots shared several gripes. They found it highly frustrating to be refused permission to bomb enemy airfields and surface-to-air missile (SAM) sites because of political constraints. The pilots wondered if Washington had ever heard the proverb "A stitch in time saves nine." Even when Washington authorized strikes, the pilots were still hamstrung by a lack of ordnance. The Crusaders of VF-51 often took off in groups of four, each carrying a single five-hundred-pound bomb and two Zuni rockets. Wayne Skaggs recalls, "One airplane could have carried the whole damned load! It was really frustrating, because they were counting sorties, and we didn't have the ordnance to put on the airplanes. We kind of resented that they were putting us in harm's way. They would launch us on these missions when there was no need to have so many pilots and planes engaged, but they wanted to count sorties! That seemed to be all that mattered."

Could they have flown half the sorties with double the ordnance per aircraft? "Absolutely, we could have done that, but that wouldn't have made

McNamara happy. He was the guy pulling the strings up there in Washington, and they were more interested in how many flights were flown than actually how much was accomplished."

The 1965–66 cruise coincided with the monsoon season in Southeast Asia. On many days throughout the cruise, thick clouds and heavy rainstorms made accurate targeting impossible. As Wayne recalls, "We did a lot of night flying as well, and they would send us out on nighttime missions or even bad-weather missions where there was no way to get to the target, and we actually dumped a lot of ordnance in the ocean that was never put on target. We'd come back, and they would count those as sorties flown, but we hadn't accomplished anything."

When the clouds cleared and the Crusader pilots could see their targets, the defenders on the ground could also see them. As Ron Evans discovered in those early months, "Any time you go over North Vietnam it's very highly defended by ground-fire, anti-aircraft fire—and there are tracers going by all the time. How in hell they miss you I'll never know. Generally, every fifth bullet is a tracer and that's all you see. The rest of the stuff, you don't see. And, you know, you're going in there and dropping bombs and strafing gun-emplacements, and this type of thing. You can't drop the bomb from straight-and-level, it'll never hit anything, so you've got to get at least a thirty-degree dive at the release point, so that when it releases off the airplane it goes where you're pointing it, according to the bombsight. So how do you live through that stuff? I'll never know. We didn't have anyone shot down in our squadron. Basically, we made it!"

In 1986 Ron was asked how he kept going. He replied, "Duty. Job. That's what you're out there for. That's what you signed up for. You pays your nickel, and you've got to take what you get."

He allowed himself a wry observation about what he got: "Anytime you're flying and people are shooting at you, it sort of changes your outlook about the enjoyment of flying airplanes!"

Not everyone who was shooting at Ron missed the target. "I picked up a couple of bullet holes in the airplane. It's a tough plane, but everything's a close call!"

On returning to the relative safety of *Ticonderoga* after one of his sorties, Ron turned his aircraft over to the maintenance staff working in the ferocious heat and humidity belowdecks and asked the men to check for any problems and to have it ready for the next day. Phil Morrison recalls the damage because Ron's Crusader was one of the first in the squadron to get shot up. "It had pretty good-sized holes in it, and I would say it was antiaircraft from the ground. I was able to put my finger in the bullet holes. They were in the wings and the forward fuselage where it had the instruments and stuff like that."

But not even being struck by enemy fire provoked Ron to record the incident in his logbook. Such an entry would have been superfluous to navy requirements, and unlike the mid-air collision, it most definitely was not an accident.

Back on the streets in the United States, something odd was happening. After the two world wars, returning American servicemen were greeted warmly as heroes. They even got ticker-tape victory parades. The Vietnam War was different. Although the armed forces and the perceived goal of pushing back the tide of communism had broad support, a growing anti-war movement had become increasingly vocal. Things would get much worse, with returning servicemen being greeted with chants of "Get back on the plane!" and "Baby killer!" But even while Ron was away on the 1965–66 combat cruise, some demonstrators were prepared to vent their spleens on the families of those who were fighting the war.

The first letter Jan received had her name and full postal address clearly written on the envelope, although the white paper was splattered with stains of a reddish-brown color. Puzzled, Jan tore open the envelope and read the enclosed letter with mounting revulsion. It, too, was heavily stained with red blotches that she later assumed were from some kind of animal blood. But the wording of the letter was what shocked her. The anonymous writer accused her husband of being a murderer and a war criminal. How did these people know her name? Her address? Were they watching the house?

It was the first of several such letters Jan received, all stained with dried blood. Initially she was shocked that some of her fellow Americans would indulge in such behavior. Then it just made her angry. "It was awful. The first time I got one I thought, 'What the heck . . . ?' And then it just pissed me off! You have to live a normal life to raise your children. So if there is something you can do nothing about it, you accept that, and you go on. And you don't let it get to you or bother you."

Fortunately, this campaign was limited to letter writing. Nobody approached the house or mounted any anti-war demonstration on Mount Tami Drive.

Nevertheless, Jan always made sure the doors were locked, and she realized how wise Ron had been to bring Flieger, the German shepherd, into the family. Nobody would get past Flieger.

Amid the increasing heat of the Vietnam conflict, the Cold War was never far away, and on 5 December 1965, it claimed the life of a *Ticonderoga* pilot. While the ship was en route to Japan for a spell of shore leave, Attack Squadron 56 (VA-56) conducted an exercise that simulated preparations to launch a Skyhawk jet with a one-megaton Mark 43 hydrogen bomb. Lt. (JG) Douglas Webster was strapped into the cockpit, canopy raised, as the ground crew rolled the Skyhawk onto the aircraft elevator for delivery to the flight deck.

According to witnesses, the ship then rolled, accelerating the aircraft's movement. Before anyone realized what was happening, the Skyhawk rolled right over the side of the ship, striking the water upside down and taking Webster and the hydrogen bomb to their final resting place sixteen thousand feet deep in the ocean.

VF-51 electrician Phil Morrison remembers general quarters being sounded, and everyone proceeding to their stations. They all quickly learned about the loss of the pilot, but news about the nuke only came later. Phil himself had received some training in nuclear techniques and realized that the unarmed bomb had a fusing mechanism with so many safeguards that it couldn't possibly detonate. Probably most of the sailors' minds grasped that, but convincing their guts was another matter. More than a few of the crew couldn't help wondering about the effects of extreme water pressure on those nuclear fusing mechanisms.

The death of Douglas Webster is recorded without explanation in *Ticonderoga*'s cruise book. The circumstances of his death and the loss of the hydrogen bomb were not acknowledged by the United States until 1989. Jan Evans never heard about it from Ron.

As Christmas approached, two of VF-51's pilots were feeling particularly wistful. The twenty-second of December was Ron's eighth wedding anniversary and his first without Jan. It was even worse for his friend and colleague Jack Allen, who had married on 26 December the year before. "That's the reason I got out of the navy! I would have stayed in, I loved it . . . but that damned war! My first anniversary—I'm in Vietnam! My second anniversary—I'm in Vietnam! And I thought, 'This is not conducive to family life!'"

But there was some consolation for Ron, Jack, and the rest of *Ticonderoga*'s crew when the legendary entertainer Bob Hope brought his group's Christmas show to the ship, five miles off the coast of South Vietnam. Addressing a huge crowd of excited sailors and airmen, ship's captain Robert Miller welcomed Hope aboard and tried to out-quip him by saying, "This is the greatest thing that's happened to the *Ticonderoga* since the last typhoon."

In one dramatic incident, as the entertainers were up on Vulture's Row watching several Crusaders of VF-53 return from operations, a junior officer came in too low and struck the ramp at the rear of the ship. The mangled Crusader bounced across the deck, but the pilot ejected just before his wrecked aircraft plunged into the sea. He was quickly recovered by helicopter, suffering only a bruised shoulder.

The next day, during the show, Bob Hope couldn't resist referring to the previous evening's crash, admitting that it was "a hard act to follow." After the jokes and the music, a young singer-actress named Joey Heatherton appeared on stage in a figure-hugging black sequined leotard and performed an energetic dance routine that wowed the two-thousand-strong audience. A dozen young sailors jumped on stage to join her. Joey's reward was to be inducted into VF-51 as an honorary Screaming Eagle.

For the weary ship's complement, it was all a welcome distraction from the grim business of waging war and a reminder that they had not been forgotten.

11. Per Ardua ad Astra

In peace there's nothing so becomes a man,
As modest stillness and humility;
But when the blast of war blows in our ears,
Then imitate the action of the tiger . . .
—William Shakespeare (1564–1616)

After Ron's phone call from Alan Shepard in 1963, he had given little further thought to a space career. A door had briefly opened, revealing an exciting new world, but it had closed again with Ron on the outside. He never contacted NASA to check if the program was going to recruit more astronauts. Instead, he enthusiastically pursued his career as a naval aviator, particularly following his early promotion. But when *Ticonderoga* was in Hawaiian waters in early October 1965, Ron received a letter, out of the blue, from BUPERS. As in 1963, the key passage read: "It is a pleasure to inform you that you possess the basic qualifications required by NASA and the Department of the Navy for astronaut training." BUPERS wanted to know by 20 October if he wished to be "listed as a volunteer."

On 4 October Jan received a call from Ron in Hawaii, telling her about the letter. There was no doubt in her mind that her husband should accept the proposal, and he duly contacted BUPERS to confirm that he wanted to be considered. Ron knew he had passed all the medical tests in 1963, but this second time around, he would have to jump through a few extra hoops that no other astronaut applicant has ever had to face.

On 12 November BUPERS wrote again. A bundle of papers was forwarded to Mount Tami Drive, San Diego, where Jan read the cover letter with a mixture of delight and despair. BUPERS had convened a selection board in late October and had recommended Ron to NASA. Attached to the letter were medical records and application forms to be completed and forwarded to NASA "prior to 1 December 1965." That left ten days, but there was a big problem: at that moment Lieutenant Commander Evans was somewhere in the western Pacific fighting a war. Jan knew there was no possibility of meeting NASA's deadline. She sought advice from Bill Doak, who was home on leave. He made some inquiries and provided Jan with a vaguely familiar name and telephone number.

When Deke Slayton's secretary told him a Mrs. Jan Evans was on the line, wanting to talk about her husband's astronaut application, Deke took the call and listened sympathetically to Jan's concerns. A combat veteran who had flown missions over Italy and Japan, he knew only too well that postal deadlines are hard to keep when you're at war. He assured Jan that if she posted the forms to Ron via the navy's mailing system, NASA would make allowances and accept a late application in these particular circumstances. In later years, Ron always said that his wife had volunteered him for astronaut selection.

The formal application to NASA was dated 7 December 1965. Despite Slayton's assurances, Ron was clearly concerned that government bureaucracies do not lightly ignore application deadlines. His cover letter noted: "The unavoidable delay in receipt and submission of the forms was due to the poor mail service connected with the operational commitments of my squadron and its aircraft carrier in fighting the war in Vietnam. It is strongly hoped that this application . . . is received in time for consideration."

However, unless and until NASA and the U.S. Navy said otherwise, Lieutenant Commander Evans had other matters to attend to. During that West-Pac combat cruise, *Ticonderoga*'s naval aviators spent a total of 116 days on the line mounting air operations, with the time divided fairly evenly between Yankee and Dixie Stations.

As a fellow veteran of the 1964 cruise, Tom Klein knew Ron very well, and they often flew together on sorties. He was aware the navy only selected the best pilots to be F-8 instructors, so it didn't particularly surprise him that Ron was in the running to be an astronaut.

"Ron was an extremely professional naval officer, an excellent pilot, and just an all-round good guy," Klein recalled. "I can see why they wanted him in the astronaut corps: he had the education, a master's degree; engineering skills; great mental aptitude—and that's exactly what they were looking for to fling out into space!"

Tom's previously stated concerns about the navy's best fighter aircraft being turned into a light bomber were borne out. The pilots of VF-51 often flew flak-suppression missions during which they would precede the main force of Skyhawk bombers. The Crusaders could each carry two two-thousand-pound bombs to "soften up" the area around the main target, often a bridge or other heavily protected structure. The aim was to silence North Vietnam's 37mm radar-guided antiaircraft guns so the Skyhawks would have a clear run.

The problem of a lack of ordnance, noted by Wayne Skaggs, eased by early 1966. The squadron now mounted "just about any kind of mission you could imagine." On Yankee Station, combat air patrols either protected the carrier and its escorting destroyers or flew farther up into the Gulf of Tonkin, always on the lookout for enemy MIGs. The Crusaders would often provide close protection for Lockheed EC-121 early warning aircraft. They attacked enemy boats along the coast of North Vietnam or supply trucks on the road network. On Alpha strikes, large waves of aircraft from as many as three carriers mounted coordinated attacks on major targets. When based at Dixie Station, they carried out multiple ground-attack missions to provide aerial support for friendly forces.

Ron had experienced the enthusiasm of soldiers on the ground shooting at attacking aircraft, but the even deadlier risk was from surface-to-air missiles, which could home in on his Crusader's radar even through clouds or darkness. Each day a special code word was used to identify the SAMs. On one mission Ron had just heard by radio that the daily code word was "telephone pole." Barely had this registered than the VF-51 pilots were alerted to expect an incoming telephone pole and to take evasive action. "So what you do," he later explained, "is you haul [the F-8] right up and stall it—upside down—so you can get the airplane pointed down toward the ground. And the 'telephone pole' went right up beside me. That's a close shave!"

Over the Christmas period, a letter bearing a Houston postmark negotiated the navy's delivery service, eventually catching up with Ron on board *Ticonderoga*. To his relief, his application had been safely received and, in spite of the delay, accepted by NASA. The key passage read: "I am pleased to inform you that you have been selected to participate in the final phases of the Manned Spacecraft Center's Pilot-Astronaut selection program." He had made the short list and was asked to attend physical evaluations at Brooks Air Force Base, San Antonio, and a final evaluation in Houston. The letter requested his discretion and "as little discussion as possible" regarding the process and was signed by Donald K. Slayton, the assistant director for flight crew operations.

Formal orders from BUPERS dated 3 January 1966 directed Ron to present himself at Brooks AFB not later than 1800 hours on 10 February. He didn't know it, but out of five thousand applications, he was one of only forty-four men being considered. But at this point, an unexpected barrier appeared in the middle of Ron's road to NASA. As a serving naval officer engaged in combat operations, regardless of the BUPERS orders, he had to make a formal request to his squadron commander for leave of absence.

McDaniel said no.

No? Ron discussed the process with his skipper, who clearly recognized that he would be losing one of his top pilots for at least two weeks and maybe permanently, depending on the outcome. McDaniel simply wasn't going to allow it. He talked about the "need of service." Given that the navy had picked Ron as a suitable pilot to represent the service in the nation's space program, why would McDaniel object? In a 1986 interview, perhaps surprisingly, there is a degree of understanding in Ron's voice: "Because he was fighting a war out there, and if I left, he's got one less body to put in an airplane to fly missions over Vietnam."

McDaniel and all the squadron commanders were under pressure to launch as many sorties as possible, ostensibly to maintain pressure on the enemy. The war was a national commitment. VF-51 was part of the bulwark against the spread of communism. Perhaps there might also have been an implied criticism of a senior pilot wanting to go stateside to be a spaceman rather than fight the good fight in Vietnam. It would be trite—and disrespectful to those who didn't have the option—to ask who would rather get shot at than fly to the moon. But anyone suggesting that Ron was seeking an easier course needed reminding that he hadn't gone knocking on NASA's door. He was doing what the navy wanted him to do.

Ron was clearly conscious of the conflicting national priorities. "I had really made a commitment to my country, because I'm a military man. I decided, hey, I'm going to stay in the navy for who knows how long, but I'm going to be in the navy; that's doing something for my country. In the process of doing that, I happened to get selected for the space program, which, by that time, had become *the* mission of our nation—to send a man to the moon and return him safely to Earth, like President Kennedy said."

In assessing where Lieutenant Commander Evans would be best placed to serve his country, Commander McDaniel might have recalled former President Dwight Eisenhower's verdict in 1963 that spending $40 billion to beat the Soviets to the moon was "nuts."

Apollo had, in large measure, been born out of the national humiliations of *Sputnik*, Gagarin's flight, and the failed Bay of Pigs invasion of Cuba. Kennedy himself had argued, "The space program we have is central to the security of the United States." Had he not been felled by an assassin's bullets on 22 November 1963, Kennedy was planning to underline his support for the moon program in a speech that evening in Dallas: "The United States of America has no intention of finishing second in space. This effort is expensive—but it pays its own way, for freedom and for America."

Renowned space historian John Logsdon has observed, "By choosing a Cold War competitive arena that did not involve military or direct political confrontation, Kennedy channeled one dimension of the U.S.-Soviet rivalry into what some have described as 'the moral equivalent of war' rather than armed confrontation." Or, to put it more colorfully, men like Gene Cernan, Dick Gordon, and (perhaps) Ron Evans could be seen as the equivalent of medieval champions put forward to settle disputes in single combat, thus sparing the need for whole armies to engage in a bloody slaughter. In place of knights on horseback, there would be astronauts and cosmonauts riding rockets.

We don't know if McDaniel couldn't or wouldn't see the bigger picture, or whether he simply felt constrained by his own orders. Whatever was discussed in his office stayed behind closed doors. It remained a professional dispute between senior officers. Two of Ron's squadron mates can confirm that he said nothing to them about it at the time, and there was no squadron scuttlebutt about the issue.

McDaniel wasn't going to budge. Ron had no alternative but to take the dispute up the chain of command to the Carrier Air Group commander Jack Snyder. Ron went in pitching, pointing out that it was a great honor and a great career opportunity. Snyder agreed and went to bat for Ron. But he wasn't going to overrule McDaniel. It was a matter for more discussion and persuasion behind those closed doors.

On 28 January 1966, with *Ticonderoga* at Dixie Station, Ron and Tom Klein launched a sortie against Viet Cong insurgents and their supply chain. Both pilots carried a two-thousand-pound bomb under each wing. When the Crusader was modified to fit bomb racks, several additional switches had to be installed, including a selector switch down near the pilot's feet. This slightly cumbersome arrangement meant that the pilot had to reach down to select weapons under the left wing or the right wing.

When they identified their target, the two pilots entered a steep dive from twenty thousand feet, using their gunsights to select the trajectory of the bombs. Having each dropped one bomb, they reached down to adjust the selector switches to arm and activate the remaining bombs. But something happened at that point, and Ron pulled up out of the dive with a bomb still hanging under his right wing. It wasn't clear if his aircraft had been struck by ground fire or had simply developed a fault, but when they rendezvoused over the sea, Ron gestured to Tom that his Crusader had suffered a complete electrical failure. He was unable to jettison the bomb. That ruled out a return to the carrier. Captains are very protective of their ships if someone tries to bring back a hung-up and probably armed bomb that could detach during the sudden jerk of an arrested landing.

Tom radioed *Tico*, and they were ordered to divert to Cam Ranh Bay, a U.S. Air Force base that was still under construction. It had no concrete or asphalt runway, but since October 1965, there was a temporary landing strip surfaced with Marston matting, which consisted of interlocking metal plates. The strip was long enough to accommodate Crusaders in reasonable conditions, but on that day, the monsoon had been doing its worst on the coast. The metal surface was slick with rain.

As they prepared to land, Tom took the lead with Ron on his wing, unable to communicate. Tom could see the sheen on the metal surface as he touched down. He tried his brakes but with no initial effect, so he decided to shut down his engine to help the aircraft to decelerate. He glanced across at Ron and gave him the "cut" signal, drawing his hand across his throat. Ron may not have seen it and did not cut his engine at that point. He was struggling to fly an aircraft with a completely dark instrument panel, not to mention almost a ton of extra weight dangling under his right wing. The two Crusaders hurtled toward the end of the metal runway with both pilots almost standing on their brakes. Tom shuddered to a halt at the very end of the runway, but Ron skidded off the metal and traveled twenty feet on the overshoot strip, finally coming to an abrupt stop with his wheels buried in sand. The bomb remained attached.

Tom explains what happened next: "You can imagine the air force was not very happy! The next thing I knew, a jeep showed up with an air force colonel, and boy, was he mad! He was the commanding officer of the base, and he started berating Ron and me as we got out of our airplanes. We didn't even look at Ron's plane. We didn't want to go near it because of the bomb . . . so we moved away, and the colonel was shouting orders to his troops to get a tow-bar to tow my aircraft off the runway. He was just irate because we had closed his airport, and he had two flights of F-4 Phantoms on a mission, and now they had to go somewhere else."

The colonel put Ron and Tom in his jeep and drove them to his office, berating them all the way and telling them that navy pilots couldn't do anything right. They ended up standing in front of his desk as he continued to excoriate them, their flying, and naval aviation in general. The colonel had every right to be angry that his vital wartime airfield was temporarily out of action, but instead of blaming enemy action, bad luck, or bad weather, he vented his spleen on the only visible targets.

The tongue-lashing was finally interrupted when an air force sergeant came to the door inquiring if a Lieutenant Commander Evans was there. Ron identified himself and was handed a message from *Ticonderoga*. The ship was sending the mail plane with specialists to dispose of the bomb and examine the aircraft.

Ron hadn't been smiling as he incurred the colonel's wrath, but on reading the final paragraph of the message, his eyes lit up, and he grinned from ear to ear. He handed the message to Tom, who read it and immediately congratulated Ron, shaking his hand and breaking into a grin of his own.

Almost forgotten, the colonel demanded to know what the heck was going on and what was so funny. Tom handed him the message. The colonel read about the plans to resolve their problem; then he saw the last paragraph: "By the way, congratulations Lt. Cdr. Evans, you're to proceed to Texas in connection with astronaut selection."

For a moment, the colonel seemed to have lost the power of speech, but then he got his second wind and started expressing his views about the quality of the people NASA was now letting into the astronaut program. By then, Ron was completely immune to his remarks. What the brief message had told him was that his skipper had withdrawn his objections and was approving his visit to Texas for the interview process. The barrier had been lifted.

Over the years, Tom Klein came to believe that the message was actually confirming Ron's selection as an astronaut. Why else would Ron have reacted with such obvious delight just to learn he was going to be lightly tortured by a bunch of doctors? Of course, Tom didn't know about the dispute between Ron and Commander McDaniel. That, and the input from Commander Snyder, had all remained confidential.

When Ron returned to the ship, he was able to read his full set of orders, dated 28 January, that required him to proceed initially to Brooks AFB for "about fourteen days." Crucially, the document was signed at the bottom by C. B. McDaniel.

But there was a sting in the tail. When discussing the arrangements, McDaniel directed Ron to take the most direct route to Texas. He was not to take a detour through San Diego to see his wife and family. As Ron explained, "He was a funny individual, the skipper. His reasoning was, you know, that the rest of his squadron couldn't 'stop by' and see their wives, so why should I be allowed to do it?"

That was true, if a little officious. Ron had a solution that didn't breach his orders or incur extra costs to the navy. He was flying by military transport via Honolulu to Travis AFB and by commercial carrier from San Francisco to San Antonio. But if *he* couldn't go to San Diego . . .

Ron had written to Jan shortly after receiving the original news from NASA about the tests and interviews, but she had no idea how it would be done. On the morning of 8 February, Jan was washing her hair when the phone rang. At first irritated at the untimely interruption, she picked up the receiver and was surprised and delighted to hear Ron's voice. He was calling from Honolulu and only had a few minutes while his aircraft was being refueled.

"You've got to meet me in San Francisco tonight!"

"I can't do *that*!"

"It's only money. Just meet me!"

At times such as this one, navy bonds of friendship made all the difference. Diane Miller and Charlotte Doak helped Jan with the arrangements, neighbors Ray and Betty Snowden agreed to look after Jaime and Jon, and Ron's old friend Ken Baldry drove Jan to the airport. This effort involved probably a more relaxed definition of "discretion" than Deke Slayton had meant in his letter to Ron, but what got talked about in Jan's circle stayed in the circle.

At the airport, Jan saw "the most beautiful sight in the world." Ron was there waiting for her. They found accommodations, spoke to the children on the phone, and flew to San Antonio the next morning. Arriving at Brooks AFB, they checked into a motel, but after Ron had reported to the Aerospace Medical Center, he returned with surprising news. Courtesy of the administration officer, they had been allocated guest quarters on the base, complete with bedroom, kitchen, bathroom, living room, and color TV—all for a dollar a day! Most of the NASA Group 5 applicants had to share living quarters. There was clearly an advantage in bringing your wife with you.

Capt. Lawrence Enders, the chief of the Flight Medical Evaluation Section at Brooks, tried to encourage the applicants by describing their imminent examination process as "pleasant and enlightening, and even exciting at times." He was fooling no one. Some of his victims had firsthand experience of the anything-but-pleasant indignities to be inflicted on them. Most of the rest had talked to people who knew what to expect.

Al Worden described the process as "brutal." Charlie Duke described being "poked, prodded, tested, and analyzed from every angle, in every conceivable position, and from every opening of our bodies." Don Lind wasn't expecting the Spanish Inquisition, but he ended up attributing many of the procedures to Grand Inquisitor Tomás de Torquemada. Ron Evans considered the process "degrading."

The Group 5 applicants were spared some of the indignities endured by the original Mercury applicants. Those pioneers all had to give themselves regular enemas until Lt. Charles "Pete" Conrad led a one-man revolt; thereafter, medical staff did the honors. But many procedures survived. In *The Right Stuff*, Tom Wolfe graphically describes some of America's finest aviators waddling uncomfortably along hospital corridors, bowels sloshing with radioactive barium. They all had plastic tubes emerging from between their legs and up from beneath their hospital gowns. The tubes led to clamps, which temporarily held back the seething tides of barium. Ron was able to confirm the report: "That's really true."

They endured all the expected tests such as heart monitoring, treadmill tests, brain function and memory tests, centrifuge sessions, balance tests, and the effects of hypoxia in an altitude chamber. Also administered were eye tests, hearing tests, blood tests, kidney tests, and liver tests. Every organ was exhaustively probed.

Then came the psychological testing, with all the stuff the group's virgins had been warned about. For the second time, Ron encountered the Rorschach inkblot test and the blotless blank sheet of paper. ("For God's sake, *don't* say it's upside-down like Conrad did!") They were questioned about their personalities, motivations, hobbies, mothers, families, faiths, likes, and dislikes. To Ron Evans, the worst part was not really knowing what his inquisitors were looking for. He knew how to be a pilot. He had a pretty good idea what was expected of an astronaut. But then came this "pot-load of stupid psychology tests." Like most of the others, Ron tried to work out whether to tell them what he was actually thinking or what he thought they *thought* he should be thinking. Ultimately, he came to the conclusion that the whole purpose of all those crazy tests was to find ways to disqualify them. Which of these would-be heroes had a glass jaw, an Achilles' heel, or an allergy to kryptonite?

Some of the latest batch of lab rats didn't make it out of the Brooks maze. That left thirty-five candidates. Deke Slayton's letters to Ron and the other forty-three finalists had assumed they would pass the medicals, but Ron's orders from BUPERS only covered "about two weeks." He was relieved when additional orders came on St. Valentine's Day directing him to report by 27 February to NASA official Jack Cairl at Houston International Airport to attend "such additional places as NASA directs." This temporary duty was expected to last "about 17 days," but someone at BUPERS was clearly looking out for Ron. He was also authorized to take up to ten days' leave to allow sufficient time to jump through all of NASA's hoops.

Ron anticipated that the message might prove unpopular in certain quarters: "BUPERS had to send some more orders to extend my previous orders. They had to send those back to the ship to get the extension, and I could just see the skipper going through the ceiling when he got that!"

Commander McDaniel's worst fears were coming to pass. The BUPERS message anticipated that Ron would now be away for up to six weeks rather than the original estimate of two weeks. McDaniel's reply to BUPERS included a request: "Unless Lt. Cdr. Evans can be returned to VF-51 in a reasonable time, request qualified second-tour F-8 pilot as replacement."

The response came the next day: "Due unavailability, requested replacement regret negative. Lt. Cdr. Evans estimated departure Houston 5 March."

On their way to the interviews, Ron and Jan arranged with their neighbors for Jaime and Jon to fly to Houston. The excited children were reunited with their father for the first time in five months. Ron organized guest accommodations at Ellington AFB, finding once again that having a family in tow opened many doors, mainly into classier quarters.

They had several days to spare and decided to tempt fate by driving around the area adjacent to the Manned Spacecraft Center to look at the houses. They saw places with inviting names like Nassau Bay, Taylor Lake, and Timber Cove, but they decided if everything worked out, the El Lago community looked as if it would be a great place to live.

The Rice Hotel, which opened in downtown Houston in May 1913, played a small but memorable role in the history of Project Apollo. Today a luxury apartment block, the former hotel once hosted President Kennedy and the First Lady, who dined there on 21 November 1963, the day before their fateful visit to Dallas. The hotel also accommodated several furtive gatherings of test pilots and military officers during some of the processes leading to the selection of three NASA astronaut groups, including Group 5.

One survivor of the medical testing, U.S. Marine pilot Jack Lousma, remembers a NASA official taking him to the Rice Hotel and down to its basement, where he had to sit for a written test. It was 28 February 1966. "It was a very technical written examination, engineering questions mostly, and test pilot kind of questions as well, and these were all written down, like a college examination. Mike Collins was the moderator, to keep us 'honest,' I guess! Deke Slayton walked into the room and whispered in Collins' ear and he left and someone else came in to be moderator. And we didn't know why until afterwards."

News spread rapidly of a fatal accident that had taken the lives of NASA astronauts Elliot See and Charlie Bassett, the prime crew of the forthcoming *Gemini 9* mission. On a flight to St. Louis, Missouri, where their spacecraft was being completed at the McDonnell Aircraft plant, See misjudged his approach in very poor weather with low clouds and snow flurries. Their T-38 jet struck Building 101, which actually housed the spacecraft, and crashed into a parking lot. The plane exploded, killing both men.

The NASA family was shocked by the tragedy, but life continued. Ron had his formal interview on 4 March. Fellow applicant Jerry Carr recalled a long table with a dark green tablecloth. On one side sat the inquisitors: astronauts Deke Slayton, Alan Shepard, John Young, Michael Collins, and C. C. Williams; spacecraft designer Max Faget; and astronaut training officer Warren North. On the other side of the table, a single chair sat under a spotlight.

Jack Lousma recalls that many of the questions were quite general, intended to gauge the personality and motivations of the individual interviewee. There were also questions very specific to each applicant's career. Remembering 1963, Ron Evans knew to expect detailed probing about his two combat cruises, the missions he had been flying, and, inevitably, the mid-air collision. His full record lay before him, and he was content to be judged on the totality of his career.

For better or for worse, the process was complete. Ron and his family flew out of Houston on 5 March, bound for San Diego. Ron spent one night in his own home before heading to Travis AFB and the long journey back to the war.

As it turned out, Ron missed only a relatively short time on the line, as *Ticonderoga* departed for Subic Bay in the Philippines on 17 February and headed then to Sasebo, Japan. By 6 March, the ship was back at Dixie Station. In February the famed TV journalist Walter Cronkite had ruffled a few feathers by suggesting that overstretched navy pilots were having to fly twenty-eight days a month. The navy replied that this was "unusual," and the average for pilots was between sixteen and twenty-two sorties per month. In January 1966 Ron flew thirty-one sorties over twenty-four flying days. He managed nine sorties in the first week of February before heading to Texas and seventeen sorties during fifteen flying days in March.

Spring was a very busy and intense time for *Ticonderoga*. On 17 March pilots from VF-51 were tasked to assist American and South Vietnamese ground troops who were facing two Viet Cong battalions inland from Cam Ranh Bay. For his actions that day, Ron was awarded the Navy Commendation Medal. Although drafted in Pentagonese, the citation does not attempt to sugarcoat the nature of war: "Lt. Cdr. Evans' skill in the devastatingly accurate delivery of bombs, Zuni rockets and 20mm ammunition in the face of intense small-arms and automatic weapons fire was a major factor in the destruction or damage to twenty-eight structures which housed food and ammunition supplies. The attacks by Lt. Cdr. Evans were major factors in twenty-seven Viet Cong being counted dead, numerous secondary explosions and the area itself being denied to the enemy. Lt. Cdr. Evans' courage, devotion to duty and professional skill were in keeping with the highest traditions of the U.S. Naval Service."

None of the surviving pilots of v f-51 can add anything to this account of Ron's actions that day, either because they were too busy making their own contributions to the attack or because there were many very similar missions in those days. Any pilot paying more attention to someone else's flight than to his own would have been unlikely to make it back alive.

A hectic time for the pilots also meant it was a hectic time for men like Phil Morrison, working long hours on the flight deck and in the stifling monsoon heat and humidity belowdecks. Jan Evans remembers Ron telling her in his taped messages of how he would descend into the bowels of the ship to give the young mechanics and engineers a pat on the back and words of encouragement for their hard work in the challenging conditions. During the lulls between aircraft checks or weapon loading, they would stretch out on top of bombs to catch a little sleep. Ron always said they were doing a hell of a good job.

On 26 March, with *Ticonderoga* on Yankee Station, Ron returned from another sortie. When everyone was safely back on board, McDaniel called all available v f-51 pilots to a meeting in the ready room at 1800 hours. When he arrived, he was unexpectedly accompanied by the ship's captain, Robert Miller, who began his remarks by congratulating the squadron on a fine day's work. But that turned out to be a secondary reason for the gathering. The *real* reason emerged when the captain produced a "twix," or teletype message, and announced that Lt. Cdr. Ron Evans had been selected as a n a s a astronaut and was to report to Houston for training no later than 2 May.

Jack Allen witnessed what happened next: "It was just 'out of the blue,' and at that point all of us gave loud cheers and clapping with everybody eager to congratulate him. We all moved round him, because the ready room wasn't that big, so we were all there, with the skipper up front, and he called Ron up. That's pretty much how I remember it happened."

Other pilots later told Jan that Ron "just floated up out of his chair" and joined the captain, at which point the ship's photographer captured the moment for posterity. There stands one of America's newest spacemen in his sweat-stained shirt and his trademark smile lighting up the whole room. It is the only known photograph of an Apollo astronaut receiving news of his selection.

As the cheers died down, Captain Miller further surprised Ron by telling him to catch the next available flight to "the beach"—slang for any air base in

South Vietnam—to phone his wife. The mail plane regularly made the three-hour flight to Cubi Point in the Philippines, so it also counted as the beach.

After Ron had departed on 5 March to rejoin his squadron, Jan endured three weeks of mounting tension. She never forgot that her husband was in harm's way, but there was also the other matter of the NASA selection. How many astronauts were the selectors going to pick? She didn't know that the NASA panel had asked Deke Slayton the same question. He had told them he would take "as many qualified guys as you can find." The magic number turned out to be nineteen, so sixteen hopefuls were about to get bad news.

On the morning of Saturday, 26 March 1966—a date burned into Jan's memory—the phone rang. Ron was calling from sixteen time zones ahead of California with the news of his selection. As Jan later noted in a written journal entry, "It is impossible to describe the delirious excitement on both ends of the line." The call cost Ron seventy-six dollars, but it was worth every cent.

He included a stern warning to Jan: the news had to be kept absolutely secret until NASA officially announced it on 4 April. Strictly wives only! Jan knew she couldn't expect their young children to keep such a secret, and though it was a real struggle not to tell Jaime and Jon, she held out.

Finally, NASA announced the news. Jan could read it in a press release that a reporter from the *San Diego Union* newspaper handed to her. There it was in black and white: among the men chosen for Group 5 was Lt. Cdr. Ronald E. Evans (USN), thirty-two, born in St. Francis, Kansas, "presently on sea duty in the Pacific."

Following the euphoria of the announcement and his brief visit ashore, Ron had to come back down to Earth. His immediate thought was that he was "good to go," but Commander McDaniel had just reacquired his experienced pilot and wasn't about to let him leave just yet. *Ticonderoga* would be on the line several more weeks with many more sorties to be flown all over North Vietnam. Whatever his private thoughts, Ron made no attempt to use his newly acquired status to achieve an early release from his squadron responsibilities. Nor did anyone who knew him expect anything different. According to Wayne Skaggs, "I do not believe Ron would have condoned any special consideration regarding flight assignments while still serving with VF-51. He was a navy fighter pilot!"

This observation is borne out by Ron's logbook, which records a further eighteen sorties from Yankee Station. Of all these combat air patrols, fourteen were "over enemy airfields or other targets."

Asked whether her husband might have thought about seeking an early release, Jan literally bursts out laughing. "Oh no, no! He would never have done that!" But she could picture Ron thinking, "Oh boy! God be with me. Gotta be careful on these flights." She could also picture other pilots telling him before each sortie, "Hey, got your back, Ron!"

But he did allow himself an uncomfortable thought: "What if I get shot down?" NASA had just told the world that one of its new astronauts was a senior naval officer currently "serving in the Pacific." It didn't take a genius to work out what *that* meant. If Ron had been shot down, captured by the North Vietnamese, and even revealed only his name, rank, and serial number, any moderately intelligent Hanoi researcher would have quickly realized the prize they had in their grasp. It would be difficult to underestimate the propaganda value to the North Vietnamese in announcing to the world that they had captured a new American astronaut. ("Hey, Yankee imperialists! Have you run out of pilots? Why are you now sending your astronauts to attack us?")

Yet no one in the U.S. Navy, or NASA, or Washington seems to have considered those implications at the time. Perhaps the wheels ground too slowly for the bureaucratic penny to drop in the space of three weeks.

Gene Cernan's autobiography tells that a similar issue arose in 1967 after the *Apollo 1* fire. With Project Apollo temporarily on hold and the war still raging in Vietnam, Cernan noted that several naval aviator-astronauts considered volunteering for temporary combat duty in the South China Sea. Deke Slayton took the following position: "You can go, but I won't guarantee a job when you come back." According to Cernan, the Pentagon hammered the final nail into the plan. "We could return to active duty if we wanted to, and even fly, but never—ever—would we be allowed into combat. Imagine the propaganda if the enemy captured an astronaut."

The penny had finally dropped.

During his final weeks at war, Ron put up with some good-natured leg-pulling by his fellow pilots. With unconscious irony, Wayne Skaggs warned him never to fly a rendezvous between two spacecraft in case he had another collision.

In a ceremony on the carrier's flight deck on 3 April, Ron was one of many pilots to receive further combat awards. Having already won the Navy Air Medal, his second, third, fourth, and fifth Air Medals were in the form of gold stars to be attached to the original medal's ribbon.

Accounts of Ron's military career often refer to him as having flown "over 100 combat missions" during his Vietnam service. Distinguishing some service flights from actual combat flights is difficult, but if the definition of "combat mission" is "one flown while on the line and ready to engage the enemy," then over his two Vietnam cruises on the USS *Ticonderoga*, Ron's total number of combat missions is 112.

As April wore on, the pilots of VF-51 seemed to bear charmed lives. Despite the enemy's best efforts, no one had been killed in action or shot down and captured. Certainly there had been accidents and near misses, but Jack Allen recalls a certain feeling of youthful bravado and even a sense of invincibility. If another pilot suffered a mishap, you knew the same thing wouldn't happen to you. And when a pilot walked away from a mishap, it just became a good war story to be told and retold.

But whether it was fate or the law of averages paying a visit or just blind bad luck, things changed on 14 April, only a week before *Ticonderoga* was due to head home. It was one of those nights the pilots dreaded and hated, with no light, zero horizon, and a pitching deck. Ron's VF-51 colleague Lt. Richard "Dick" Hastings was the duty landing signal officer as a young ensign struggled to bring his Crusader back on board. After two aborted attempts, he tried again but was coming in too low. Hastings stepped away from his LSO platform to gain a clearer view as he advised the pilot by radio. Jack Allen, a good friend of Dick Hastings, was watching on a TV landing monitor when a flash of light filled the screen. The incoming pilot had suffered a ramp strike. He managed to eject and was recovered safely, but debris from the shattered Crusader sprayed across the flight deck. Men standing on deck, including fellow LSO Ralph James, threw themselves over the side into the safety nets, but Dick Hastings was standing in the path of the debris.

Jack Allen was first to reach his friend, who had been bowled a long way down the flight deck by heavy aircraft parts. Dick had a hideous open wound to his head. Jack was pretty sure Dick was dead, but he still ran to the sick bay to get help. No one could do anything for Dick with the ship's limited medical facilities. Despite the deteriorating weather, a young helicopter pilot volunteered to fly him to a hospital ship in Da Nang harbor. What most people would have considered an act of bravery above and beyond the call of duty was probably seen by the pilot himself as being all in a day's work.

Lt. Richard Hastings clung to life for a month before succumbing to his injuries on 14 May 1966. The accident cast a dark shadow over the final stages of the cruise. In San Diego the shocked and saddened squadron wives rallied round Vi Hastings, who had to endure not just the news of the accident but the extended agony of her husband's fight for life as well.

On 21 April Ron Evans returned from a two-hour combat air patrol. He slammed hard onto the deck of the ship and felt the familiar brutal deceleration as he caught the three wire. Climbing out of his F-8E Crusader, he must have realized he had just made his last flight in the remarkable aircraft. A new career and a new life beckoned.

12. El Lago

The way to get started is to quit
talking and begin doing.

—Walt Disney (1901–66)

When news of the Group 5 selection broke on 4 April 1966, Jan was greatly relieved. At last she was able to phone her parents, who were so excited they both shed tears. Her mother exclaimed, "Bless Ron's heart, we're so happy!" Meanwhile, Jaime and Jon were running around the front yard with capes made from towels pinned around their necks, playing Batman. Jan took them aside and did her best to explain to her six-year-old daughter and four-year-old son that Daddy was going to be an astronaut. They both jumped up and down and wanted to know if Daddy was going to the moon, and when she said she certainly hoped so, they cheered in delight. Whether they really understood is anyone's guess.

While waiting for the story to hit the news, Jan had been burning off nervous energy gardening. Then the phone starting ringing, one call coming after another. First in line was a Mr. Wolfe from KOGO-TV News Channel 10 who wanted to come to the house with cameras. He was followed by a procession of newspaper, radio, and TV reporters, not to mention friends and well-wishers. Jan only had time for a quick tidy-up and change of clothes before the onslaught began.

Other reactions came from farther afield. From Topeka, a letter of congratulations arrived from William Avery, the governor of Kansas, expressing the state's pride on Ron's selection. From the House of Representatives in Washington DC came a similar letter from future presidential candidate Bob Dole, who represented the St. Francis area.

A letter from BUPERS passed on NASA's request that Ron attend "normal processing" at Room 139, Building 2, Manned Spacecraft Center, Houston, on 2 May. It was signed by a friend of Ron's from Miramar days, former VF-53 Crusader pilot Everett "Ev" Southwick, who was on temporary shore duty. He had added a handwritten note: "Congratulations Ron, you fortunate dog! All of us who know you will be following your new career with interest. NASA made a good choice!" The following year Ev Southwick was back flying over North Vietnam when he was shot down. He survived 2,122 days as a POW.

Lt. Fred Dale, one of Ron's VF-51 colleagues, thought his selection was "really interesting" and even a little surprising because Ron clearly wasn't a hotshot or a show-off. He wasn't aggressive; he didn't talk or act "big." Fred had never, ever heard Ron lose his temper about anything. To Fred, Ron "was a good pilot, for sure, but he wasn't particularly concerned that you noticed he was good." He wondered if that really fit the template of the "typical astronaut." To Fred, Ron Evans was the kind of guy you would pick to handle a difficult situation. He was always very much in control, the archetypal "safe pair of hands."

On 21 April 1966 *Ticonderoga* came off the line and began the long journey home. Ron's own return was rather quicker: he traveled by mail plane to the Philippines, where he caught a military flight to Lindbergh Field Airport, San Diego, and arrived on Sunday, 28 April. Jan, Jaime, and Jon were there to greet him, but they had no hope of a private reunion. The news media turned out in force, filming Ron scooping the children up in his arms and embracing Jan. Then came the interviews. Jan had seen how the nation had idolized the early astronauts, and with America—and possibly her husband—now on course for the moon, she felt it was good practice for what might lie ahead.

Four hours later, the family made it back to Mount Tami Drive. Over the next two days, they enjoyed a friendly invasion of old navy colleagues and their wives, but somehow Ron and Jan also planned the family's move to Houston. It occurred to Jan that navy men had plenty of uniforms but not many ordinary clothes beyond casual stuff. Personnel did not wear uniforms at NASA, so she insisted on a shopping expedition to enlarge Ron's wardrobe. Over the coming years, he also received many fashionable outfits at discount prices from his father-in-law.

Ron flew from San Diego to Houston on Sunday, 1 May. On arrival he bumped into another new astronaut Maj. Jerry Carr (USMC). Whatever else the two fighter pilots discussed on the eve of a new career, the interesting coincidence that they had both been on the USS *Ticonderoga* at the same time in October 1964 never emerged.

Carr recalled: "Ron and I got to Houston at the same time. We got in at midnight the night before we were supposed to report, and according to navy tradition, we were supposed to report at noon the next day. Ron and I were staying that night at the bachelor officer quarters at Ellington AFB, and neither one of us was aware that anything was going to happen before noon. But it turned out that while we were traveling, an announcement apparently went out saying they were going to have an eight or nine o'clock press conference with all the new astronauts. Of course, Ron and I totally missed it! We showed up at noon and were told we had missed it, so we both felt pretty embarrassed. And that's when Ron and I first met!"

The men who turned up at Building 2 were an interesting bunch. The ironically self-styled "Original Nineteen" hailed from twelve different states, more or less equally split between "townies" and "country boys." There were four sons of Colorado, while Kansas, Texas, Michigan, and Pennsylvania claimed two each. All were less than six feet tall except Jack Lousma, who was "five feet thirteen" if measured first thing in the morning. Seventeen were married; Ken Mattingly and Jack Swigert were bachelors. William Pogue and Jim Irwin were the oldest at thirty-six years old, while Bruce McCandless, at twenty-eight, was the youngest. Significantly, thirteen members of the group were (or had been) test pilots, and two of them—Ron Evans and Paul Weitz— were Vietnam combat pilots. Pogue had flown combat missions at the end of the Korean War. Uniquely in the group, Joe Engle already had U.S. Air Force astronaut wings for flying the X-15 rocket plane higher than fifty miles. In those early days, anyone looking around the group to determine who would be his chief rivals for future missions had plenty to consider.

Just making it to Houston didn't turn Ron and his eighteen colleagues into Columbuses overnight. More than half a century later, Vance Brand, the last of the group to retire from NASA, wryly observes that he and the others "kind of knew what to expect in our astronaut career." But he concedes they all realized they had a long, hard road ahead. "Well, at that point nobody knew the future. Everybody there was working hard to make Apollo a success. We didn't know if we would make it a success—it was such a difficult project—so I don't think any of us had a very clear picture of the future. We all hoped we could fly to the moon but there was . . . well, now it's easy to see what happened, but back then there was uncertainty. It seemed to me almost like an impossible task at that time—to get to the moon. But perseverance and smart people made it happen!"

The long journey began at what had once been a thousand-acre cow pasture twenty-five miles southeast of Houston. According to new arrival Al Worden, the Manned Spacecraft Center was "spartan but functional" and "not designed to impress anyone." But the deeds done and controlled from there would certainly make an impression.

The nineteen men were allocated the trappings of any bureaucracy: identity badges, offices, secretaries, and an orientation briefing. Most early contacts were with NASA's administration staff. They didn't see much of Deke Slayton at first. They saw rather more of Al Shepard, chief of the astronaut office. A general training program was laid out for them, with academic studies in classrooms that inevitably concentrated on subjects such as orbital mechanics and spacecraft systems. They would visit the Morehead Planetarium in North Carolina while learning about navigating a spacecraft by pinpointing star positions. The new astronauts would also spend a year on geology training both in the classroom and on multiple field trips to some of the most spectacular wounds on the face of the planet, including the Grand Canyon and Meteor Crater in Arizona and the volcanic landscapes of Alaska, Hawaii, and Iceland.

As a skilled test pilot, Fred Haise hadn't ever expected to study geology, much less poke around volcanoes and canyons. Now he was picking up rocks in an effort to understand the geological peculiarities that had brought them to these places. In the classroom they would examine thin slices of the samples through microscopes. But Fred understood the goal. "They didn't want to make us lab technicians; they wanted us to be good observers in the field. The real hope was for us to become good field geologists."

Perhaps to avoid, or at least cool, overheated expectations among those eager new astronauts, Slayton and Shepard held an early briefing at which they put things into perspective. None of the Original Nineteen could expect to fly in the early stages of Apollo, up to and including the initial landings. Jerry Carr recalled being told NASA already had enough astronauts in training from earlier groups; many had previous spaceflight experience and were all ahead in the pecking order. The best that Group 5 members could expect would be mission support roles, doing all the dirty jobs the prime and backup crews wouldn't have time to do. But it would be a start and give them a chance to shine and get noticed, ideally leading to a future backup role or even a prime crew assignment.

At least all of Group 5 knew they were good at one activity from day 1: NASA maintained a fleet of jet aircraft at Ellington Field to allow their astronauts to travel quickly to outlying sites or simply to maintain their flying skills. Ron was back in the air on 11 May for a local familiarization flight with another new astronaut, Charlie Duke, in the rear seat. As a former Crusader pilot, Ron was slumming it in the NASA T-33A (better known to him in the navy as the TV-2 Shooting Star), but his second and third flights that day introduced him to NASA's primary runabout, the T-38A Talon. This supersonic jet trainer with fighter variants could achieve Mach 1.3 at thirty thousand feet and had a ceiling of fifty thousand feet. Compared to the brute force of the Crusader, the Talon was a small, nimble vehicle that was deceptively robust and versatile. Ron loved flying the T-38. "Oh, it was beautiful! A beautiful airplane! An actual fighter airplane, it's got an afterburner . . . kind of short legs compared with a lot of fighter airplanes, but you could power it back, you could get on top of the thunderstorms most of the time—it's a neat airplane to fly!"

Over the rest of the month, he flew with several other Group 5 members: Jim Irwin, Stuart Roosa, Al Worden, and Joe Engle. It is difficult to imagine such a high-performance bunch making contrails over Texas without some effort to demonstrate their personal flying skills. Even the pilot who didn't show off might have been tempted, on those occasions, to make an extra effort. Bearing in mind how many astronauts were ahead of them in that daunting pecking order, one point was becoming clear to the group: moon-bound Apollo spacecraft would not have nineteen spare seats.

During his first weeks of astronaut training, Ron somehow found time to go house hunting. Jan and the children were going to stay in San Diego until Jaime finished school in late May, but the family had already decided to find a home in El Lago, a new development close to Taylor Lake, one of a network of brackish inlets leading to Galveston Bay. Ron and Jan had both liked the easygoing atmosphere in El Lago. As for their choice of a house, they had a list of must-haves. The houses were mostly single-story buildings in the Spanish style, and Ron wanted a bathroom just inside the back door. They preferred a quiet cul-de-sac location, and it was important to be near the schools and a swimming club. Ron found pretty much the perfect property: 1310 Woodland Drive ticked all their boxes and was a brand new, four-bedroom house just waiting to be occupied.

On 16 May *Ticonderoga* returned to San Diego, where it was met by the crew's cheering families, including those of the VF-51 pilots. Jan and her friends had mixed emotions, as they were still coming to terms with the death of Dick Hastings. As always, they drew strength from each other while awaiting their individual reunions. It seemed strange for Jan to be greeting a ship her husband had left weeks earlier, but she had to offload a lot of accumulated property in Ron's old quarters and pile it in the back of their station wagon. There was even a tandem bicycle he had bought in Japan. The other pilots were delighted to help Jan and shoulder the burden.

As the big move grew nearer, Ron kept dropping pieces of advice into his daily phone calls to Jan. First, he suggested she might want to think about getting an air conditioner installed in the station wagon. Jan wondered if this astronaut thing had gone to his head. Then he begged her to do it. Finally, he *demanded* that she do it. Jan was a Kansas girl who had acclimatized to the hot, dry summers of San Diego. Her only experience of Houston had been three months earlier, in wintertime. Just how bad could the Houston summers be? With some skepticism, she paid to have air conditioning installed.

Ron flew back to San Diego to help with the move. On 6 June the family loaded their essential possessions into the Rambler station wagon, a vehicle Ron had dubbed the "Gutless Wanderer." It took two days to reach El Paso and two more to reach El Lago. Long before then, Jan was forced to admit Ron had been absolutely right about the air conditioning. She couldn't believe the brutal heat and humidity of what was still only early summer in south Texas.

The climate wasn't the only problem. The house was new but wasn't *quite* ready for occupation. It had no electricity, no telephone, and no furniture. Somewhere between California and Texas, all their household goods had been misplaced, and nobody seemed to know where they had all gone. The family set up temporary quarters at Ellington AFB, a lingering benefit of being a military family.

Day after day, Ron dropped his family at the empty house on his way to the Space Center; then he picked them up again after work. Jan, Jaime and Jon sat in the powerless, phoneless house waiting for the utility companies to arrive. At least it gave the children the opportunity to explore the neighborhood and make new friends.

One of those days, Ron drove by to take them to lunch, and they accepted the risk of missing a call. Sure enough, when Ron left them at the house after lunch, Jan did find a note pinned to the door. It read: "Neil and I are having a 'pot luck' dinner tonight at our house. Please come as you are." It was signed Jan Armstrong. There could hardly have been a better way for the new kids on the block to meet the neighbors.

The furniture arrived at last. All electrical connections to the outside world had been made. Moving-in day was 6 July. Once the Evans family had finally settled into the new home, Jan had a chance to reflect on the major changes to all their lives in the past few months. She was relieved to find that one thing had not changed. Although Ron never told her exactly what being at war had meant, Jan was not naive; she had a fair idea of what had been required of her husband as a naval aviator flying combat missions over North Vietnam. But the man who had come back to her was the same man. Jan never saw any changes in him. She got back the same old Ron.

In the early months of his training, a postscript to Ron's combat career must have helped get him noticed by NASA management. Catching up with the final months of *Ticonderoga*'s cruise, the U.S. Navy arranged for NASA to present Ron with a silver star in lieu of his sixth Air Medal and gold stars in lieu of his seventh and eighth Air Medals. All were awarded for combat missions he had flown between 2 February and 18 April, and they were presented in a small ceremony in early July.

On 31 August Commander McDaniel, Ron's former skipper, forwarded the Navy Commendation Medal to NASA. The accompanying citation referred to Ron's actions on 17 March and authorized him to wear the medal with the Combat V, a device indicating heroism in direct contact with an armed enemy.

The executive secretary of the Manned Spacecraft Center's Awards Committee sent a memo to Dr. Robert Gilruth, the center's director and one of the most highly regarded management figures in NASA. The memo suggested that as the combat award for meritorious service was being made to a member of the new astronaut group, Dr. Gilruth might want to make the award himself. Gilruth needed no further encouragement. Early in September, in a ceremony attended by Deke Slayton and a very proud Jan Evans, Gilruth solemnly presented Ron with his final navy decoration. It was a transitional moment for Ron as his former supersonic career gave way to a career that he hoped would allow him to fly much higher and faster.

1. One of the earliest known photographs of Ron Evans, probably taken in 1936 in St. Francis. Courtesy Kelly Evans Sears.

2. An Evans family portrait, probably taken around 1943, showing Jimmy in his navy uniform and the three boys in sailor hats (*from left*: Jay, Ron, and Larry). Courtesy Kelly Evans Sears.

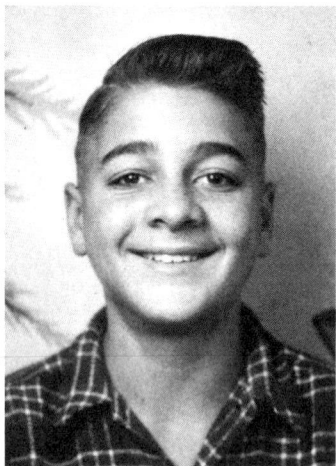

3. A portrait of Ron, possibly a school photograph. The date is unknown but likely around the time he started attending Highland Park High School, Topeka, in the fall of 1947. Courtesy the Evans family.

4. Navy cadet Ron Evans, bound for Europe, on the deck of the battleship USS *Wisconsin* during his first midshipman cruise in the summer of 1952. Courtesy the Evans family.

5. In the summer of 1953, Ron and his midshipmen colleagues were offered a glimpse of life in the U.S. Marines at Little Creek, Virginia. Courtesy the Evans family.

6. At NAS Memphis, on 12 April 1957, Rear Adm. Frank Akers presents Ens. Ron Evans with a certificate designating him a naval aviator. Jan Pollom stands by to pin Ron's wings of gold on his chest. Courtesy the Evans family.

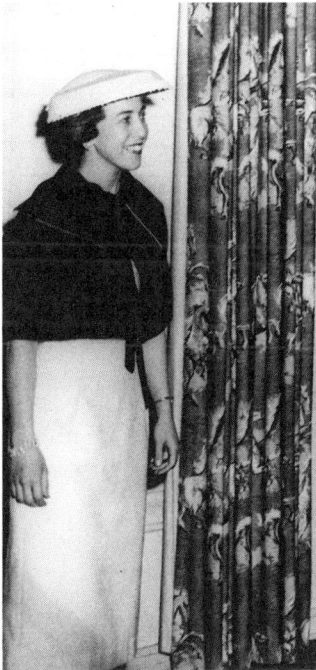

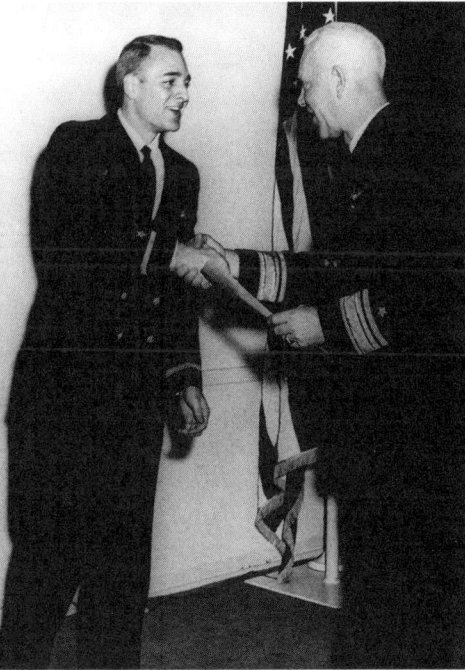

7. (*opposite top*) "Just married": Ron and Jan shortly
after their wedding in Topeka on 22 December 1957.
Courtesy the Evans family.

8. (*opposite bottom*) Lt. (JG) Ron Evans pictured
at NAS Miramar with an F-8 Crusader of Fighter
Squadron 142 (VF-142), probably in 1958. U.S. Navy
photograph, courtesy the Evans family.

9. (*above*) "Top Guns": VF-142's team in the Fourth
Annual Naval Air Weapons Meet included two
future admirals (Jimmie Taylor and "Moose" Myers)
and future Apollo astronaut Ron Evans. U.S. Navy
photograph.

10. Family man: now an F-8 Crusader instructor at Miramar, Ron Evans holds his infant son, Jon, in the family's home in San Diego in late 1961. Courtesy the Evans family.

11. Capt. Robert Miller of the USS *Ticonderoga* shows Lt. Cdr. Ron Evans the teletype message confirming his selection as a NASA astronaut. But many combat missions still lie ahead. U.S. Navy photograph, courtesy the Evans family.

12. Dr. Robert Gilruth, director of the Manned Spacecraft Center, Houston, presents Ron Evans with the Navy Commendation Medal in September 1966, observed by Deke Slayton and Jan Evans. Courtesy NASA.

13. Buzz Aldrin, with pipe, working with Ron Evans and Jack Schmitt on 21 January 1969 in preparation for the *Apollo 11* mission. NASA photograph, scan courtesy Ed Hengeveld.

14. *Apollo 14* backup crew members Joe Engle, Gene Cernan, and Ron Evans (*all standing, center*) monitoring the plight of *Apollo 13*. Courtesy NASA.

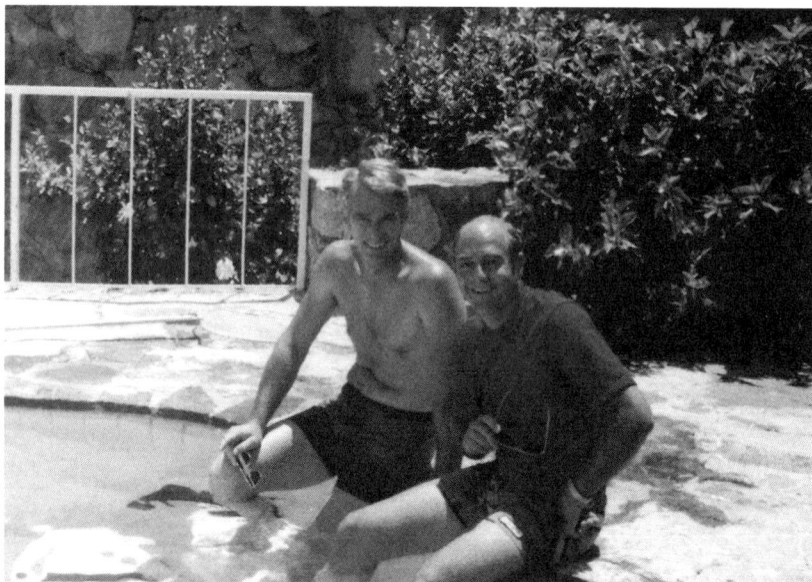

15. Relaxing in Acapulco in early August 1966, Gene Cernan and Ron Evans look as if they might have received good news about the *Apollo 17* crew selection. Courtesy the Evans family.

16. Ron Evans studying charts of the moon with his trainer, geologist Farouk El-Baz, in preparation for *Apollo 17*. Courtesy NASA.

17. Wearing false mustaches, the crewmen of *Apollo 17* pose for a lighthearted portrait with their backup crew (*center*) and support crew (*rear*). Courtesy NASA.

18. (*opposite top*) Jack Schmitt, Ron Evans, and Gene Cernan are dwarfed by the enormous bulk of their Saturn V launch vehicle. Courtesy NASA.

19. (*opposite bottom*) Dressed for the moon, Ron Evans undergoes spacesuit pressure checks in preparation for the launch. Courtesy NASA and J. L. Pickering.

20. The sleeping giant awakens and launches *Apollo 17* to the moon on 7 December 1972.
Courtesy NASA.

21. "Captain America" on his way to the moon: an onboard portrait of CMP Ron Evans. Courtesy NASA.

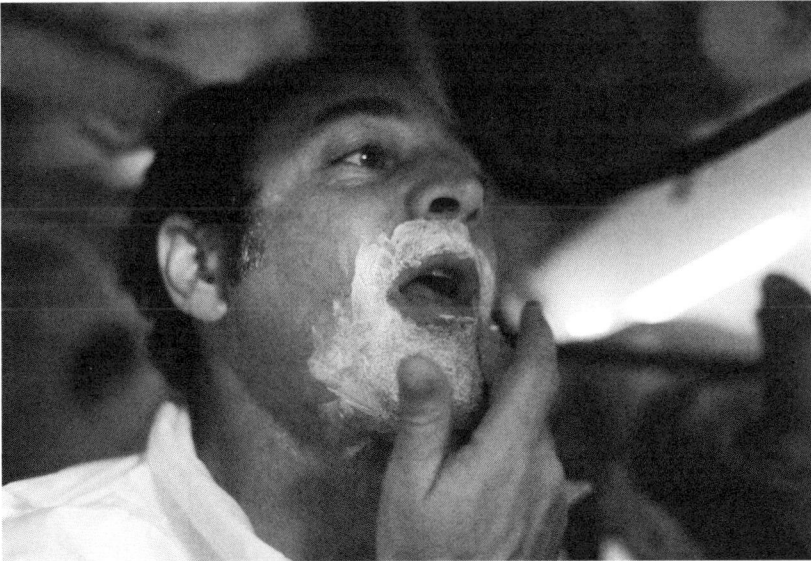

22. Ron Evans preparing for a shave in weightlessness, disappointing his son, Jon, who wanted him to grow a beard in space. Courtesy NASA.

23. The so-called chowhound of the *Apollo 17* crew pictured with a "square meal" fit for a spaceman. Courtesy NASA.

24. (*opposite top*) In preparation for docking with the returning lunar module, spaceship *America* crosses the lunar horizon while piloted by CMP Ron Evans, the last man to orbit the moon alone. Courtesy NASA.

25. (*opposite bottom*) The mesmerizing sight of a crescent Earth rising above the rugged lunar horizon. Courtesy NASA.

26. (*opposite top*) A real spaceman: in the void between worlds, Ron Evans retrieves the mapping camera's film cassette from the service module's SIM bay. Courtesy NASA.

27. (*opposite bottom*) Ron Evans and Jack Schmitt smile for the camera on their return to Earth. Courtesy NASA.

28. (*above*) An overhead helicopter captures the moment of splashdown, marking the successful end of *Apollo 17* and of Project Apollo. Courtesy NASA.

29. A picture is worth a thousand words: Jan
Evans; her children, Jaime and Jon; and her sister,
Marian, celebrate the splashdown of *Apollo 17*.
Courtesy the Evans family.

30. (*opposite top*) Spacecraft *America* awaits
recovery by Ron Evans's old ship USS *Ticonderoga*.
Courtesy NASA.

31. (*opposite bottom*) Sporting a startling new
hairstyle, Ron Evans and his *Apollo 17* crewmates
visit Downey, California, to thank the workers
of North American Rockwell for building them
a superb spacecraft. Courtesy North American
Rockwell and the Evans family.

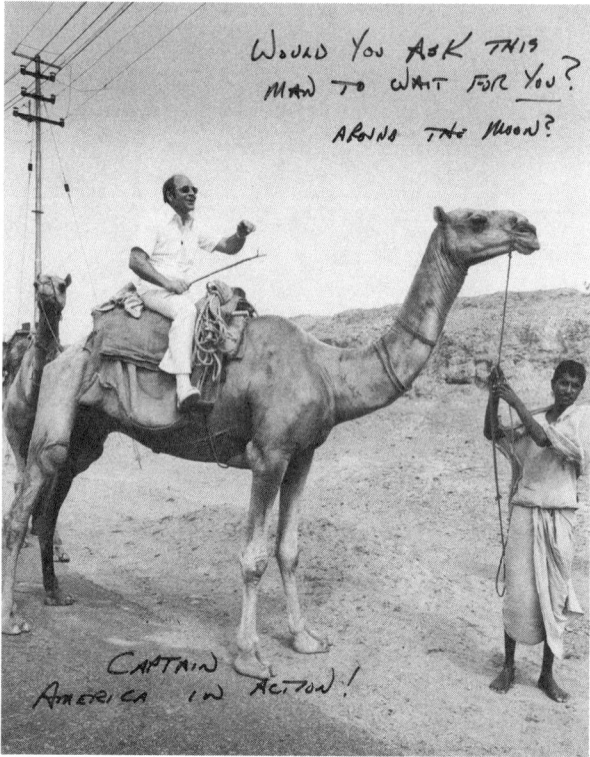

WOULD YOU ASK THIS
MAN TO WAIT FOR YOU?

AROUND THE MOON?

CAPTAIN
AMERICA IN ACTION!

32. (*opposite top*) The crew of *Apollo 17* with President Richard Nixon at the White House on 3 March 1973. Courtesy the Richard Nixon Presidential Library and Museum.

33. (*opposite bottom*) Outside Agra, India, Ron Evans swaps a spacecraft for a slower mode of transport on the *Apollo 17* presidential goodwill tour. Gene Cernan later captioned the picture "Captain America in action!" and added: "Would you ask this man to wait for *you* around the moon?" Courtesy the Evans family.

34. (*above*) Ron Evans, Anatoly Filipchenko, Jack Lousma, Alan Bean, and Nikolai Rukavishnikov of the U.S. and Soviet ASTP backup crews at the Baikonur Cosmodrome in April 1975. NASA photograph, scan courtesy Ed Hengeveld.

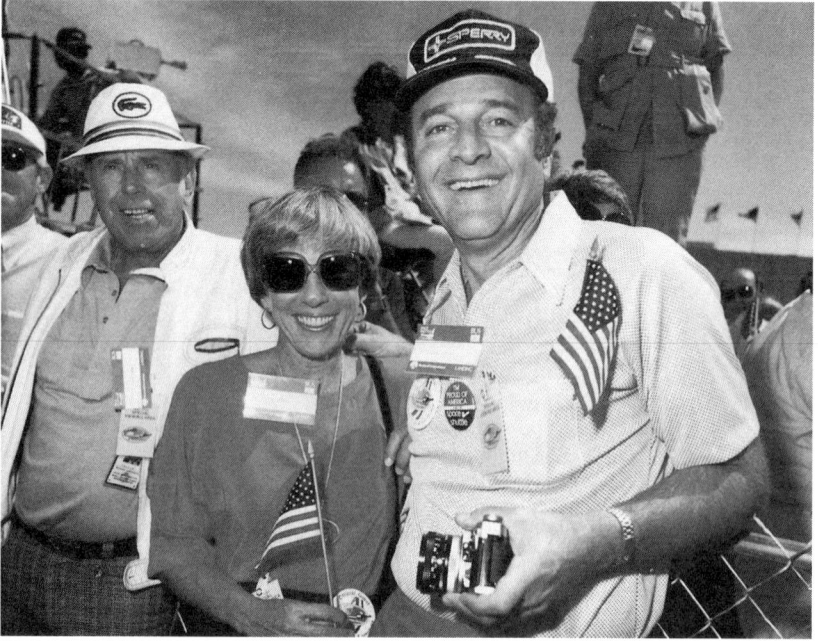

35. Ron and Jan Evans at Edwards A F B, California, on 4 July 1982 for the return of space shuttle *Columbia* on mission S T S-4, commanded by Ron's former Group 5 colleague Ken Mattingly. Courtesy the Evans family.

36. Ron Evans giving a talk at the U.S. Space and Rocket Center, Huntsville, Alabama, in September 1986. Photographer unknown.

13. Modules

Competition has been shown to be useful up to a certain point and
no further, but co-operation, which is the thing we must
strive for today, begins where competition leaves off.

—Franklin D. Roosevelt (1882–1945)

Dee O'Hara (nobody ever calls her by her given name, Dolores) was a surgical
nurse before being persuaded to join the U.S. Air Force "to see the world." In
May 1959 she had only gotten as far as Patrick A F B in Florida when the commander of the base hospital proposed her as a suitable person to work with
the newly elected Mercury astronauts. Moving to Cape Canaveral in January 1960, she set up the Aeromed Lab and carried out the preflight preliminary medical checks on each astronaut.

When Project Mercury concluded, Dee was persuaded to choose N A S A over
her air force commission. She moved west to the newly established Manned
Spacecraft Center, where, as a civilian, she established the N A S A Flight Medicine Clinic. Inevitably Dee became known as "the astronauts' nurse." The
term always embarrassed her a little because it wasn't a title she ever applied
for, but as it was difficult to describe her position, she came to terms with it.

Dee's role could have been portrayed as serving as a buffer or an intermedi-
ary between the astronauts and the flight surgeons. Pilots are generally wary
of doctors, who have the power to ground them, and as noted previously, they
did ground Deke Slayton because of a minor heart irregularity. The astronauts
found it easier to relate to Dee, who would measure their temperature, blood
pressure, preflight weight, and the like. She was in awe of these national icons
but was made to feel at home in their all-male world and, in turn, befriended
each of them. The astronauts knew that if Dee could not resolve a medical

issue, she was duty bound to refer it to the flight surgeons, but the men always trusted her to do the right thing.

As the drive to land men on the moon accelerated, Dee's clinic was not only looking after the astronauts and their families but also the air force T-38 pilots at Ellington Field. When the Group 5 astronauts and their families swept into Houston in May 1966, Dee was under pressure and thinking, "If I see one more astronaut I don't know what I'm going to do. They're coming out of the woodwork!" By 1967 she had around five hundred patients in total.

Among the Group 5 intake, several arrived in a swirling cloud of testosterone, their boisterous high-flying personalities hard to miss even in an office already housing the likes of Pete Conrad and Dick Gordon. But Dee also noticed Ron Evans, ironically because in contrast to some of his fellow trainees he was so calm and quiet. Her early assessment of Ron—from which she was never given cause to depart—was that he was "obviously a very, very nice man, very quiet and such a gentleman." While some of the new astronauts seemed to be jumping up and down, saying, "Look at me!" Ron would sit there and let them have their say and when—or *if*—it became appropriate, he spoke up. Dee appreciated his quiet manner. She remembers with a laugh, "There were some who would talk endlessly, and you wanted them to shut up, but that was never the case with Ron!"

She was even more impressed once she had met Jan, Jaime, and Jon. Dee notes, "I think one of the things I admired and loved most about Ron was that he really loved his family, and he was very much a family man. Not that some of the others weren't, but Ron definitely was. And I think I always appreciated that about him, because he was just a really, really good guy, and he did care so much about his family."

Dee always insisted that the real heroes of Project Apollo were the wives. While the astronauts oversaw the development of the spacecraft or gathered volcanic rocks in Iceland or Hawaii, the wives stayed behind and looked after the home, fixed the broken dishwasher, mowed the lawn, and cared for the children, wiping away their tears when they scraped their knees and holding their hands when they were ill. They attended to a hundred and one other tasks, weighty and trivial, that most wives in the 1960s resolved with the plea, "Honey . . . ?" For most astronaut wives, their honeys were away so often that they learned to fix the problems themselves.

Dee sums it up succinctly: "By golly, the wives came through!" But she could see it wasn't easy for them. Often they felt lonely and isolated, and that problem was much worse in the program's early years. Later, when missions were flying to the moon, the wives left at home had a very effective support network of friends and relatives who visited and took care of many of the little jobs.

Inevitably, some wives coped better than others. Most navy wives were used to their husbands heading off on seven- or eight-month-long cruises with only limited communication opportunities. Their experience, of course, wasn't quite as bad as that of the wives of eighteenth-century sailors. The great English explorer Capt. James Cook was away from home and out of contact for years on end. But even in the twentieth century, wives of naval aviators knew they would have long partings. They understood this when they got married, and for Jan Evans, Ron's absences as an astronaut were rather easier to cope with than his absences as a navy fighter pilot.

For one thing, in her private thoughts, she at least assumed nobody would be shooting at Ron while he was working for NASA. For another thing, she was well used to the network of navy squadron wives banding together for mutual support, practical help, and just honest-to-goodness friendship. She didn't know what they did in the air force, but she was sure the navy wives did it better. In El Lago in 1966, Jan found herself surrounded by astronauts and NASA engineers, but in many ways, it seemed as though they were part of another big squadron. She was particularly good at reaching out, making connections, and forging friendships that in many cases proved to be lifelong and even survived divorces, widowhood, and relocations.

The squadron atmosphere was never headier than when a crew returned safely to Earth. Parties erupted around Houston to celebrate the achievement. The most select were the "pin parties," which were strictly limited to astronauts and their wives. On arrival at NASA, Ron and the other eighteen recruits received silver astronaut pins to wear on their lapels. Once an astronaut made a successful spaceflight, he was awarded a gold pin at an exclusive pin party. This singular honor was not earned painlessly, as the backup crew would also "roast" the recipient; fortunately, the victim usually had the opportunity to retaliate in kind. Ron and Jan attended every pin party from *Gemini 9* onward and were warmly welcomed into the circle. Jan remembers each party as a celebration and "just a fun, fun thing to do."

The year 1966 was a crucial time in the drive to land men on the moon, even though the first manned Apollo mission was not expected until the following year. Project Gemini was launching a pair of astronauts roughly every two months to practice the techniques that would be required on Apollo. El Lago seemed to be front and center in the effort, providing the commanders of four consecutive missions: Frank Borman, Neil Armstrong, Tom Stafford, and most recently John Young on *Gemini 10*.

On 10 August Ron flew to Patrick AFB with Fred Haise on a two-day familiarization visit to the Kennedy Space Center. They toured the Gemini launch facilities, where preparations were being made for the next mission. The Group 5 astronauts knew they would not be flying Gemini missions, but in early September, Ron returned to Cape Canaveral with several of his colleagues on another orientation tour. As their visit coincided with the launch of *Gemini 11*, they were invited to join astronauts Pete Conrad and Dick Gordon at the traditional prelaunch breakfast. The flight was delayed a day, but on the morning of 10 September, astronauts John Bull, Evans, Brand, Carr, Haise, and Mattingly sat round the breakfast table in the crew quarters. Jerry Carr recalled the superlative double act of Pete and Dick being in good spirits and ready to go.

Unfortunately, Murphy's Law applied for the second time. After the astronauts had suited up and were heading to the launchpad, a fault developed in the unmanned Atlas-Agena rocket that was going to place their docking target vehicle in orbit. The launches were postponed another two days. While this was frustrating for Ron and his colleagues, who had to return to Houston and miss both launches, it was rather more frustrating for Conrad and Gordon.

As a fighter pilot, Ron had experienced high g-loads during training and combat. Those loads were usually only brief spikes, but a launch into space and the ensuing reentry into the atmosphere would be very different, with sustained g-forces lasting many minutes. Preparations required training sessions in NASA's centrifuge in the space center's Building 29.

In November Jerry Carr endured two centrifuge sessions with Ron Evans and Bill Pogue, simulating a three-man Apollo crew. Carr found the experience "incredible" and not necessarily in a good way. He added, "I describe it as like having an elephant sit on your chest. The worst case we ever expected was about 7 or 8 g's I think." But NASA studies referred ominously to a human endurance limit of 16 g's during certain abort scenarios, and the astronauts were given a brief taste of 15 g's.

According to Jerry Carr, "I think they ran the centrifuge up to a 'spike' of that at one time, just to let us know what it was like." With measured understatement, he added, "I think you could probably have a few little rib-separations there as a result of it. That would take you a little while to get over!"

Project Apollo has been described in numerous ways. Many people saw it as the largest national effort since the Manhattan Project. It was not lost on some critics that Apollo was at least a peaceful enterprise, although it did raise issues about fighting a proxy "war" in space. It was easy to think of the nationwide phenomenon—at its peak employing over four hundred thousand Americans spread across every state of the union—as a giant beginning to stir, stretching its limbs, and rising to its feet. That image is particularly apt when contemplating NASA's hatchery for moon rockets, the 525-foot Vehicle Assembly Building with its sixty-five thousand cubic yards of concrete and nearly ninety-nine thousand tons of steel. Or the enormous Saturn V rocket itself, rising to 363 feet and producing, at launch, some 7.6 million pounds of thrust. At the tip of that giant arrow rested, quite literally, the point of it all—the Apollo command and service module (CSM) housing the crew of three and, just below it, the surprisingly flimsy and ungainly lunar module (LM).

As an army of dedicated workers strove to complete the machinery and infrastructure of the Apollo Program to previously impossible engineering tolerances, the men who hoped to fly to the moon were trying to familiarize themselves with the modules that would make the journey possible. The CSM, or "mothership," was built by North American Aviation at Downey, California. Simultaneously the Grumman Corporation of Bethpage, Long Island, New York, was piecing together—often by hand—the ascent and descent stages of the lunar module. Depending on how you counted them, the combined Apollo spacecraft had two, three, or four sections.

In July 1966 the Group 5 astronauts (now joined by the earlier Group 4 intake of scientist-astronauts) underwent a familiarization course in the systems, controls, and instrument panel displays of both vehicles. It must have occurred to each man that much depended on which module he would work on. They all understood that the mission commanders would be drawn from the ranks of the more experienced earlier groups. But what of the other two crew members? One would be entrusted with piloting the CSM solo in lunar orbit and with achieving a safe docking with the returning lunar explorers. The command module pilot (CMP) would be second in command on the flight and might later command his own mission. The junior crew member, the lunar module pilot (LMP), would not actually pilot the lunar module. Al Worden, probably to the displeasure of some colleagues, always considered the LMP to be "essentially a systems engineer" who monitored the vehicle's performance while the commander flew it. Although the CMP had the prestige of piloting the mothership, the LMP, the so-called junior guy, got to walk on the moon!

One of the unsolved mysteries of Project Apollo is how individuals were allocated either to one spacecraft or the other. In assessing how Ron Evans, or any other Group 5 astronaut, could expect to progress toward the holy grail of an Apollo lunar flight, it might be helpful to examine the men's motivations, rivalries, and qualifications. Deke Slayton often remarked that he could have matched any three astronauts in a crew because he knew they were all capable of making a flight (or NASA wouldn't have selected them). But that didn't mean crew positions were all interchangeable. In the nine years following their selection, nine members of Group 5 would make it to the moon. Three of that select group would walk on its surface. Another five would orbit the earth. Others would have to wait much longer, and two would never make it into space.

When Ron was presented with the opportunity to become an astronaut, he never made any bones about his motivations. He wanted to serve his country in the great national commitment that was no longer just about beating the Soviets but also was something of a debt of honor to the memory of a slain leader. On a personal level, he now realized that he, Ron Evans, from a tiny little town in rural Kansas could do something he had never dreamed possible as a child. "I had an opportunity to explore. I had the feeling you could no longer be a Lewis or Clark and run up to the end of a river to find out what's at the end of the river. Somebody else had done that! And I happened to be born in a period of time, and happened to get selected for a program, where I was going to get the opportunity to leave the earth and—probably—go out to the moon. At least, I was going to be associated with a new program that was going to explore space, and it really didn't make any difference where I was going to explore, but I was going to be part of that program."

Jan had no doubt what form Ron hoped that exploration would one day take. There were craters and mountains and valleys to be explored. Ron had an expression he was fond of using: as an astronaut he hoped he could one day "get his boots dirty."

But that raised the question of Ron's place in the pecking order. Deke Slayton had asked previous astronaut groups to conduct peer reviews of their colleagues, and he did the same for Group 5. Each man had to suggest the order in which the others should fly a mission. The only rule was you couldn't vote for yourself. The winner who emerged was Lt. Cdr. Edgar "Ed" Mitchell (USN), who held a doctorate in aeronautics and astronautics from the Massachusetts Institute of Technology, no less. For good measure, he had graduated at the top of his class at the U.S. Air Force Aerospace Research Pilot School. Winning the peer review offered no guarantees, but it meant that Mitchell became the unofficial spokesman for the Original Nineteen. No one now seems to remember the rest of the results, although future events suggest high rankings for Fred Haise and Ken Mattingly.

By their very nature, astronauts are competitive and goal orientated, even the quiet ones. All nineteen men knew they already had some two-dozen pilot-astronauts from earlier selections above them in the pecking order. Inevitably they would have to assert themselves within their own group to get noticed. But could you still be "one of the guys" if you were simultaneously trying to clamber over the others to get to the top? Would too obvious an effort to shine actually be counterproductive if Deke was looking for team players who got on well with the others? Fred Haise notes, "There wasn't really a way to be 'rivals' in terms of being graded, or whatever. We all kind of split in different ways, and you just tried to do the best at what you were assigned to do. In other words, there was no common ground where you came together and you took a test, and by the results of the test, it would give you a score. There was no direct way to—if you want to call it that—have a 'rivalry' and show yourself better than the other person in some full-up way like a test. It just wasn't built into the program."

They had no direct way, but maybe an obvious indirect way was available. Jerry Carr was happy to offer his two cents: "I think for the most part we were all very competitive men. Nobody seemed to have an idea how the process for selection of crews was made, and it was kept, really, a 'dark secret' away from us. Our form of competition was to do the best job we could and stand out for doing the best work. In other areas, socially, we didn't compete. There was a lot of camaraderie there, and I don't remember any other types of competition or activities to get ahead. . . . Everybody just kind of put their noses to the grindstone and worked hard. Certainly, that was my credo!"

Standing out by doing the best work implies a level playing field for all the lunar hopefuls. But the widespread perception was that test pilots had a distinct advantage over "ordinary" fighter pilots. Could Jerry Carr or Jack Lousma really measure up to crack test pilots like Ed Mitchell, let alone X-15 pilot Joe Engle?

Michael Collins had something to say on the subject shortly before his historic *Apollo 11* flight. He told *Life* magazine that he liked fighter pilots and that they were good people. But the fighter pilot could afford to be impetuous. The test pilot generally had to be older, smarter, and steadier to avoid making a wrong judgment on a new airplane that might later lead to some pilot getting killed. No white silk scarf trailed in the slipstream for the test pilot. His job had more to do with engineering and studying charts and graphs to establish an aircraft's limits.

But there are fighter pilots, and then there are *combat* fighter pilots. In one of the last interviews he ever gave, Neil Armstrong reflected on the challenges and compensations of being a combat fighter pilot: "The risks in combat are substantial and I think in general they are higher risks than I faced in my test pilot work or in my astronaut work. And the consequences are severe. There's a good side and a bad side. The bad side is you lose colleagues, and that's painful. The good side is you create very strong bonds with your colleagues who survive. And those bonds exist throughout your lifetime. And I value those experiences very highly because they build a lot of character, you build a lot of back-bone, and you are a better person for having learned to endure that environment, that situation and those risks."

Jack Lousma is pretty sure being a test pilot trumped being a fighter pilot, even a Vietnam combat veteran. "I think the test pilots had a clear advantage. Not all of us were test pilots. I wasn't, and neither was Ron. We were military pilots, but we hadn't had any test pilot school. And there were a few other guys in our group as well: Paul Weitz, Jerry Carr, Bruce McCandless, Don Lind. And when you look back you can see, well, [Deke] assigned the test pilots first, with the exception of one guy, Bill Pogue."

But Lousma possibly overlooks a factor that must have been very hard to measure or to weigh. Fred Haise recalls Ron in familiar terms, but he goes further: "Ron, in the kind of things we did together as a group, was very easygoing, good sense of humor. He was a likable guy. I was frankly somewhat envious in a way—because I had missed the chance—but also admired the fact that he had seen combat. He had been in Vietnam, and I had not. . . . I felt guilty that I had not been exposed to combat. I admired Ron for that, as well as Paul Weitz."

Dee O'Hara had no flying hours but was sufficiently close to the astronaut corps to pick up the vibes and opinions of her charges, voiced and unvoiced. "I think those who served in Vietnam were definitely accorded great respect," she observed.

Witnessing Dr. Gilruth's presentation of the Navy Commendation Medal to Ron, World War II combat veteran Deke Slayton may have shared the opinion of Ron's pilot friend Ev Southwick that NASA had "made a good choice." Deke may even have felt an extra sense of camaraderie with Ron. We can speculate, but Deke didn't record such views for posterity. And he had a lot of test pilots on the books.

Nor did Deke leave any explanation as to why some Group 5 astronauts were chosen as CMPs and others as LMPs. As a military man, he understood the lure of command. Those who remained in orbit around the moon would miss out on getting their boots dirty, but they might go on to command future lunar landings or space station missions. By 1966 the flood tide of funding for Project Apollo was starting to ebb, and considerable doubt arose about whether any more Saturn V moon rockets would be built after the initial order of fifteen. It is unclear to what extent, if any, this issue influenced the spacecraft assignments made in 1966 that would set the candidates on one path or the other. There might be a clue in what Slayton wrote after the Group 3 selection in October 1963: "In my tentative crew plans carrying through the last Gemini missions into Apollo, I figured it would be the test pilots I would count on for the more immediately difficult work—training as command module pilots for future Apollo missions, where they would, in effect, be solo pilots."

Slayton also noted that before the budget started to get squeezed, NASA had plans extending beyond Project Apollo that might include "a series of more ambitious manned lunar landings . . . giving a crew the ability to stay on the surface for two weeks." With the benefit of hindsight that now sounds fanciful, yet by 1968 Ron Evans, Al Worden, Vance Brand, and others were learning to fly helicopters, a seemingly rather unnecessary diversion unless Slayton thought that some of his command module (CM) guys might one day find the experience useful for landing on the moon. The navy conducted helicopter training at Ellyson Field near Pensacola, Florida. The course, spanning several weeks, was designed for pilots already experienced in flying fixed-wing aircraft. Ron recorded over 170 hours in helicopters during Project Apollo.

After the initial weeks and months of classroom studies, NASA site visits, and geology training, the new astronauts began their more detailed familiarization with the constituent parts of the Apollo spacecraft. Instructors from North American Aviation arrived from Downey in the fall of 1966 and took the men through ninety-five hours of intensive analysis of the CSM subsystems. Instructors from Grumman turned up and did the same over eighty-two hours for the LM.

Later, individual assignments were handed out that involved direct experience on the spacecraft production lines. Two men of the group, Duke and Roosa, were initially assigned to development work on the Saturn rockets at Huntsville, Alabama. Lousma had an early assignment on the Apollo Applications Program. The others were sent on assignments to North American or Grumman. It would have been easier to understand their ultimate designations as CM guys or LM guys if the process had been completely consistent, but it wasn't. For instance, Jan Evans recalls Ron visited Grumman several times early in his career and at least once with Jack Swigert.

Edgar Mitchell wrote in his 1996 autobiography that on arriving in Houston, one of his first major decisions was "choosing to work on the lunar module ... as my technical assignment." Set against the accounts of other Group 5 astronauts, it seems a little unlikely that Mitchell was given a choice, but, of course, he had topped the peer review process. Perhaps Slayton felt that entitled Mitchell to choose his module. Jim Irwin similarly claimed that he "chose" the lunar module to improve his chance of walking on the moon. But he also suggested that if NASA hadn't picked as many as nineteen Group 5 astronauts, he "wouldn't have had a chance," so in his case the claim seems a little implausible. Unfortunately, Slayton's records provide no corroboration or rebuttal.

What Slayton does tell us in his autobiography is that by late 1966, by some process or another, the Group 5 astronauts had been "broken down into command and service module (CSM) and lunar module (LM) specialists." How did he decide who would do what? The question continues to puzzle the surviving Group 5 astronauts, and there is no real consensus. But comparing their answers perhaps allows a plausible conclusion to emerge.

Watching the development of the Saturn V at the Marshall Space Flight Center in Alabama, Charlie Duke and Stuart Roosa worked under veteran astronaut Frank Borman on propulsion systems. It therefore made sense to Duke that he was later assigned to work on LM propulsion systems. However, the same logic didn't apply to Roosa, who was sent to North American to work on the command module.

Slayton may have mentally divided up Group 5 by late 1966, but he certainly didn't tell Jack Lousma. "Trying to understand Deke Slayton's rationale was something that I think all of us spent a lot of time doing. We tried to figure out what Deke was thinking, and I didn't know that Deke had said that he had already broken the entire group down into CM and LM specialists. I wasn't aware of that."

If Deke had a master plan, it was also a mystery to Al Worden. "I never figured it out, and I don't think anyone else has either!" With everyone concentrating on their own detailed assignments, Al didn't get to know Ron Evans particularly well, but he really liked him. He found him to be "a charming guy, very friendly, very low key, very unassuming, but an excellent pilot." Al was assigned to work on the command module at Downey. He suspected his engineering background and experience had a lot to do with it, although it is worth noting that sixteen of the Original Nineteen had some type of engineering degree.

Fred Haise spent much of his early NASA years working on the lunar module. In one nine-month period, he was away from home seven months. As with future crew assignments, the division into CM or LM specialists was "something of a mystery, frankly." Deke had not set out written criteria to allow the men to determine in which direction each of them might be headed. Fred wonders if Slayton and Alan Shepard made use of the peer reviews, but he accepts that still wouldn't explain how the division was made. However it was done, Fred wasn't complaining. "I mean, I guess you would say the lunar module was kind of a prize thing because you *might* get to land on the moon! In terms of the training requirements, what you had to accomplish from a piloting or technical standpoint, there wasn't much difference." Whether the LMP was a proper pilot or a systems engineer, he certainly had to be capable of flying the LM solo into lunar orbit in case the commander died on the lunar surface.

Jerry Carr had an interesting suggestion: "I never, ever, felt like I knew why I had been selected for the lunar module section. I suspect it was random. I don't have any other indications that there could have been any other reasons for that decision."

A random choice seems a little unscientific, but if everyone was good enough for either module, why not? As Carr noted, "It was the only thing I could think of. Those other fellows working with me on the lunar module, they were all kinds. Of course, we were all pilots, but some were engineers, some were not engineers. Pretty much the same for the command module group. They didn't know us very well at that time because we had just arrived. So my guess is they just randomly chose to put us where we were."

Vance Brand has a slightly different take on the subject. "I was never told. Al Shepard used to say: 'If you get a lunar flight, there's no bad seat in the spacecraft!' Early on, we were assigned jobs by the chief of the astronaut office, and yeah, I was assigned to work on the command module. I don't think any of us, at least at our level, had insight into why we were selected for one or the other."

But Brand doesn't think it was a random process. He prefers to think there was some matching of test pilot experience or educational experience and expertise with particular tasks. "No, I should think they would have looked at our work histories and just made decisions on that basis. Or primarily on that basis." Perhaps.

Joe Engle suggests a slightly more mundane explanation. "I can't answer why the initial assignment to different tasks and activities was made. Probably availability! If you had been working on a certain program, project, or experiment and in the middle of it they needed someone to fill a slot to support either testing or development at one of the manufacturers, that probably had more influence on the initial activity that you were assigned to than anything. But once you *were* working on a particular vehicle, if you were doing a particularly good job at it, then I think that is when the strategy was that you were earmarked for either command module or lunar module."

Perhaps it was, after all, the random element of an individual's being available at the right moment. That individual is sent to Downey, puts his heart and soul into the task, and gets noticed. North American says, "Wow! This guy knows his stuff! Let's get him back to help with that other CM problem next month." Eureka!

In the specific case of Ron Evans, would his specialty of *electrical* engineering have been a particular asset when dealing with the miles of wiring snaking throughout the command module? Probably not. Joe Engle believes an electrical engineer would have been an asset working on either vehicle. "I think it was very applicable for both vehicles, because we were kind of entering a new era in flight control systems, and understanding flight control systems and computers was beginning to surface as important."

As 1966 drew to a close, Project Gemini had given way to Project Apollo. NASA was preparing three early flights, tentatively identified as *Apollo 1*, *Apollo 2*, and *Apollo 3*. The second, third, and subsequent flights would use an advanced Block 2 CSM with a docking mechanism for the LM, but the first mission would fly the older Block 1 spacecraft as a test of flight systems, particularly of the main engine and the ability of the environmental control system to keep three men alive for two weeks in a tin can circling the earth.

The prime crew consisted of Mercury and Gemini veteran Gus Grissom, *Gemini 4* spacewalker Ed White, and Group 3 rookie Roger Chaffee. NASA was aiming for a launch in February 1967.

Preparing to fly a completely new spacecraft was a daunting business. Jim McDivitt, who was slated to command *Apollo 2*, suggested having more astronauts involved to help with tasks that the prime and backup crews didn't have time to handle. Deke Slayton accepted this good advice and formally introduced the concept of the support crew, thus raising the number of astronauts on each mission team to nine.

Slayton selected "three CM guys"—Ron Evans, Ed Givens, and Jack Swigert—as the *Apollo 1* support crew. If there was a ladder to the moon, Ron was now on the first rung. The announcement was made on 22 December. Being their ninth wedding anniversary, Ron and Jan had two good reasons to celebrate.

14. "Go Fever"

There will be risks, as there are in any experimental program,
and sooner or later, inevitably, we're going to run head-on
into the law of averages and lose somebody.

—Virgil I. "Gus" Grissom (1926–67)

Jim McDivitt was frustrated. He and Ed White were lying on their couches inside the *Gemini 4* spacecraft, perched on top of a Titan II launch vehicle, on 3 June 1965. But for a problem with part of the launch support structure, they were ready to go. The erector, which had raised the Titan to its vertical position, was now supposed to pivot down to a horizontal resting position. Each time it reached a certain angle, a circuit breaker popped out, and the erector stalled. The launch crew raised the erector and lowered it again, and again the circuit breaker popped. Finally, they scouted around and found two wooden brooms. With one man on each side pressing the brooms against the circuit breaker to hold it in place, they were finally able to lower the erector and launch *Gemini 4*. With a healthy dose of right-stuff understatement, McDivitt noted, "We were significant risk-takers back in the early days. . . . It was a different philosophy to what we see today."

Addressing a group of space-enthusiasts in England in 2008, McDivitt acknowledged that NASA was less risk averse in those early days. Unorthodox though the broom technique might have seemed, it was a *calculated* risk taken by experienced engineers who knew what they were doing. What no one in NASA seemed to understand in 1965, however, was that some pretty unacceptable risks were being taken in the rush to prepare the new three-man Apollo spacecraft for its early flight tests. One procedure in particular was completely discounted as being "nonhazardous," not because an accident was completely unforeseeable, but because it had been carried out safely many times before. Nobody seemed to notice the false logic in that.

As a member of the *Apollo 1* support crew, Ron Evans had no illusions about the nature of his new assignment. He was a gofer. No task was too small or too menial for Ron, or Ed Givens, or Jack Swigert. If the prime crew didn't have time to do it and if the backup crewmen weren't around, the gofers were there to handle it. Ron was delighted to be involved in the first manned Apollo mission and embraced the task with typical zeal. A support crew gofer was the lowest form of animal life on an Apollo mission, but gofers who did well got noticed.

Jan Evans again knew what it was like to have a husband who was away a lot. After his long visits to the contractors' sites and the geology field trips, Jan didn't really notice any difference when Ron went on support crew duty. All too often he would leave for work on Monday and not return until Saturday evening. Sometimes he left on Sunday and got back on Friday. She wryly felt it was like having an affair with her own husband when he managed to get a whole day and night in his own home.

Jan had met the three *Apollo 1* crew members at various NASA functions but couldn't pretend to have gotten to know them well. However, Jan and Betty Grissom hit it off and quickly discovered they shared a passion for bowling. They both joined the local league and went bowling most Thursdays for several years. Ed and Pat White lived beside the Armstrongs, and Jan knew Pat from conversations at the local swimming pool. Martha Chaffee was closer to Jan in age, and they soon established they had both been Kappa Alpha Thetas in college. The Thetas held an annual charity fundraiser in downtown Houston, and Martha had no difficulty persuading Jan to do some volunteer work for the event. It seemed to Jan that the old squadron camaraderie was indeed alive and well among the astronaut wives.

Even before his appointment to the support crew, Ron had been flying around the country with Gus Grissom and Ed White. By January 1967, he was often flying around the country *instead of* Grissom's crew. Exactly as Jim McDivitt had suggested to Deke Slayton, the gofers were turning up at far-flung locations where the presence of an astronaut, not necessarily an experienced astronaut, would help.

On 25 January 1967 Ron and Ed Givens flew to Patrick AFB in preparation for a fairly routine prelaunch test that the prime crew would conduct inside the *Apollo 1* spacecraft, which was on top of the launch vehicle at Pad 34. The so-called plugs-out test was essentially a dress rehearsal for the flight. It had been done many times on Mercury and Gemini. First, the pressure-suited crew would be shut into the capsule with the minimum number of wires, pipes, or plugs connecting them to the pad. After the cumbersome inward-opening hatch was bolted shut, the capsule would be pumped up with pure oxygen in excess of atmospheric pressure to check for leaks. Finally, the crew would run through the simulated countdown with the Launch Control Center.

On the evening of 26 January, the backup crew of Wally Schirra, Donn Eisele, and Walt Cunningham conducted a "plugs-in" test with the hatch open and power provided from outside. At a later debriefing session, Schirra warned his old friend Gus he wasn't happy with the spacecraft. Although there was nothing definitive he could put his finger on and everything had checked out, Schirra felt in his gut that something wasn't right. He told Gus that if he didn't like it either, he and his crew should evacuate the capsule.

The backup crew had done their jobs and weren't needed for the plugs-out test. The routine preparations were left to the support crew. Before the prime crew arrived, Ron's job was to climb gingerly into the *Apollo 1* command module and ensure all the subsystems were ready and that all the switches were correctly set. Ron was on board *Apollo 1* for a couple of hours, methodically running through the checklist. With the hatch open, the spacecraft was slumbering on external power, cocooned inside the environmentally controlled "white room." Nothing seemed amiss.

The prime crew entered the spacecraft shortly after 1:00 p.m., two hours later than planned. The test suffered further delays, and Grissom found to his frustration that communications with the Launch Control Center were decidedly scratchy. "How are we going to get to the moon if we can't talk between two or three buildings?" he inquired pointedly. The air he was breathing was pure oxygen at more than sixteen pounds per square inch.

With the test running late, the three backup crew members decided to fly back to Houston. On landing at Ellington AFB, they were met by two familiar but grim-faced air base personnel.

At the Los Angeles International Airport, Jerry Carr and several other astronauts were returning from a visit to North American Aviation at Downey. Waiting for their flight to Houston, they started hearing rumors that some kind of accident had happened at the Cape. Jerry telephoned the CBS newsroom, identifying himself as an astronaut, and asked what the reporters knew.

The news that evening sent shock waves across the United States and around the world. Just after 6:31 p.m. in Florida, as best as could be determined, a short circuit created a spark somewhere in the miles of wiring within the *Apollo 1* spacecraft. A tiny smoldering flame, fed by high-pressure oxygen, quickly became a ravenous inferno. Sitting at the capsule communicator's console, Stuart Roosa heard unusually animated voices across the radio: "Fire! We've got a fire in the cockpit! We have a bad fire! We're burning up!" Astronauts aren't supposed to scream, but the last sound from the spacecraft would haunt Roosa and his colleagues for the rest of their lives.

Jan Evans doesn't remember exactly how she first heard of the tragedy. NASA didn't want astronauts and their families to hear about it from the TV news or even from inquisitive newspapermen, so it arranged for available colleagues to break the terrible news in person. When Ron was away from home, he always called Jan in the evening. The call that night was uniquely somber. Inevitably they shared expressions of shock and great sadness at the loss of their friends and colleagues. Jan then learned for the first time that Ron had actually been inside the spacecraft that morning. The revelation sent an eerie shiver down her spine.

When Ron finally got home the next day, he and Jan talked further about what had happened or at least as much as he knew. Shocking though the events had been, Jan detected no sense of despair or despondency in her husband, nor would she have expected it. While NASA and the nation mourned, and preparations were made to honor the memories of the three men in solemn ceremonies at Arlington National Cemetery and the U.S. Military Academy at West Point, New York, Jan never once heard anyone in El Lago express any doubts about the program continuing. Some newspapers may have asked such questions, but after a respectful pause, the NASA family just wanted to find out what had gone wrong, to fix the problems, and to get the Apollo Program back on track. Ron was anxious to return to the astronaut office and help with the inevitable investigation. The military mindset kicked in for Ron and Jan. Accidents happen, and you have to try to get past the pain and carry on.

Following the fire, all flight assignments were canceled, and NASA deputy administrator Robert Seamans established the Apollo 204 Review Board. As the mission had not flown, NASA was not even calling it *Apollo 1* anymore, but this rankled so many people that it officially restored the designation in April. No future flight would usurp the name.

One astronaut, *Gemini 7* veteran Frank Borman, was appointed to the nine-member board, and twenty-one panels of specialists were set up to carry out detailed investigations. Among those reporting to Borman were the seven members of Panel 20, which included Chairman Jim Lovell and fellow astronauts Ron Evans and Jack Swigert. Walt Cunningham was on Panel 13, investigating launchpad emergency procedures. Those men were the only five astronauts directly involved in the inquiry. Many years later Lovell wrote, "While Borman and the NASA brass heading up the fire investigation became something close to media darlings, Lovell, Swigert, Evans and the rest of the men on the other panels toiled in near obscurity."

Jan Evans wouldn't have presumed to enter such a debate, but she privately felt from her own observations that El Lago resident Frank Borman was just the sort of guy the investigation needed. She found Borman stern, handsome, and businesslike, with a very military bearing. There was no nonsense with Frank, and he didn't smile much—not that the task would have given anyone much cause to smile.

Jan could also vouch for Lovell's comment about Ron and his fellow panel members "toiling in near obscurity." Their brief was to review existing plans for in-flight fire emergency procedures and to recommend any necessary changes. One widespread concern was that if the fire had happened to *Apollo 1* while in orbit, the cause might never have been established. Panel 20 found that a fire in pure oxygen at five pounds per square inch in space would be less severe and spread more slowly than it had on Pad 34, but it would still generate considerable smoke and could kill the crew. Existing firefighting procedures in space were found to be next to useless. To extinguish a significant fire, the crew would need to vent the atmosphere. That would take around a hundred seconds. But first the astronauts would have to don their pressure suits, a task requiring at least nine to twelve minutes, all while keeping clear of the fire. The movements of three men trying to climb into their suits, however, would literally fan the flames. Then, assuming the fire had been extinguished by venting the atmosphere, the astronauts would have to hope there was no damage to their suits or their oxygen feed, as an emergency repressurization of the cabin to minimum acceptable pressure would take *thirty-five minutes.*

The panel didn't need to belabor the obvious point: prevention is better than cure, so NASA must not have a fire in space—ever! Others could argue about why the fire occurred, but the crux of Panel 20's report was that the best way to prevent any fire catching hold and spreading was to remove the numerous combustible materials from the spacecraft.

When the review board's final report landed on the desk of NASA administrator James Webb, it cited many instances of sloppy workmanship by the prime contractor, North American, including frayed and bare wires among the miles of cabling snaking around the capsule. An abandoned socket wrench found among the wiring bundles in no way contributed to the fire but was indicative of lapses among a workforce consumed by "go fever"—that is, the push to meet President Kennedy's deadline for landing a man on the moon by the end of the decade. The malady had infected everyone, from the contractors to the astronauts and all the way up through NASA.

In the end, the precise ignition source was never identified. Somewhere below Grissom's couch an arc had jumped, and the pressurized oxygen and combustible materials had done the rest. When Jim McDivitt told his mainly English audience in 2008 that NASA people in those days were "significant risk-takers," he had been talking about calculated risks—namely, the risks inherent in riding a huge rocket with the explosive potential of a small atom bomb or in flying a flimsy LM down to the lunar surface. Those risks were real but had been assessed and accepted. It was *not* acceptable to place a crew in an electrically operated capsule pumped up on the launchpad with pure oxygen. It should have come as no surprise when the review board found that to be "extremely hazardous." Yet no one had thought to call a halt to the practice, and three good men had died.

Having been very publicly hauled over the coals, NASA and its contractors were allowed to get on with the chastening business of fixing their moonship, their internal procedures, and their safety standards. When the next Apollo crew was ready to fly on a mission redesignated *Apollo 7*, the astronauts would be boarding a much-improved Block II vehicle. Walt Cunningham, one of the men who would fly the mission, summed up the situation in a way that probably mirrored most people's thoughts, even if some might have been squeamish about voicing them: "And out of the whole mess, North American was to bring forth one of the greatest machines ever built by man. I am convinced that it would not have been possible to reach the moon in only five missions had we not gone through this rebuilding process, which was the inescapable result of the fire on Pad 34."

Even while mourning the loss of the three astronauts, including his best friend Gus Grissom, Deke Slayton never stopped planning ahead. In early March 1967, even before the review board published its report, he told Schirra, Eisele, and Cunningham they would have the next mission. The backup crew would be Tom Stafford, John Young, and Gene Cernan, a talented trio with five previous flights to their credit. Observers of how Deke operated were aware of the unofficial crew rotation system in which a backup crew skipped two missions and then became the next prime crew. If that rule of thumb held true, Stafford's crew would be in line for *Apollo 10*. For good measure, Slayton put together two additional crews led by Neil Armstrong and Pete Conrad. Initially they would back up the second and third Apollo crews, but they would be hoping for future prime crew slots and lunar landings. All the men were drawn from Groups 2 and 3.

Deke also appointed Ron Evans, Ed Givens, and Jack Swigert as the *Apollo 7* support crew. Essentially, they were reprising their role on *Apollo 1*, which they had barely had time to embrace. Their assignment was neither a promotion nor a demotion; it was an expression of continuity.

For one of the support crew members, it was a short appointment. Highly rated by Slayton, who saw him as a future CMP on a lunar flight, Ed Givens died in a road accident on 6 June 1967. The thirty-seven-year-old Texan was driving to Ellington AFB from an aviation society meeting with two air force friends. That dark, wet night he missed an unlit and unmarked right-angle bend. The car tumbled into a deep drainage ditch, resulting in Givens's death and injuries to his passengers. Jan Evans recalls Givens as a fine, dedicated family man. At the funeral in his hometown of Quanah, Ron and Jack joined the prime and backup crews to serve as pallbearers. Deke Slayton later appointed former test pilot and Korean War combat pilot Bill Pogue to the support crew in place of Givens.

Allowing for the fact that Slayton had nineteen available astronauts ahead of the Group 5 rookies in the pecking order, Ron's role as a support crew member on *Apollo 1* and *Apollo 7* gave his career as big a boost as he could reasonably have expected in 1967. But he did have a concern. "So here I am in a real good position where I can get some operational experience, and still be 'capcom' [capsule communicator] and gain other experience which evidently helped out in the long run. But every time a crew would be selected and they would announce the crew, you would try to figure out, 'Where am I going to be able to get in there?' It actually became somewhat discouraging around the beginning of Apollo because it looked like you could not fly on Apollo if you hadn't flown before. And then eventually some of the guys out of our group, the 'Original Nineteen,' were selected on backup crews so we thought, 'Maybe there's going to be a chance here, one of these days!' So you keep trying, you get better, you keep going after it, and doing what you're assigned to do. And doing the best you can in whatever assignment you get."

But in 1967, "one of these days" seemed a long way off for men who were still learning the ropes. In July sixteen Group 5 astronauts traveled to Iceland for extensive geological training. They were joined by Neil Armstrong and several scientist-astronauts, including a geologist named Harrison Schmitt.

Joe Engle recalls the early part of his astronaut career as "a very intense time." He didn't have many chances to socialize and get to know everybody else, but he remembers the geology field trips as having opportunities to unwind a little in the off-duty hours. He had already found Ron Evans to be "one of the warmest, most sincere and most friendly people that you could hope to meet," and on the trips to Iceland and elsewhere, Ron was "just as much fun and just as warm." If they had an extra degree of bonding, it probably helped that they were not just fellow Kansans but also fellow Jayhawks.

Jack Lousma remembers his Monterey colleague Ron Evans as a "good guy" who was easy to get along with and very sociable with a good sense of humor. He particularly recalls the geology trips to Alaska in August 1966 and to Iceland in July 1967 when the endless daylight of the arctic summers allowed time for a little off-duty sport. He and Ron would go fishing together before breakfast and after they had hammered their last rocks for the day. The dramatic scenery, the luxury of a little time off, and the chance to chew the fat while catching dinner made for a mellow experience.

For NASA's latest astronauts, their training wasn't always about hunting for rocks or understanding the spacecraft. Once in a while, each man had to spend a week "in the barrel." This period usually involved meeting the taxpayers and explaining what NASA was doing with their tax dollars. Sometimes they met with the sons and daughters of the taxpayers at schools and colleges. The astronauts would frequently meet leaders of industry and commerce, or they would appear as guest speakers at town festivals. Many of the men saw this duty as an unwelcome distraction from their work. Others detested public speaking, but some, including Charlie Duke, enjoyed meeting people and made the best of it. Ron found himself warming to public speaking, but he quickly discovered the difference between silver and gold. Eager citizens would line up to greet the space hero. When the word went around that this hero was wearing a silver pin and hadn't actually blasted off yet, the smiles on the faces would alter subtly. They were still glad to see him, and though nobody actually said anything, he knew what they were thinking: "Aw, shoot! He ain't even flown in space yet!" At such times, it was hard to avoid thinking about that ladder to the moon and how many more rungs it had.

The Block II Apollo spacecraft that Walt Cunningham praised so highly took shape during 1967 and 1968. It had a new outward-opening hatch, an oxygen-nitrogen atmosphere during ground tests, and an interior stripped of flammable materials. Grumman's LM was behind schedule but would not be needed for *Apollo 7*, which was planned as a rigorous eleven-day test of the CSM. It would also test the mettle of the crew. Many future Apollo astronauts would get the chance to stretch their legs on a mission with extravehicular activity (EVA), but *Apollo 7* would be three men in a tin can with nowhere else to go.

During the long buildup to the mission, Ron and Jan got to know the prime crew quite well, particularly the commander, Mercury and Gemini veteran Wally Schirra. A Korean War combat pilot (who had shot down two MIGs) and a crack test pilot, Schirra flew probably the most technically perfect of the Mercury missions. He was also renowned for his jokes and puns, and had a particular penchant for "gotchas." Many other astronauts were adept at practical jokes, mostly aimed at their fellow astronauts, but Wally had honed the skill to perfection.

A few days before his Mercury flight, Dee O'Hara gave Wally a small glass bottle and asked him for a pre-mission urine sample, which he was to leave on her desk. When she returned to her office, she found a huge glass receptacle filled with gallons of pale amber fluid with a slight head. She was actually feeling the sides of the jar to see if it was warm when Wally appeared with a cry of "gotcha!" Dee got him back by giving him a four-foot-long plastic bag, telling him it was the urine collection bag for his nine-hour flight. But Wally had the last laugh when he attended his preflight medical with the bag dangling below his bathrobe and along the floor.

To unenlightened outsiders, the idea of grown men—high-ranking military officers and space heroes—playing practical jokes on one another might have seemed a little odd. Dee had no difficulty rationalizing the astronaut office culture of gotchas: it provided a safety valve, pure and simple. These men lived in a high-pressure, high-risk environment. They were regulated up to their eyeballs about what they could and couldn't do. Letting off steam by having a laugh at a colleague's expense was very cathartic, and in the world of the gotcha, what goes around comes around. Everybody would eventually share a laugh at somebody's expense.

Unfortunately, "Jolly Wally" seemingly left his sense of humor on the ground for *Apollo 7*. Ron had flown to the Cape to help with launch preparations and immediately flew back to Houston with Gene Cernan to take up capcom duties. Over the course of the eleven-day flight, the redesigned Apollo spacecraft passed every test with flying colors. If there was a problem, it was with the crew. After a day Schirra developed a head cold, followed by Eisele. According to Cunningham, Schirra felt he had caught "the cold they were saving for Judas." It made him irascible and intolerant of any changes to the crew's flight plan tasks. He summarily canceled the first planned TV transmission and declared himself the "onboard flight director" who would veto any "crazy tests we never heard of before." Eisele followed Schirra's example, complaining about a "hastily conceived" navigation test. One reporter at a NASA press conference asked whether the spacecraft was occupied by "a bunch of malcontents."

Ironically, although renowned for his sense of humor, Schirra thus ended his NASA career with a successful mission remembered mainly for its commander's bad-tempered crabbiness. In Wally's case, it didn't matter so much as he had already indicated his intention to retire, but the careers of Eisele and Cunningham seemed blighted. Neither flew in space again. Walt Cunningham suggested that the backup and support crews working in mission control "also suffered during the mission for defending Wally's actions in some pretty indefensible positions." Referring specifically to the three support crew members, who all served as capcoms during *Apollo 7*, he added, "For guys like Jack Swigert, Bill Pogue and Ron Evans it may have had a carryover effect on their future career assignments." He wrote those comments shortly after the end of the Apollo era, but it is difficult to see any evidence of a "carryover effect." Swigert and Evans each made a lunar flight, and Pogue flew on the final, eighty-four-day *Skylab* mission. Some others in Group 5 would have been happy to have their careers blighted like that. Jan Evans never once heard Ron lamenting any perceived negative impact from his time on the *Apollo 7* support crew. On the contrary, he had relished all the operational responsibilities.

Conceivably, Cunningham was hinting that Ron might have expected his next assignment to be on a backup crew. The problem, as Ron had already suspected, was being stuck in the vicious circle that particularly affected command module specialists. It didn't matter so much for the LM guys who would fly alongside seasoned commanders, but to be chosen to fly solo as the CMP, Deke Slayton's rule was that you had to have flown before. For the first five Apollo missions, that rule held even for backup positions. Slayton eventually relaxed the rule when he selected Al Worden as the *Apollo 12* backup CMP. After that, the men of Group 5 came into their own.

Even as NASA geared up for the final push to land men on the moon by the end of the decade, the busy men still needed to take an occasional break. On 15 November 1968 several NASA T-38 jets from Ellington AFB flew to Kansas, landing at Forbes AFB near Topeka. According to his logbook, Ron Evans was the copilot in one of the aircraft. The pilot in charge for most of the flight was Deke Slayton, whose medical status allowed him to fly provided an unrestricted pilot was on board. This way he could check on the flying skills of his astronauts while also keeping himself as sharp in the air as his medical restriction allowed.

The five-day visit to Kansas was carefully timed: the quail season had begun on 10 November, and Ron was hosting a hunting trip to the St. Francis area. Vance Brand remembers being on a quail hunt in Kansas with Ron and Deke. He clearly recalls socializing with the residents of St. Francis and realizing how proud they were of their very own astronaut.

Slayton likely discussed Ron's next assignment with him during the trip. Ron was going to be a capcom on *Apollo 9*, the first manned test of the much-delayed LM. He was not on the support crew, which had both advantages and disadvantages. On the one hand, supporting the first flight of the complete Apollo spacecraft would have been very good experience. On the other hand, he did not have to fly around the country playing the gofer again. How would he have prepared for his task as an "unattached" capcom outside the nine-man crew structure?

Future lunar motorist David Scott, the CMP on *Apollo 9*, recalls Ron as "a great guy who made excellent contributions to the programs, Apollo and ASTP [Apollo-Soyuz Test Project]." The two men didn't know each other particularly well, simply because they were never part of the same flight rotational cycle. They were therefore seldom in direct contact, except during Ron's stints as a capcom on *Apollo 9*. In terms of his future prospects, Scott feels Ron's lack of test pilot credentials put him at something of a disadvantage compared with his test pilot peers.

As for a capcom's preparations for a mission, Scott points out that nothing was formal or structured. Capcoms didn't punch time cards. There were no tests or exams. NASA in the days of Gemini and Apollo was not regimented in the same manner of today's bureaucracy. Ron and Stuart Roosa, who was also an unattached capcom on *Apollo 9*, were responsible people who knew what was required or could figure it out for themselves. A detailed knowledge of the final flight plan was an obvious minimum requirement.

Ron's debut as the capcom toward the end of the mission's first day was not particularly auspicious. He could have recited large chunks of the flight plan, but that didn't stop him greeting the crew with the words, *"Apollo 7*, Houston . . ."

Cdr. Jim McDivitt, who by then probably could have recited the *whole* flight plan, retorted, "That's *Apollo 9*."

"Sorry about that," Ron replied.

But McDivitt wasn't going to let him off too easily. "That's all right. New guys are that way."

Ouch.

Gerry Griffin was one of the key participants who made Project Apollo such a success. Born in 1934 in Athens, Texas, Gerry obtained a degree in aeronautical engineering before flying in supersonic air defense fighters in the U.S. Air Force. Leaving active duty in 1960, he worked with Thor and Atlas missiles in the aerospace industry before joining NASA in Houston in June 1964. After serving as a flight controller throughout Project Gemini, he was appointed an Apollo flight director during the period of recovery following the *Apollo 1* fire. He would later serve as lead flight director on *Apollo 17*, and in that role, he had a very close working relationship with Ron Evans and the rest of the crew.

Well before that experience, however, Ron was a known quantity to Gerry. "I got to know Ron very well because he passed through the control center as a capcom and worked with us on *Apollos 7, 9,* and *11.* I know he got pretty busy after that because he got assigned to a backup crew first, then got very busy with [*Apollo*] *17.* But when astronauts came through mission control as a capcom, we got to know them very well, because once they were assigned into mission control, they became a little more like flight controllers than astronauts for that period. So they sat in on a lot of our meetings and all of the simulations we did, which were hours and hours on every phase of the mission. When they were on duty, they were there, so you not only got to know them in a professional sense, but you had a lot of breaks, or time off for a cigarette—in those days!—so we got to know each other and talk to each other."

Asked to describe the Ron Evans he knew, Gerry is very forthright. "Irrespective of being an astronaut, Ron was one of the nicest guys I ever met. He was always friendly, a smile on his face—he could find the light side of almost any situation and not dwell on all the negatives. He was a very capable guy, and he was an excellent communicator, which I got to see quite a bit of in mission control because he was speaking as the capcom, talking to the crew. Of course, that's where his naval aviation background paid off. Like most of the guys, they were good communicators. But I would describe him as one of the calmest of all the astronauts. He was kind of unflappable. He wasn't a 'back-slapper.' He wasn't real loud or anything like that. He was just a steady, steady guy. . . . You know, the astronaut corps was full of different personalities, and he wasn't the only guy I would put on the 'calm side,' but he had to be one of the coolest guys I knew."

Over the years of the Apollo era, Gerry saw that the men of mission control, who were mostly in their twenties and thirties, worked so closely together and forged such bonds of friendship and cooperation that they could essentially function as a single entity. "You know, in all of Apollo, we did have a very strong 'team concept,' and the astronauts fit into that world, and Ron did especially. One thing I always noticed about the years of Apollo: you seldom ever heard the pronoun 'I.' It was usually 'we,' . . . and Ron fit right into that. He was definitely a team guy. You know, we could have our opinions, and we could express them. In fact, we were all kind of brought up that way in mission control, to let everybody speak out. But once the decision was made on which way to go, then everybody got behind it, and that's where the team swung into action and said, 'Okay, here's what we're going to do. The decision's been made, no more "ifs" or anything like that, just get on and get the job done!' And Ron fit that mode extremely well."

In those early years of the Apollo Program, it is impossible to say whether or how Ron Evans featured in Deke Slayton's forward planning. If he said anything in confidence to Ron, Ron didn't tell anyone, not even Jan. Future crew planning was affected by numerous factors, not the least being the loss of other astronauts. We can speculate endlessly about where the late astronauts See, Bassett, Grissom, White, Chaffee, and Givens—as well as rookie astronauts Ted Freeman and C. C. Williams who died in T-38 crashes—would have been slotted. Alan Bean got Williams's seat on *Apollo 12* and walked on the moon. In April 1968 Slayton lost another of his most promising LM guys but not through a fatal accident. Group 5's John Bull developed chronic pulmonary disease, and when the navy grounded him, he could not remain an astronaut. Slayton had seen him as a potential moonwalker.

In July another of those random bolts from the blue made major changes to Slayton's plans and to the historical record. Mike Collins should have been the CMP on the first lunar flight, *Apollo 8*, but he had developed serious neurological symptoms, including numbness and a collapsing left knee. A spur of bone in his neck was pressing against his spinal column, and he needed urgent surgery—from the front, via his throat. Collins recovered but not in time to fly on *Apollo 8*. Jim Lovell took his place, and Slayton assigned Fred Haise as the backup LM pilot . . . on a mission without a LM.

Ron Evans and other Group 5 astronauts saw this as the first sign that they might, after all, get a seat to the moon. But flight experience still won out. If the flight rotation system had applied normally to the rookie Haise, he would have rotated into the *Apollo 11* crew and could have been the second man on the moon. That would have been unfair to *Gemini 12* veteran Buzz Aldrin, so on *Apollo 11* Haise was again the understudy. Looking on the bright side, Fred must have felt well placed to make an early lunar landing, but it underscored Ron's concern that the best way to get a flight was to have had a previous flight.

This concern was probably not alleviated in early 1969 when Ron was appointed to the *Apollo 11* support crew. Careerwise, he must have felt that a third support crew assignment was something of a backward step. Still a gofer after three years! But he could hardly have expected to beat his neighbor Bill Anders, the man with the shiny new gold pin, to the backup CMP slot on *Apollo 11*. That made a support crew role the best option available. As his friends and colleagues all knew, Ron Evans would not dwell on what might have been; he would look on the bright side and make the best of it. Above all, he knew that *Apollo 11* was being tipped as the first mission to land on the moon. It would be an honor to participate—in any capacity—in such an exciting and historic undertaking.

15. Two Lives

A career is wonderful, but you can't curl
up with it on a cold night.

—Marilyn Monroe (1926–62)

On 21 December 1968 Ron and Jan watched El Lago residents Frank Borman and Bill Anders set off on a voyage of exploration that would surpass the achievements of Columbus, Magellan, and Cook in scale if not in duration. Poised at the tip of the first Saturn V to carry a crew, Borman, Anders, and Jim Lovell were breathtakingly thrust into the early morning Florida sunshine on humanity's first voyage to another world.

Ron had jetted in with Jack Swigert two days earlier. They had no direct involvement in *Apollo 8*, but one of the perks of the job was getting to see their colleagues blasting off into the wild blue yonder. Jan was offered a flight to the Cape by contacts in North American and a return flight courtesy of Grumman. The Holiday Inn, Cocoa Beach, always had room for astronauts and their wives. For Ron and Jan, it was a special break, particularly as the day after the launch was their eleventh wedding anniversary.

Five months later, Marie Evans joined them at the Cape for the launch of *Apollo 10*. Ron gave his mother a guided tour of the launch facilities and accompanied her to a couple of pre-launch parties. Marie later wrote to her friends the Millers in St. Francis, describing her experience as "quite a trip," but her letter is most noteworthy for the absence of any description of the launch itself. Marie was perhaps not relishing the prospect of her elder son riding on top of one of those thunderous pillars of fire.

Project Apollo was in top gear, with near-flawless missions every two months. All elements of the Apollo hardware had been tested in space, and the goal of landing men on the moon by the decade's end seemed within NASA's grasp. It was a heady time to be working at the sharp end of the space program, and Ron was loving every minute of it.

"Well, the way I look at it is this," he later enthused, "we all had the best job in the world. The *best* job in the world! We really did! All the time we were in the astronaut program down there!"

Ron's former colleague Jerry Carr wholeheartedly concurred: "Yes, those were very heady years, all the Apollo missions that we worked on and participated in. Those were wonderful times, we called them the 'glory years.'"

But one time during the Apollo years the normally unflappable Lieutenant Commander Evans actually "saw red." Very few people can recall seeing Ron genuinely angry. Marv Miller, his old friend from St. Francis, remembers a hunting trip when a pheasant flew up in front of Ron so close that his shots missed. Marv heard Ron utter "a few words the preacher wouldn't want to hear on a Sunday," but they were hardly volcanic.

Shortly before *Apollo 11*, on a whirlwind break from being a gofer, Ron flew with Jan and others to Detroit as guests of Shirley "Murph" Murphy and his wife, Dorothy. A wealthy hotel businessman, Murph was also a keen pilot, hunter, and motor-racing enthusiast with contacts in the world of politics. Accompanied by Governor Edgar Whitcomb of Indiana, they all attended a dinner and rally addressed by Vice President Spiro Agnew.

After the rally, Ron, Jan, and five or six others decided to walk through the crowds to their hotel, only two or three blocks away. Turning a corner, they found themselves in the path of an anti-war demonstration. The ragtag bunch of protesters was holding up a torn and stained North Vietnamese flag. As Ron's group and the protesters converged, the flag fell right across Ron's face. Jan felt Ron's shoulders hunching and saw his fists balling. She grabbed his arm and yelled to one of their companions, "Get Ron!" She was afraid her husband's instinctive reaction might be to confront the protesters, and she had visions of the prospective headline: "NASA Astronaut Arrested in Riot." Their friends bunched around them as they all walked on. Although Ron was incandescent with fury, he managed to contain himself. He later told Jan that seeing "that flag" being flaunted in an American city had made him so mad. It just broke his heart that people could turn against their own country in that way.

Ron Ammons and his family had moved into number 1242 Woodland Drive, El Lago, shortly before the Evans family moved into number 1310 in 1966. Ammons had a degree in aeronautical engineering from the University of Illinois and had joined what was still the Douglas Aircraft Corporation in 1957. He later spent two very enjoyable years working in Huntsville, Alabama, with Wernher von Braun's team evaluating the third stage of the Saturn V rocket (s-iv b). When he arrived in El Lago, Ammons was still working on the s-iv b, sweating blood for the moon program but thinking that one day he would like to set up his own business. He was sitting at one of the contractor consoles at mission control in Houston when fire took the lives of the *Apollo 1* crew. After that experience, something inside him changed. The following year, rather than accept a move to California, he set up his own equipment rental company.

The Ammons and Evans families lived a few houses apart on either side of Tallowood Drive. As Ron Evans was away a lot, the wives and children met first. The El Lago swimming pool was on the far side of the cul-de-sac, so most of the children along Woodland Drive had to pass the Evans home to get to it. This made it easy for Jaime and Jon to make friends in the area, and it helped that the Ammons children were close to them in ages. Once their families had bonded, the two Rons soon found they had a great deal in common and became the best of friends.

Contrary to popular belief, El Lago was not populated with "wall-to-wall astronauts." The astronaut residents were probably outnumbered by NASA engineers and contractors. The Evans family shared their cul-de-sac with the Puttnams and the Krauses. The Anders' house backed up to Ron and Jan's side-front yard, and the Dukes were behind them. Out of the cul-de-sac and around to the right were the Carrs. Straight along Woodland were the Ammons and Lorenz families, and then the Hurds. Beyond that were some more familiar names: the Haises, the Armstrongs, the Whites, the Lousmas, the Staffords, the Youngs, and the Bormans.

The closest friendships of Ron and Jan turned out to be with the non-astronauts. Five or six families became the kind of close friends who could pop in unannounced and were always ready to babysit. Ron and Jan could fly off to the Cape for a launch knowing that Jaime and Jon would be well looked after. It might seem strange that their closest friends weren't astronauts, but those who work together don't always automatically play together. There were plenty of astronaut parties and barbecues, pin parties, and other NASA functions where they all got together, but the long absences of each individual astronaut meant that two neighboring astronauts might not see each other for weeks. By contrast, the other residents kept more normal hours and were always around when Ron was home. And children of different families playing together create bonds among their parents, whatever their occupation.

The two Rons looked at life the same way and enjoyed each other's company. They didn't actually talk a lot but would spend time tinkering on some piece of equipment that needed fixing or adjusting. Ammons knew that his astronaut friend had been a fighter pilot in the Vietnam War, but they never discussed it much. Ammons now wishes they had. They didn't talk much about the space program either. It consumed them both during the week, so they didn't talk shop while they contentedly passed the time on their pet projects on the weekends.

Ron Ammons understood his friend's dedication to his career, and he saw for himself how much Ron enjoyed his time at home as a husband and a father. In fact, he could see that Ron Evans was a man who lived two lives. When he was off duty in El Lago, he was a fully involved, fully integrated member of the community and not in the least bit aloof. At a time when most citizens regarded the Apollo astronauts with great respect, folks in the neighborhood greeted and treated Ron as one of their own because that was what he was. He just wanted to enjoy life, to drink deeply from the wine of life. Whether out with his family or out with his friends, for Evans, the important thing was to enjoy the experience and always look for new experiences.

NASA profiles of Ron Evans usually list golf, hunting, fishing, boating, and swimming among his hobbies and recreational pastimes. Reciting this list to Ron Ammons provokes a bemused laugh. "Well, I don't know where you got that from. He didn't have those kind of interests, really." Evans's golfing skills had not improved much since his letter to Jan about his efforts in 1957. Ammons pulls no punches: "Ron was a *terrible* golfer." In fairness, Ron Evans described *himself* as a "lousy golfer." Although trained to locate a lunar module from hundreds of miles away, then rendezvous and dock with it, the man had difficulty applying thrust to a small white ball and making it rendezvous with a hole a few hundred yards away.

Ron did love the camaraderie of playing occasional rounds of golf with fellow astronauts. He was genuinely delighted to turn up at celebrity golf tournaments and launch a few divots along the fairway. For him, it was all about companionship and socializing. If it meant some starry-eyed celebrities got to meet the astronauts, and the astronauts (and their wives) got to meet the celebrities, all well and good.

To hear Ron Ammons apparently dismissing his old friend's interests in fishing and hunting is a little more surprising, but he is referring to how seriously Evans took them. "Yeah, he was more than happy to do those things, he enjoyed being in those situations, whether it was hunting or fishing. I have sons that are fishermen. I have sons that are hunters. That's what they *do*. But not Ron." Again, for Ron, it was a matter of participating and sharing an experience with friends and colleagues. At least in the cases of hunting and fishing, he was a skillful and capable participant, but these activities did not give meaning to his life.

As for water sports, Ron never owned a boat of any great size, but he loved to go sailing in the ponds and streams around El Lago with his children. It was more of a case of "dad time" than personal recreation and usually involved as much time in the water as on it. It helped that Jaime and Jon both swam like fish.

Surprisingly, the NASA profiles never mention Ron Evans's genuine love for woodworking. He would sometimes drive into the desert and bring back old wind-blasted tree trunks, put them on his lathe, and turn them into pieces of furniture or ornaments. This was his *real* hobby, an activity in which he was highly skilled.

To Ammons, Ron Evans was a down-to-earth guy who was fun to be around, a firmly planted family man, a man who really wanted to make each day count. It was almost as if he somehow knew he wouldn't enjoy a long life and was determined to make the most of his time. He had to be serious and meticulous in his work, but that didn't mean he couldn't have fun doing it—for instance, while keeping up his flying hours. Ammons remembers one of many weekend trips to Lake Livingstone when he was sitting on a balcony with several friends, drinking a beer. Their astronaut buddy usually tried to join them for this bonding ritual, but that day he was flying. As the group gazed at the lake, they saw a dark spot in the distance getting bigger and bigger, rushing silently toward them. As Ron's T-38 flashed four hundred feet overhead, the sound waves caught up and rolled over the small audience. Ron knew where they would be and wanted to let them know he hadn't forgotten them.

Ammons tries to paint a verbal picture of his friend, toying initially with the idea of a leprechaun, but the image of a little Ron Evans sitting cross-legged on a toadstool is going a tad too far. Perhaps "puckish" conveys the right idea, particularly when Ron wore that impish grin his El Lago friends knew so well. He was an enigma. He could be really comical. He had a lot of mischief in him. Everything he went into had to be a challenge, to add spice to his life, but he also wanted it to be *fun*. That was why the two Rons got along so well.

When asked if Ron Evans was a "typical astronaut," Ron Ammons answers emphatically: "Noooo! No, no, no, nuh-uh!" He knew—and really liked—most of the El Lago–based astronauts, referring to one as "a prince of a guy," then expressing concern that this might appear to damn the rest with fainter praise. He didn't dislike any astronaut he ever met, but he did notice that some astronauts took themselves a little seriously at times. Ron Evans never did that. He was quintessentially down-to-earth.

Ammons noted similar characteristics in Neil Armstrong. Of course, both astronauts were engineers, naval aviators, and combat pilots with the special bond of having been Screaming Eagles. Armstrong and Evans liked each other and often gravitated toward the quiet end of astronaut get-togethers. Years after both men had retired from NASA, Ron would sometimes answer the phone and hear the familiar voice: "Hey, Ron. It's Neil. I'm doing a talk here in Phoenix. Can I drop by to say hello?" Sure enough, he would arrive at the door, relax on the couch, drink a beer, and talk about old times. Theirs was a lifelong friendship.

But Neil Armstrong was unlikely to have gotten involved in some of Ron's off-duty escapades. Ron was fascinated by the question, "What will happen if I do this?" Photographs show him standing beside the family's Chevrolet Suburban, trapped up to the axles in mud someplace off the beaten track. Jaime Evans is convinced her father was testing the limits of the vehicle's ability to power its way out of the mud. Sometimes his calculations were wrong, and only a tow truck could save the day.

The Suburban was useful for off-road driving in the countryside beyond the expanding southern suburbs of Houston. In the area were old abandoned houses succumbing to the elements before the developers sent in demolition teams. In one case, the two Rons got in first, pulling down sections of a wall and taking piles of bricks home. Ammons built a patio with them. He still remembers Evans carefully chipping the old mortar off the bricks and using them to pave his side yard.

In a similar vein, Evans would sometimes load up the Suburban with his own children and their friends on foraging expeditions to the local dump. They sometimes found useful or interesting items that could be converted into toys or garden furniture or the like. It was probably the bizarre and slightly grungy nature of the exercise that so delighted both the children and Ron's inner child. Nowadays it might be called recycling. On one occasion, their salvage expedition averted a small tragedy when they rescued a family of kittens abandoned in a large cardboard box.

One escapade straight out of the Ron Evans "What if?" manual occurred shortly after the families all moved into El Lago. The developer, Mr. Vick, who knew most of the residents very well, had left a large earthmover on empty ground in the cul-de-sac. The local children were fascinated by this monstrosity; so, too, were the neighborhood engineers, who couldn't resist examining it to see how it worked. Finally, scientific curiosity got the better of Ron, who climbed aboard and figured out how to start the engine. The roar from the exhaust provoked screams of delight from the children, who naturally wanted to see the earthmover work. Standing at the side of the road, Jan exchanged glances with Ron, who assured her he *definitely* wasn't going to start it moving! It was aimed right at the Krauses' house. A local police patrol stopped to find out why a crowd had gathered but saw nothing to concern the officers. The only problem was none of the engineers could determine how to switch off the contraption's engine. Finally, their only option was to cut the fuel line.

The next day, Ron had an early flight to the Cape, so Jan joined a slightly sheepish delegation of engineers and NASA contractors to explain what had happened and to offer payment for the repairs. Jan gathered from Mr. Vick's laughter that he wasn't too bothered about it.

Most of these incidents fell into the "you had to be there" category and reveal the impish side of Ron's nature, but another, more fundamental aspect of his character defined both his working life and his home life: Ron Evans did not like to fail at anything. He approached everything with a positive attitude, determined to be successful. He was very serious about his career and equally so about his home life. Ron Ammons understood. "Not being successful in his family life would just not have been Ron. He would have done anything he could to make that successful."

Many astronaut marriages broke up under the triple threats of the long absences from home, the allure of starry-eyed young groupies, and the risk (real and perceived) of spaceflight-induced widowhood. Jan's experiences as a navy wife gave her a head start in the astronaut marriage survival stakes, but what Ron and Jan mainly had in their favor was the obvious fact that they were crazy about each other. As Ron had noted in one of his 1957 letters to his wife-to-be, "A happy and enjoyable home life is the first thing to strive for."

Jaime Harner, who grew up in El Lago as Jaime Evans, always had a complex relationship with the space program, but there is nothing complicated or hard to understand about her relationship with and her memories of her late father. Quite simply, she adored him. She says he was "the best Dad ever." She still misses him, and her contributions to this book have come at the cost of a few tears. Mostly happy ones.

Born in 1959, Jaime has only fleeting memories of her father's naval career. She can remember playing on a swing set outside their home in San Diego and knowing that if Daddy was on a cruise, he would be gone a long time. Her mother kept her and Jon busy until their dad came back to play with them. Apart from his lengthy absences, she has no recollection of her father's time on board *Ticonderoga*.

Nearly seven years old when the family moved to El Lago, Jaime settled in quickly and easily. She recalls the new neighborhood had "lots of kids our age." As the local swimming pool was right behind their house, Jaime spent much of her free time with the swim team. She laughs that swimming "kind of kept me out of too much trouble all the way through high school."

Jaime was acquainted with several of the nearby astronaut families, although she really only thought of them as neighbors and people her father hung out with because of his work. Charlie and Dotty Duke's boys were much younger (she thinks she babysat them once). Even though Gayle Anders was closer to Jon's age, she and Jaime played in each other's houses, but the Anders family moved not long after *Apollo 8*. Jaime mostly remembers growing up with a very close-knit group of friends from the Evans, Ammons, Hurd, Lorenz, and Puttnam families. They all shared her own parents' laid-back and easygoing approach to life. She pictures them as adults who never minded all the local children running in and out of their houses, playing hide-and-seek, shouting, and doing all the other things happy children do.

The homes of other astronauts who were also, mostly, in the navy or air force seemed a little less approachable and lacked the same playful atmosphere in young Jaime's eyes. When she later went to college, a friend of hers whose father was in the military didn't mention "the astronaut thing" but did mention "the military thing." She wondered whether Jaime had experienced "a real regimented life" with her father's being a naval officer. Coming from a most unregimented home, this question greatly surprised Jaime. There was no military-style discipline in the Evans household, and she and Jon certainly never had to call their father "sir."

In fact, the real organizer in the Evans household was Jaime's mom. With Ron away so much, Jan assumed the roles not only of housewife and mother but also of Sunday school teacher, Girl Scout leader, Little League assistant, Parent-Teacher Association member, and more. Ron Ammons considers Jan a remarkable woman who was "very well suited for that life," and Jaime is clearly still in awe of the strength her mother brought to the marriage. Her parents were "definitely a perfect match for each other." If her father had a free weekend and said, "Hey, let's go!" Jaime knew he wasn't suggesting some tame road trip but most likely an off-road adventure in the Suburban. And if the heavens opened, her mom would be the one to produce the umbrellas and the coats. Even today, Jaime is torn between her father's preference for "just going" and her mother's way of making all the preparations first.

Adults don't always perceive their friends in the same way their children do, but Jaime strongly identifies with Ron Ammons's observation that her father seemed to live two lives. One of her most abiding childhood memories is of astronaut Ron Evans arriving home from work and taking his ten-minute nap. Ron could sleep anywhere. He would lie on the couch or even on the floor, snoring loudly. Then he would wake up, full of energy. Astronaut Ron would bound to his feet, transformed into her daddy, all ready to play with her and Jon, and do all the "dad things" they loved.

One of Jaime's fondest memories of dad time around the house sounds suspiciously space themed. At breakfast or snack time, Ron would break up a pile of cookies and mix them into a cup of milk, producing a paste of "cookie goop" that Jaime loved. "It was a silly thing I remember," she reminisces. "It was just a fun thing we did. He did it, I watched him do it, and it was part of his whole 'thing.'" Jaime enjoyed it enough to introduce it to her own children. And on *Apollo 17*, astronaut Ron Evans had plenty of opportunities to mix up his own food goop with hot water for consumption in zero-g.

With his engineering zeal, Ron was an inveterate handyman. Throughout their life together, Jan never needed to bring in a tradesman to do work around the house. Jaime remembers the "tree fort" her father built for her and Jon in the backyard. Not actually in a tree, it was the size of a small room and mounted on stilts. They accessed it by a trapdoor and a dangling rope, which was later replaced by an easier ladder. As a finishing touch, Ron planted bamboo all around it to give the fort a jungle look. It was a perfect spot for hiding out and holding sleepovers. Sometimes if Ron was out of town, Ron Ammons, in the role of "deputy dad," would have to clamber up to the fort and look for the missing offspring of several families.

Jan has many fond memories of Ron spending as much time as Project Apollo allowed to play with the children. When Jon was a little boy, if a toy broke, Ron would take his son into the garage, and together they would sit on stools and assess the damage. Ron would select the right tools for the job, carefully explain their functions, and show his son how they could fix the toy. When Jon was able to do it himself, he glowed with pride in the accomplishment, and over the years, this "good with his hands" expertise passed from father to son.

With Taylor Lake and a network of waterways nearby, Ron and the children spent a lot of time "messing about in boats." They enjoyed many a voyage in an old wooden sailboat, which was later replaced by a little fiberglass Dolphin sailing dinghy. Ron also took a basic wooden boat and applied his creative skills to add pontoons for extra buoyancy and paddles turned by bicycle pedals for motive power. The Evans navy had another vessel, a flat-bottomed "Jon boat," which had room to fit a motor. When the children were older, Ron used it to take them partway up the Houston Ship Channel, which brings to mind the famous line in *Jaws* about "needing a bigger boat."

Around the house, Ron built a cedarwood room divider, a king-size bed with storage drawers, bedroom cabinets, coffee tables, and, more ambitiously, a whole extension at the back of the house. He let his beard grow during the construction, as if to say, "astronaut on vacation."

For the children of the El Lago community in the days of Apollo, it must have been exciting to grow up right in the heart of the space program, mustn't it? All over the world, through TV coverage and in newsprint, people followed the stories of the astronauts and waited in keen anticipation for the footprints that would signal the realization of an age-old dream.

One day, probably around 1968, Jan Evans heard the front door opening and watched as her young son, Jon, walked into the house followed by about a dozen complete strangers talking in animated Spanish. They only knew a little English but warmly greeted a shocked Jan with multiple hellos. It turned out they were a Mexican tour group lured to Texas by their fascination with the space race. On arrival in El Lago, they had driven into the cul-de-sac and asked the only visible resident where all the astronauts lived. Jon had told them his daddy was an astronaut and invited them all into the house.

But if you're in the eye of the hurricane, you don't actually get blown by the winds. Jaime Evans wasn't the only "space brat" to feel a little underwhelmed by the whole space experience. After all, most fathers in El Lago were connected in some way with the space program, but they all were just ordinary guys who happened to have extraordinary jobs. It wasn't that Jaime lacked pride in her father's hard-earned status, but she didn't want to know all the details, such as how the Apollo docking mechanism worked. Jaime turned ten years old in the summer of 1969 and became a teenager shortly before her father flew to the moon. She just wanted to hang out with her friends, go swimming, and try to get noticed by boys, sometimes all at the same time. If her father hadn't been an astronaut, Jaime might not have paid much attention to all the worldwide hullabaloo about Apollo.

Over the years Jaime has tended to shy away from interviews about Apollo, and her heart sinks a little, even now, if a friend introduces her to someone new as "the daughter of an Apollo astronaut." Quite apart from being judged on her own merits, she would prefer to talk about her husband, her children, or her job. This is certainly not through any lack of pride in her father's achievements. She is immensely proud that her father made a successful flight to the moon. Her concern is that she lacks the technical knowledge about *Apollo 17* and her father's work in lunar orbit, and discussing it all with strangers would reveal her lack of technical detail. She doesn't want to be embarrassed or made to feel that she has somehow let down her father's memory. After all, if Ron Evans had been a Nobel Prize–winning theoretical physicist, Jaime could be genuinely proud of his achievements without people expecting her to understand relativity or string theory.

Growing up in El Lago, Jaime really only saw the "dad" side of her father. When he set off for the space center, that was outside Jaime's experience, and she could hardly imagine her father's other life. She can appreciate his calm unflappability and his complete lack of anxiety in the buildup to *Apollo 17*. She never sensed any concerns, let alone fears, on her father's part. He didn't make it a big deal. It was his job, and he tried not to let it impact too much on their home life.

Only in later years did Jaime get a real inkling of what had been involved in her father's work life. With much concentration and effort, she learned to fly a light aircraft. "If it was that hard to fly a little Cessna," she often wondered, "what must it have been like slamming a supersonic jet down on the deck of a carrier in a war zone?"

Then she recalls a trip to Germany and Austria some years after *Apollo 17*. Jan was recovering from minor back surgery, so Ron invited Jaime to accompany him on the trip. As she enjoyed the comfort of first-class travel across the Atlantic and a big, brash welcome by space fans at Frankfurt Airport, it dawned on her that other people saw her father in an entirely different light. She remembers thinking, "Wow! This is, like, crazy!" She was finally glimpsing her father's other life and realized, "Okay, this is interesting! He's more than 'just Dad'!" They were accompanied on the tour by a group of people Jaime struggles not to call an "entourage," but clearly it was. The visit included trips to the top of Germany's highest mountain, the Zugspitze; to Oktoberfest in Munich; and to Mozart's birthplace in Salzburg. At the public meetings, Jaime saw her father being greeted as a celebrity and feted like a hero. Complete strangers, some barely able to speak English, lined up just to shake the lunar explorer's hand. The two lives of Jaime's father came together right in front of her. It was a real eye-opener.

16. Footprints in the Dust

This operation is somewhat like the periscope of a submarine. All you see
are the three of us, but beneath the surface are thousands of others, and to
all of those I would like to say, "Thank you very much."

—Michael Collins (1930– 2021), *Apollo 11* CMP, 24 July 1969

Deke Slayton had always hoped the first man on the moon would be one of
the original Mercury astronauts, probably Gus Grissom or Alan Shepard.
Then in 1963 Shepard developed alarming symptoms diagnosed as Meniere's
disease, an inner ear condition causing bouts of vertigo, nausea, and tinni-
tus. He was grounded by the navy and by NASA, and lost his position as commander of the first Gemini mission. A cruel fate for any pilot, it was a private
purgatory for America's first astronaut.

NASA persuaded Shepard to remain at the heart of the space program as
its chief of the astronaut office and work alongside Deke Slayton, the director
of flight crew operations. Neither man could fly without a qualified second
pilot. They couldn't even fly the T-38 together. As Dr. Charles Berry memo-
rably but cruelly put it, "Two half pilots don't make a whole."

If Gus Grissom had not died in the *Apollo 1* fire, he would probably have
been the first man on the moon. The reality in late 1968 was that Deke Slayton
had always expected to lose men from the program for one reason or another.
The total by then was nine, or twelve if he counted three scientist-astronauts
who had left, realizing they would never feature in Project Apollo. The history of what Jerry Carr called "the glory years" would be written by and about
the men who were still there and drawing NASA salaries.

The men of Group 5 who were climbing the long ladder toward the beck-
oning moon recognized that it would be very difficult to get past someone

above them on that ladder unless he fell off or jumped off. Or got pushed off by someone higher up.

As in any bureaucracy, NASA had internal politics, but most of it took place above the pay grade of men like Ron Evans. Although the benefit of test pilot credentials had become increasingly obvious, none of the rookies could work out how or why Deke Slayton selected particular individuals for missions. Some of the group might have been tempted to ingratiate themselves with Slayton. Inevitably such efforts would have included participating in a lot of hunting trips, but even those who were not dyed-in-the-wool hunters gained enjoyment from a few days away from the office, bagging a few quail and drinking a few beers with their colleagues. There was a fine line between camaraderie and ingratiation, but Ron seemed to understand on which side of the line to hunt. As he pointed out in 1986, "There was no way to 'brown-nose' Deke. No way! Some people may have thought they could, but they didn't know Deke well enough!"

Jan Evans always liked and respected Deke. She knew exactly what role he was performing: the astronaut corps was similar to a big squadron, and Deke was their skipper. Of course, that made Marge Slayton the skipper's wife. Marge tried to disavow the term, but the more she insisted, "I'm no skipper's wife," the more the other wives teased her about it.

Strictly speaking, the real skipper was Al Shepard. Compared with Slayton, Shepard was notoriously moody. One morning he'd turn up at the office as "smiling Al"; the next morning he'd be "the icy commander," chilling the blood of anyone within his orbit. Some people around NASA whispered that Shepard spent too much time in the business world, making himself a millionaire, but Ron Evans respected Shepard's efforts as the chief of the astronaut office. The men in his charge were all highly motivated and basically leaders in their own right in one capacity or another.

"And then," Ron later noted, "some guy had to pull all this together—which was Al Shepard. And he had kind of a tough problem, leading leaders." Ron felt Al got the job done by letting the men find their own way. Shepard didn't need to hold their hands while they trained. He wasn't standing outside the gym or the simulators, checking that the guys turned up on time. But if they didn't, he knew.

In the space of around fifteen months, beginning in May 1968, the fates of several astronaut careers were sealed, and the shape of Project Apollo after the first moon landing began to snap into focus. Unknown to all but a few in NASA, Al Shepard had undergone pioneer surgery to correct his inner ear problem. It took a year before he was officially restored to flight status and eligibility for flight crew selection. Even before the announcement, Shepard was pulling strings to secure the first available flight.

Gordon Cooper had commanded *Gemini 5*, served as the backup commander of *Gemini 12*, and then was appointed the backup commander of *Apollo 10*. As a Mercury veteran, he may have felt slighted by that sideways move, but as the usual rotation system would have put him on the moon as the commander of *Apollo 13*, it isn't difficult to see why he bit his lip. But whether he realized it or not, his star was on the wane. Being a great natural pilot wasn't good enough anymore. Deke Slayton was by no means alone in judging that Gordo just wasn't putting his heart into the training and seemed more interested in getting his kicks from motor racing.

If Ron had any qualms about a third support crew assignment, he must have felt some sympathy for Gordo when he heard that the Mercury veteran being proposed as the *Apollo 13* commander was not Cooper. Slayton had accepted the recommendation of the chief of the astronaut office: the mission commander should be the chief of the astronaut office—namely, the new, improved Al Shepard. The man with fifteen minutes' spaceflight experience! This choice completely ignored the usual rotation convention, a development that Ron kept in mind over the next two years. Gordo Cooper himself was furious, later claiming he had been offered—and had refused—the sop of backup commander. Slayton later wrote that he hadn't had any plans for Gordo after his *Gemini 5* flight.

While Slayton had too many applicants for the commander's job on *Apollo 13*, he had the opposite problem seeking a LM pilot for Shepard. He tried to sell the idea to Jim McDivitt, who was just back from commanding *Apollo 9*. McDivitt refused, telling Slayton that Shepard wasn't ready to jump straight into a commander's chair after so many years out of the loop. That was the polite explanation. Four decades later, in addressing a meeting of space enthusiasts in England, McDivitt expressed his feelings about the proposal in robustly salty terms.

The whole issue may be academic because Dave Scott recalls that McDivitt indicated during *Apollo 9* that it would be his last mission. Be that as it may, the proposition is hard to see as anything other than a rare lapse of judgment by Slayton. How could the lunar module's first test pilot have been expected to swallow his pride and accept a demotion to systems engineer? He would have been junior not only to Shepard but also to the likely command module pilot, Group 5 rookie Stuart Roosa. This wasn't a mere matter of ego; it was a matter of self-respect. For McDivitt, it meant turning down the chance to walk on the moon, but some prices are too high to pay. Besides, what would it have said about Shepard's suitability for command if he needed a flight-hardened veteran to look after him? Instead, McDivitt applied his considerable skills and experience to administration as the Apollo Program manager, deftly overseeing the missions from *Apollo 12* to *Apollo 16*.

Other veterans from Groups 2 and 3 had either decided to retire or were already on other assignments. Shepard and Slayton mulled over their options and picked two of the Group 5 rookies to join the crew, but it would be misleading to suggest that Shepard had to "make do" with the men who were left. The Apollo machine was purring smoothly, and Slayton no longer required a lunar-orbiting CMP to have flown before. He was confident that his Group 5 troops were ready to fly. One day in April when most of NASA was thinking ahead to *Apollo 11*, astronauts Stu Roosa and Ed Mitchell answered a summons to Al Shepard's office. They were stunned when he told them, "If you guys don't mind flying with an old re-tread, we're the prime crew of *Apollo 13*."

When the dust stirred up by this selection had finally settled, cynics still smiled knowingly and hinted that it conveniently solved any concerns Shepard might have had about being overshadowed by greater experience. Fifteen minutes' flight time beats zip! But that was unfair to both rookies. Roosa, an accomplished test pilot, had gained glowing reviews in his previous assignments, and Ed Mitchell had not only topped the Group 5 peer review but also was one of the leading experts in the operation of the lunar module.

But NASA said no. Specifically, George Mueller, the head of the Office of Manned Spaceflight, vetoed the proposal. For the first time, one of Slayton's selections was not rubber-stamped from above. Apparently, others agreed with Jim McDivitt that Shepard needed more training time before grabbing a command.

There was a relatively simple solution. Jim Lovell was backing up Neil Armstrong on *Apollo 11*. If he was good enough to be just a broken leg away from stepping into Armstrong's shoes—and the history books—then he was good enough to command his own mission. Slayton was lining up Lovell to command *Apollo 14*, so getting him to agree to switch to *Apollo 13* wasn't too difficult. What could possibly go wrong? Shepard's crew would now fly *Apollo 14*, allowing Shepard the time he needed to master his brief.

Selection problems weren't limited to the *Apollo 14* prime crew. Before George Mueller quashed the Shepard selection, Slayton tried to put together a backup crew. He suggested the unflappable John Young for backup commander and Jack Swigert as the CMP. He also asked the twice-flown veteran Gene Cernan to be Young's LMP. That would probably have allowed Cernan to follow Young down the ladder on *Apollo 16*. To Slayton's surprise, Cernan politely declined the offer. He wanted to hold out for his own command. He felt he had done enough to lead his own crew to the moon. Slayton knew only too well that NASA's budget was decreasing and that a successful *Apollo 11* ironically might accelerate that problem. He made it clear to Cernan that NASA might not have enough future missions to accommodate his lofty ambitions, but Cernan was adamant and left Slayton again scratching his head in puzzlement.

While not the same as the Shepard-McDivitt situation, Cernan's decision reveals a lot about the military mind. Cernan recognized Young as the more experienced man (Group 2 with three flights versus Group 3 with two flights). He would not have suffered a bruised ego nor any loss of self-respect by playing Robin to Young's Batman, but as a military man, Gene Cernan had ambition. Had he remained an active naval officer, he would have worked hard to get his own squadron and eventually his own ship. As an astronaut, the nearest equivalent was to command his own Apollo mission. Whether ballsy, brave, or deluded, his decision had consequences that would affect several careers, including that of Ron Evans.

Meanwhile, in spite of McDivitt's and Cernan's declining roles on *Apollo 13*, it would not have escaped Ron's attention that seats to the moon were filling up quickly. After *Apollo 12*, eight more lunar missions were planned, needing only eight more command module pilots. It already looked as if Ken Mattingly, Stu Roosa, Al Worden, and Jack Swigert would fill four of those seats. All were test pilots.

On the eve of *Apollo 11*'s leap into the history books, Ron Evans sat down with nine friends and colleagues for what Deke Slayton called "a quiet dinner" in the NASA crew quarters at the Kennedy Space Center. To Ron's right sat Bill

Anders, Neil Armstrong, Buzz Aldrin, and Bill Pogue. On the other side of the table were Ken Mattingly, Mike Collins, Jim Lovell, Deke Slayton, and Fred Haise. It was a low-key affair, something of a thank you to the backup and support crews for their help during the preparations for this most anticipated mission.

They dined on broiled sirloin steak, mashed potatoes, tomato puree, cottage cheese, bread, and butter. If anyone was tempted to refer to it as "the last supper," it was not reported. Photographs of the event show mainly somber, pensive expressions, but despite the task ahead, Armstrong managed some unposed cheerful grins. He did have one gripe about the meal and later confided to his wife, Jan, "I'm sick of steak."

For Ron, the dinner marked the culmination of a busy six months on support duties, which had included frequently flying around the country with Pogue, Swigert, Mattingly, Aldrin, and Collins. He flew into Patrick AFB with Jim Lovell on 7 July as the preparations built up to a climax and managed to fit in some helicopter training a few days before the launch. On the morning of Wednesday, 16 July 1969, Ron joined the million-strong crowd of people who had arrived from every state of the Union, and far beyond, to witness the beginning of the journey that—if successful—would mark the fulfillment of President Kennedy's pledge. Ron's viewing point near the Vehicle Assembly Building was rather more exclusive than for most of the visitors, but from wherever they watched the launch, no one who was dazzled and deafened by *Apollo 11*'s fiery ascent would ever forget it.

Back in El Lago, Jan watched the launch on TV. Busy wives and mothers can't always drop everything, even to witness history being made.

After the launch, the world's focus switched from Florida to Texas. Ron made the two-hour flight back to Ellington. He was not under any time pressure as his first stint as capcom did not start until fifteen hours into the mission, and by then the crew of *Apollo 11* was asleep. Not for nothing was flight director Glynn Lunney's shift called the "Black Team." Ron and his colleagues had a quiet night, but like all night watchmen, they had to be ready to spring into action if something went wrong.

Ron was back on duty thirty-eight hours into the flight. Black Team's shift was sandwiched between a TV tour of the command module and the first TV views inside the lunar module. During the intervening hours, there was silence, broken only by the occasional snore.

At exactly seventy-five hours, forty-one minutes, and twenty-three seconds into the flight, *Apollo 11* disappeared behind the moon. All contact was lost for just over thirty-four minutes before Capcom Bruce McCandless was able to resume communications. The engine burn on the far side of the moon had been close to perfect, and the spacecraft was in orbit.

It had not gone unnoticed at mission control that Ron's opportunities to talk to the crew had so far been very limited. Toward the end of another quiet night shift, NASA's public affairs officer couldn't resist announcing, "Standing by for Ron Evans's big moment as he makes his call to the spacecraft as, being the sleep-watch, his job has been rather easy. . . . Here we go!"

Ron duly greeted the crew as the men emerged from a deep sleep and confirmed that the CSM *Columbia* was in good shape. He added, "Black Team has been watching it real closely for you."

A still-groggy Collins replied, "We sure appreciate that. 'Cos I sure haven't."

On 14 December 2013 the People's Republic of China earned international headlines as it successfully soft-landed a spacecraft on the moon. The vehicle that had caught the world's attention was named *Chang'e 3*, and shortly after landing, it deployed a self-propelled wheeled rover named Yutu (Jade Rabbit). Much of the media coverage seemed more interested in an ancient Chinese myth concerning Princess Chang'e and her pet rabbit than in the impressive scientific results beaming back to Earth. What none of the media seemed to recall was that the world had already been introduced to this Chinese myth forty-four years earlier.

To ease the *Apollo 11* crew gently back to wakefulness, ready for the challenge ahead, Ron Evans continued a capcom tradition of reading out a summary of world news and sports headlines, both serious and funny. One wacky story had caught his attention. "Among the large headlines concerning Apollo this morning is one asking that you watch for a lovely girl with a big rabbit. An ancient legend says a beautiful Chinese girl called Chang-O [*sic*] has been living there for four thousand years. It seems she was banished to the moon because she stole the pill of immortality from her husband. You might also look for her companion, a large Chinese rabbit, who is easy to spot since he is always standing on his hind feet in the shade of a cinnamon tree. The name of the rabbit is not reported."

Collins later suggested that the "steady chatter" from Houston might not have been the best idea as the crew ate breakfast and began to select items of equipment for transfer into the LM. At the time, although still a little grumpy and groggy, he managed a witty response to Ron's story: "Okay. We'll keep a close eye out for the bunny girl."

Ron was well aware he was only a bit-part player lost in the immensity of the drama that was about to unfold. He didn't mind. Whatever the future might bring, whatever his opportunities, he was thrilled to participate in *Apollo 11* in any capacity and to watch the culmination of the national commitment to which he had applied himself body and soul. At least he had one of the best seats in the house.

While Ron mainly listened to the snores of the crew, Jan made her own contribution as an El Lago resident, joining the small army of friendly faces calling on Jan Armstrong and providing both nutritional and moral support. Jan knew her namesake well enough to understand the weight of expectations piled on her. While Neil was away, an ordinary flesh-and-blood man doing an extraordinary job, Jan Armstrong had to face the world and acquit herself in a manner befitting the wife of a reluctant international celebrity. When major mission events took place, a phalanx of family, friends, and neighbors cocooned her. Jan Evans recognizes that the full impact of what was being attempted only sank in later. History was not uppermost in their minds as they followed each step of the mission. And yet they could hardly forget the mission was unique. "You know," Jan muses, "you wanted it so bad, and you knew that it *was* different from any other flight!"

Jan Armstrong was as strong as any of the wives, but her friends knew she was harboring understandable concerns. The lunar landing was the one part of an Apollo mission that had not previously been carried out for real. She didn't dwell on it, but she couldn't help that some apprehension seeped into her thoughts. Luckily, when she had a question or a concern, she had plenty of experts to consult. Ron was one of many astronauts who popped in to lend her support.

The landing of astronauts Armstrong and Aldrin, riding their flimsy little *Eagle* down to the Sea of Tranquility, has been written about, analyzed, and eulogized for five decades. Those who followed the event on TV or radio will never forget it. Those who were not alive to experience it will never truly understand. For all but the most plugged-in NASA specialists, the landing approach seemed to be a confusion of static, harsh and clipped radio messages, and baffling alarm calls. Almost no one understood what "12-01" or "12-02" meant, but the words "program alarm" sounded bad, *really* bad. Hundreds of millions of human hearts were pounding. As the clock ticked past the predicted landing time and fuel warnings were called, we all just sat riveted to our TV sets and hoped for the best. Then came an unmistakable call: "Contact light!" Then a babble of jargon that few understood. Then at last, clearly over the airwaves, the voice of Neil Armstrong saying: "Houston, Tranquility Base here. The *Eagle* has landed."

Delight. Relief. Ecstasy. All over the world, from Sydney to Tokyo, Cairo to London, Buenos Aires to El Lago, family and friends turned to each other, slightly bemused by what we had heard. But we were all now clear about one thing, and the words we uttered, in many languages, all had the same meaning: "They made it! They're down."

Did anyone watching at home notice that the self-effacing Armstrong gave top billing to the machine, built by the dedicated workers at Grumman, over the men—or the man—who had actually made the landing? No matter. The *Eagle* and its crew rested safely in the dust of what the ancients called the Mare Tranquillitatis. An age-old dream had become reality.

It was a glorious and satisfying moment for Jan Armstrong and everyone packed into her house, but it was still too soon for champagne. Neil and Buzz were on the moon, but never had two of us so completely cut themselves off from the rest of us. The first part of Kennedy's commitment had now been fulfilled. Someone even placed a handwritten note beside the eternal flame on John F. Kennedy's grave that read, "Mr. President, the *Eagle* has landed." Now the second part of the commitment, the rather crucial part about "returning…safely to the earth," would occupy everyone's minds. But first the astronauts had important work to do.

Charlie Duke had been in the spotlight as the capcom during the lunar landing. Now it was Bruce McCandless's turn to add his own words to an indelible page of history as he talked to the first men to walk on the moon. It is tempting to speculate that McCandless did not feature in Deke Slayton's Apollo plans and that this high-profile shift as "the Voice of Planet Earth" was something of a consolation prize. Whether true or not, McCandless did a fine job. His words will live forever alongside those of the men who made the first footprints in the lunar dust.

In the early hours of 21 July, Armstrong and Aldrin were safely back inside the LM. They had provided the world's largest TV audience with a starkly surreal view of their ungainly, bouncing movements across the rocky moon-scape. For Jan Evans, as for everyone else, the experience had been unforget-table. Now it was time for her to walk the few blocks home alone. Like most of the astronaut corps, Ron was at mission control, playing the passive role of proud and delighted off-duty spectator.

Jan found it shocking to emerge from a house where everyone had been staring at the men on the screen into a place where the center of attention was the Armstrong house itself. Neil Armstrong was away on business, so the world's press was besieging his home with a blockade of vans, trucks, genera-tors, and "horrendous, enormous big spotlights like you see at fairgrounds." There were enough cars to stop the traffic for two blocks around. The press couldn't be on location with the astronauts, so the next best thing was to put their families under the microscope. Visitors like Jan Evans had to run the gauntlet simply to walk home.

It was a hot, humid, still summer night. As she walked along Woodland Drive, Jan turned and looked back at the Armstrong home, bathed in the glare of the lights. But it was also surrounded by an eerie glowing blue radi-ance. Almost every member of the besieging army was chain-smoking, and the rising columns of smoke were coalescing into a cobalt haze that enveloped the spotlit house and spread out along the street, with its subtle hues gradu-ally fading through indigo to black.

When the newspapers appeared that day, only one story made the head-lines and most of the inside pages. The drowning of a young woman at Chap-paquiddick was mostly relegated to page 2. Amid the adulation for Armstrong, Aldrin, and Collins, the astronauts who lost their lives in pursuit of John F. Kennedy's goal were also remembered. *Chicago Today* printed an achingly poignant cartoon showing Armstrong raising the U.S. flag, Iwo Jima–style, aided by the spectral figures of Grissom, White, and Chaffee.

It was time to go home. After a little more than twenty-one hours on the moon, Armstrong and Aldrin were preparing to launch the *Eagle*'s ascent stage back into orbit to rejoin the *Columbia* and Michael Collins. Five days earlier, they had needed thousands of engineers and technicians to launch their Saturn V rocket. Now they were alone, linked to mission control by the voice of Ron Evans, who was back on duty as the capcom.

The fate of the astronauts and the mission rested on one small rocket engine.

Weighing only fifteen pounds more than Armstrong, the ascent propulsion system produced 3,500 pounds of thrust, enough to counteract the moon's low gravity and to propel the spacecraft into orbit. Engineers had made it as simple as possible to limit its risk of failure. Its only moving parts were the valves to release the fuel and oxidizer. When these hypergolic liquids mixed in the combustion chamber, they would ignite spontaneously and continue to accelerate the *Eagle* as long as the liquids flowed. Ever the engineer, Armstrong had wanted to simplify the system even further by including "one big handle to open the valves," but NASA had insisted on a more sophisticated system of "electronics and solenoids and valves."

As the time for liftoff approached, Ron's contacts with the *Eagle* were mainly clipped and businesslike. The men exchanged essential guidance and navigation data and checks on the health of the spacecraft. Ron was briefly joined by backup commander Jim Lovell, who radioed his congratulations on a job well done and his prayers for what was to come.

With sixteen minutes to go, Ron advised the crew, "You're cleared for takeoff." Buzz Aldrin then "spontaneously injected a touch of humor" to relieve the tension by replying, "Roger, understand. We're Number 1 on the runway." After four seconds, Ron acknowledged with a "Roger," and Aldrin assumed the rookie, unfamiliar with the LM pilot's "sardonic sense of humor," had taken time to process the remark.

Finally, Ron reported, "You're looking good to us." Aldrin began a short countdown. Armstrong armed the engine. Aldrin pressed the proceed button.

If there are infinite parallel universes besides our own, there must be at least one in which the *Eagle*'s ascent engine failed to ignite, stranding Armstrong and Aldrin on the moon. Ron Evans never discussed with Jan what he would have done in such a uniquely public emergency. He must have given at least a passing thought to how he should conduct himself while acting as the conduit between NASA's problem solvers and the two men in need of urgent solutions. In the parallel universe where the *Eagle*'s engine ignited but failed a few seconds later, Ron might have found himself having to break the news to Michael Collins before sharing the airwaves with President Richard Nixon. The president had a pre-written speech ready that began, "Fate has ordained that the men who went to the moon to explore in peace will stay on the moon to rest in peace."

Happily, in the only universe we know, it was never necessary to test Ron's mettle, or that of the others involved, in such bleak scenarios. The full vigor of the ascent engine launched the crew along a near-perfect trajectory into an egg-shaped orbit. The *Eagle* was neatly positioned for later retrieval by Collins in the *Columbia*.

A few minutes after entering orbit, Armstrong formally reported, "Roger, Houston. The *Eagle* is back in orbit, having left Tranquility Base."

It is a fair bet that Ron Evans wore one of his trademark smiles as he responded, "*Eagle*, Houston. Roger, we copy. The whole world is proud of you."

Thursday, 24 July 1969, was splashdown day. For more than eight years, NASA had been working in overdrive to meet the challenge set by John F. Kennedy. This was the day of reckoning. The high-profile capcom shifts by Charlie Duke and Bruce McCandless had been associated with the moon. The period of reentry and splashdown was rather less glamorous but hardly less crucial. Ron Evans would conduct the final capcom shift of the mission.

In 2003 Neil Armstrong was asked which landing meant more to him— the landing on the moon or the "landing" back on Earth. After a surprisingly long pause, he acknowledged that for a pilot, the first landing on the moon was very satisfying. But NASA had been tasked to land a man on the moon *and* return him safely to Earth. For that reason, Armstrong chose the splashdown.

Earth was now a giant looming crescent, but as the command module accelerated toward the thin layer of atmosphere, the crewmen had their backs to the direction of travel. As anyone with a sardonic sense of humor knew, it was always a good idea to hit the atmosphere blunt-end-first.

Twelve minutes before the entry interface, Ron advised the crew, "You're still looking mighty fine here. You're cleared for landing."

"We appreciate that, Ron. Thank you," Collins replied.

Continuing his line of military humor, Aldrin added, "Rog. Gear's down and locked."

A few minutes later, Ron advised, "And Eleven, Houston. You're going over the hill there shortly."

With reasonable expectation, Neil Armstrong replied, "See you later." And after that transmission, the *Columbia* entered the upper traces of the atmosphere. Soon it resembled a man-made meteor heading toward a rendezvous with the aircraft carrier USS *Hornet*.

In El Lago, Jan Evans joined the Armstrong family, Jim and Marilyn Lovell, and many others waiting for news. They never got to see the spacecraft descending under its parachutes because of the murky weather, but when one of the recovery helicopters radioed the report of a successful splashdown, everyone in the Armstrong home erupted in an outpouring of relief, joy, and pride. Half a century later, Jan recalls the moment and says, with obvious emotion, "They did it! They *did* it!"

That was the moment to bring out the champagne to celebrate the mission's successful accomplishment. An elated Jan joined in the toasts. Even though she never much cared for champagne, at *that* moment it tasted like nectar.

When the TV pictures showed the first man on the moon emerging from the bobbing spacecraft in the water, Jan pulled a bunch of small American flags from her handbag. Soon everyone was waving them, cheering, and quaffing champagne.

A live picture from Houston showed mission control jammed with astronauts, NASA officials, and support staff. When—and only when—the astronauts arrived on the deck of the *Hornet*, almost everyone, probably including many nonsmokers, lit up the traditional celebratory cigars. With so many waving flags, the TV picture from mission control was a blur of red, white, and blue. A large screen flashed the wording of President Kennedy's 1961 pledge, followed by a picture of the *Apollo 11* mission patch and the message: "Task Accomplished. July, 1969."

Among the hundreds of faces and bobbing figures pictured amid the euphoria, several bald and balding heads can be seen near the front of the room. Shown a photograph of the scene, Jan Evans has no problem identifying one of the celebrating figures.

"That's my Ron!"

17. Backups

Anyway, that's why you do it. Not to be famous, just to be good.
To do good work. Find the thing you really love doing, and
do it to the best of your ability.

—David Nicholls (1966–)

After the last champagne bubble had popped and the last reveler had returned home, Houston still basked in the euphoric glow of *Apollo 11*'s success. A major national goal had been achieved, and the United States was riding high on a wave of international praise and esteem.

For Ron Evans and other members of Group 5 who had not yet heard their names associated directly with future missions, it was time to take stock. Perhaps, with the greater goal now achieved, it might even be a time for personal ambition. Only the best should expect a seat to the moon, but if you didn't believe you were one of the best, what were you doing as an astronaut?

Certainly Ron's expectation was that to get a prime crew seat, he would first have to be selected as a backup command module pilot. In July 1969 the crews for *Apollo 13* and *14* had not been officially announced, but the next free backup seat appeared to be on Al Shepard's *Apollo 14* flight. Unbeknownst to Ron, an important decision in the month before *Apollo 11* would have a major knock-on effect on his prospects.

When Michael Collins took to the air with Deke Slayton in June 1969, he was looking forward to *Apollo 11*. On the flight to Patrick AFB, Slayton told Collins that he wanted to plug him into the flight rotation after the mission. If Collins would back up Shepard on *Apollo 14*, he could walk on the moon as the commander of *Apollo 17*. Collins turned it down. Assuming *Apollo 11* went well, he wanted to retire from the training grind. After all, even as the "forgotten man," could he really top the crowning glory of the first lunar landing mission?

Shortly after *Apollo 11*, Slayton called Gene Cernan to his office and offered him the position of *Apollo 14*'s backup commander. Cernan's gamble had paid off, and he happily accepted. He later wrote that if he had known that Collins was Slayton's first choice, he would have had second thoughts about turning down the chance to land on the moon as John Young's sidekick. Cernan also wrote that he and Slayton "quickly settled on a crew of two space rookies." The backup CMP would be "my old comrade from Monterey, Ron Evans the Vietnam veteran." Completing the trio would be X-15 test pilot Joe Engle. It was a good day for Kansas. Cernan added that he "liked and respected both men and was determined to make ours the best crew ever."

Whether Slayton or Cernan suggested Evans and Engle is not known. Perhaps a number of names were proposed and considered, leading to early agreement. We will never know for certain.

Barbara Butler, then Gene Cernan's wife, doesn't know how the selection was made. Gene never discussed it with her, but she is certain he must have had some input "because you want to be sure you're flying with the right people!" Gene Cernan admired Ron for his combat service and always felt he should have "done his bit" alongside so many of his contemporaries. Barbara recalls the period after the fire when Gene and other navy astronauts considered returning to active duty. She talked it over with him and told him, "Gene, everybody has a place in life and a thing to do. Yes, you're qualified to go to Vietnam and do all those things that I know you've been trained to do. But now you've been trained to do something else, and you'll have to make that decision."

Barbara realized the choice was hard for Gene because he had "the fighter pilot instinct." In the end, she told him, "The good Lord puts a path in your way, and you have to follow it." The Pentagon's attitude regarding the astronauts' flying combat missions clearly helped Gene choose the right path.

Cernan didn't tell his wife how his first crew was selected, and he didn't tell Ron Evans either. Asked in 1986 whether Cernan had personally picked him, Ron's reply tells us a little more about his attitude to life: "Gene had something to do with it, sure. Oh yeah, you bet! I know Deke had asked Gene. I don't know if he was saying, 'Hey, would you go with this?' or 'Who do you recommend?' I don't know how it was done, and it's probably just as well I don't.

And it doesn't make any difference! If you didn't get selected on a crew, it's just as well you didn't know why you weren't selected. And if you were selected, it doesn't make any difference, because you're there! So why bother worrying about it? And I never have worried about that type of thing!"

Nevertheless, Ron certainly felt great satisfaction and undoubtedly some measure of relief with his selection. He had climbed another rung up the long ladder and was entitled to think that his years of hard work and relentless training were finally paying off.

On 6 August 1969 NASA announced the prime and backup crews for *Apollo 13* and *14*. It was official. No one had made any promises, but the usual rule of thumb suggested that if all went well, Ron would one day be flying to the moon on *Apollo 17*. Assuming nothing went wrong.

Ron was still a serving naval officer, although now NASA was submitting the regular reports on his performance. In the fall of 1969, he received more good news, this time from the U.S. Navy: the letter from BUPERS informed him of a promotion. As from 1 October 1969, Ron Evans would be a full commander.

By January 1970, the future of Project Apollo was shrouded in doubt. Along with budgetary cutbacks and the slashing of its already reduced workforce by another fifty thousand employees, NASA had to accept that production of the Saturn V rocket would end after fifteen vehicles had been delivered. On 4 January the *Apollo 20* mission was canceled to make the last Saturn V available to launch the planned *Skylab* space station. After the prime crew of *Apollo 15* was announced in March, only four more Apollo CMP seats were available.

At least the previous concerns that there would be no room on lunar flights for Group 5 astronauts had been laid to rest. By early April 1970, Group 5 members Mattingly, Haise, Roosa, Mitchell, Worden, and Irwin were in training for forthcoming flights. The established pattern now seemed to team a spaceflight veteran as commander with two Group 5 rookies. Ron could only hope that pattern would continue.

Shortly after 9:00 p.m. on 13 April 1970, the party was beginning to liven up. If Ron wasn't away on duty, he and Jan always opened up their home after a successful launch, inviting neighbors, astronauts, and any NASA contractor friends who were in town. As *Apollo 13* headed moonward, guests were spreading through the roomy interior of the house and sampling the chili.

Drowned out by the babble of many voices, the telephone rang in the kitchen. Tom Stafford happened to be standing beside the phone and took the call. His party smile quickly disappeared as he strained to hear the voice from mission control talking about "a problem" with the spacecraft. He listened intently, then replaced the receiver, and began to move discreetly among his colleagues. Whatever the news had been, it quickly passed from the astronauts to the contractors who had built the Apollo hardware, turning smiles to frowns of concern.

At first unnoticed by Jan and the other women, the front door kept opening and closing, and the numbers at the party began to dwindle. Eventually they looked around the half-empty room and realized all the guys had gone. None of the astronauts had said a word to their wives, partly so as not to cause unnecessary concern but mainly because they didn't know the scale of the problem. Ron phoned Jan later to confirm that *Apollo 13* had suffered some kind of malfunction, which they were all investigating, and he had no idea when he would be home. After Ron's capcom sessions, Jan knew enough about mission control to be confident the "smart guys were thinking of ten million ways to take care of the problem."

What happened next has been written about extensively and etched indelibly in popular culture. What the press had written off as a humdrum repetition of history suddenly erupted onto the front page of every newspaper and led every news bulletin from Houston to Timbuktu. The *Apollo 13* mission failed to achieve its goal of landing on the moon, but it shined a spotlight on NASA's response to a life-threatening crisis. Future historians seeking to understand the overall success of Project Apollo in an era of slide rules and primitive computers need look no further than NASA's dedicated teamwork. When calamity struck, the people who built the hardware, wrote the software, and molded mission control into a precision mechanism all rose magnificently to the occasion.

But no machine built by the hand of man is perfect. Without warning, oxygen tank number 2 exploded in the bowels of *Apollo 13*'s service module almost fifty-six hours into the mission. Later investigations revealed that a series of errors and omissions dating back several years had doomed the mission. Unknown to anyone, *Apollo 13* was launched with a damaged thermostat and damaged Teflon insulation inside the tank. For the second time, an electric arc, flammable material, and high-pressure oxygen changed the course of Project Apollo. From that moment, the lives of the crewmen depended on how well they and the NASA team on the ground could work together to create solutions to problems that had mostly never been anticipated.

Jack Swigert wasn't even supposed to be on this flight. Two days before the launch, he replaced Ken Mattingly, who had been exposed to German measles and might have become unwell and feverish at a crucial stage of the mission. The backup, by definition, is not as well trained in the minutiae of the mission as the prime crewman, but he must be trained well enough to win the confidence of the mission commander. Swigert was a superb pilot who knew the Apollo spacecraft inside out, but his substitution was made official only after several days of intensive simulator work convinced both NASA management and the commander, Jim Lovell, that the understudy was ready.

Ron had often worked with Jack Swigert, both at North American and in Houston, so he and Jan knew him well. Jack had the reputation of being a real ladies' man, allegedly with a girl at every airport. Jan Evans and Barbara Cernan knew that his reputation, while not unjustified, tended to misrepresent or even trivialize Swigert. Both women liked him a lot. To Barbara, he was "a really nice man, kind and polite and attractive." Jan remembers his bringing dates to various NASA events, but she never saw Swigert as a "lady-killer." She thought he looked more like a college professor who could have used a little wardrobe advice. To Jan, Jack was "a very nice man who would have made the right girl a good husband." The idea that the wives might have felt nervous if their husbands spent too much time in the company of NASA's "playboy astronaut" makes both Jan and Barbara laugh. The thought never crossed their minds. As Barbara points out, "The ones that 'went astray' didn't need any influence. They did it on their own!"

The late substitution of Jack Swigert was a real eye-opener for the NASA family and brought home to Ron, and Jan, that backup astronauts were not just for show. At first they were definitely surprised that rather than postponing the mission or substituting the whole crew, NASA would replace a single crew member so close to the launch, but that was the way the NASA system had been designed. Once the deed was done, Jan could only feel sorry for Ken but happy for Jack.

A quirk of fate had set Jack Swigert on course for the moon, and now another dashed his hopes of a smooth and successful mission. With their spacecraft all too clearly dying around them, the crewmen knew they could not engineer their own survival. But help was at hand. On Earth, news of the emergency brought more and more astronauts to mission control to help in any way possible. Tom Stafford, Ron Evans, and other party guests began to arrive. It was perhaps fortunate the explosion had not occurred a few hours later, or many of NASA's astronauts would have been unfit to drive a car, let alone an Apollo simulator.

In space, still traveling away from home faster than a rifle bullet, Lovell's crew shut down the command module *Odyssey* but not before activating the lunar module *Aquarius*. The LM would now have to serve as the men's lifeboat.

With the CSM's main engine out of action, future course corrections would have to be made with the LM's descent engine. Gene Cernan led his *Apollo 14* backup crew to Building 5, where he, Ron, and Joe began to run through procedures to test the LM's ability to steer the combined spacecraft and to identify problems involved in navigating by the stars from the LM's cabin. Other astronauts, including Tom Stafford, Dave Scott, Vance Brand, and the *Apollo 14* crew, took their turns in the simulators, and in Florida Dick Gordon operated the Kennedy Space Center's command module simulator, which was linked by radio with Houston. John Young, as the *Apollo 13* backup commander, never seemed to take a break and later worked with Charlie Duke in the LM simulator to test the procedures for the crucial reentry.

Only Ken Mattingly himself knows how he felt during the crisis, but he later joked in his NASA oral history that it didn't take him long to forgive the NASA doctors for grounding him. That was only human, but no dedicated astronaut in his position could have banished the equally human feeling that he should have been up there. Mattingly was a constant presence during the recovery efforts and was heavily involved in troubleshooting the multiple problems and in planning the procedures that had to be communicated to the astronauts. But Hollywood's portrayal of Mattingly as virtually living in the command module simulator employed a great deal of dramatic license.

As Fred Haise points out, "We had to make the mid-course [engine firings] without a computer. When those procedures were written up on the ground, various astronauts were dispatched to the simulators to test those procedures and maybe find errors to correct and improve them before they were ever sent up to us to utilize. And I'm sure that included Ron and other command module experts and others at the site—but, unlike the movie, not Ken Mattingly!"

Because he was so involved in writing the procedures, they didn't want him in the simulator testing them. That would have been 'cheating'!"

On 17 April Jan Evans was among thirty or so friends and family who joined Mary Haise in her El Lago home for the crucial reentry and Pacific Ocean splashdown. Jan got to drink more champagne, but this time the sense of triumphal achievement was replaced by a profound sense of relief at the crew's safe return. Ron was one of the many weary astronauts crammed into mission control to see the culmination of their efforts rewarded in the only way that mattered.

Ron's part in the team effort to save his fellow astronauts earned him the Manned Spacecraft Center's Superior Achievement Award. In fact, along with dozens of other astronauts, flight controllers, and contractors, all three *Apollo 14* backup crew members received the award. All were cogs in the greater mechanism. President Nixon awarded Jim Lovell, Jack Swigert, and Fred Haise the Presidential Medal of Freedom for their courage through adversity. Jim Lovell was quick to draw attention to the teamwork that had brought them safely home, and the president also awarded the Medal of Freedom to the whole *Apollo 13* Mission Operations Team in Houston.

In an unguarded moment while his spacecraft was limping back to Earth, Jim Lovell suggested that *Apollo 13* might be the last mission to the moon for some time. There was certainly no question of launching *Apollo 14* until an inquiry established what had gone wrong and what remedial steps would be necessary. The mission would eventually be delayed until the end of January 1971, providing extra time for the prime and backup crews to hone their skills.

During the extended gap between flights, rumors of further cuts to Apollo proved correct. In 1970 NASA's budget, measured as a slice of the American pie, was less than half its thickness in 1965. On 2 September, faced with additional cuts, NASA administrator Tom Paine announced the cancellation of *Apollo 15* and *19*. Ostensibly it was to save $42.1 million, but some also speculated that Jim Lovell's close call in April had rattled NASA management enough that it had no stomach to fight to retain the two missions. The last three missions would be redesignated *Apollo 15*, *16*, and *17*.

For Ron Evans and the other CMPs, the number of available seats to the moon now had dropped to just two. Ken Mattingly seemed a shoo-in for *Apollo 16*, and no one would begrudge him his seat. To all intents and purposes, only one seat was left. Although Cernan, Evans, and Engle seemed to be in pole position for *Apollo 17*, rumors persisted that more missions might be canceled. Meanwhile, lurking in the wings was the *Apollo 15* backup crew led by veteran Dick Gordon, who was itching to walk on the moon just as his good friends Pete Conrad and Al Bean had. Now that *Apollo 18* had been consigned to oblivion, Dick's only chance would be to snatch *Apollo 17* from under Gene Cernan's nose.

The only way the men of Cernan's team could prove themselves worthy of *Apollo 17* was to do the best possible job as backups. It was crucial for them to work well together not only among themselves but also with the *Apollo 14* prime crew. Cernan was quite aware that his ambitions could yet founder on the granite reef that was Alan Shepard, but the two men actually established a very good relationship, both professionally and personally.

Gene's former wife, Barbara, remembers how Shepard molded the six men into an efficient unit, almost a "mini squadron." She recalls, "Of course, Alan Shepard carried a big stick. When the guys first came to NASA, everybody was kind of scared of him because he was a 'big honcho.' So then, to get to work with him when he was able to fly! He and Gene became friends, really nice friends, and Gene kept saying, 'You'd better be ready because if you're not—I'm there!' He kept telling him that! 'If you can't do it, if anything happens, I'm stepping into your shoes!' And they formed, I thought, a really nice, nice friendship. I know he always said he and Al worked really well together."

It was a rather odd mismatch in terms of experience. This squadron's skipper had not yet orbited Earth, while its XO had spent two weeks in space, walked in space, and orbited the moon. Shepard was going to be the oldest moonwalker, and Cernan had been the youngest American in space. Seven years later, some people might have compared them to Obi-Wan Kenobi and Luke Skywalker.

The spirit of friendship was shared among the whole team, and Ron found that Al Shepard could be very gracious when he got to know you well. It wasn't always dawn-to-dusk training for *Apollo 14*. Jan remembers Ron getting several calls from Shepard that ran along the following lines: "Hey Ron, I've got tickets for the football game. Why don't you and Jan meet us in town, and we'll have dinner and go to the game."

Shepard seemed to have shed his old icy commander persona, and Ron didn't think it took a genius to work out why. "Al Shepard changed totally, changed personality, when he got selected for the prime crew of *Apollo 14*. He had been chief of the astronaut office and finally he was back on crew status! The change was he was a lot more of a likable guy. He changed because in my mind he no longer had to be the strict administrator. Now he was in the position where—hey!—he was finally getting a chance to do what he came into the program to do, and that's fly! He'd been grounded for all that time, so now he's got ungrounded. Now he no longer has the responsibility of being chief of the astronaut office; he has the responsibility of being commander of his crew. And it pretty much goes without saying: to get your own command is great—an operational command! That's what everyone strives for in the flying game, or the space game, or whatever. To be your own commander!"

Jan was of two minds about Shepard. She certainly admired him as a pilot and as a national hero, but she found him "a cocky guy." By the time of *Apollo 14*, Al had become a very successful businessman. His wife, Louise, was independently wealthy. The Shepards lived in "a very ritzy, expensive part of Houston," but they were always very sociable, turning up at most NASA events and astronaut parties.

Jan had no ambivalence about Louise Shepard. "She was the most gracious, beautiful person that you could ever hope to meet." Louise was the real deal: she was as friendly and as genuine as she was elegant and always enjoyed her get-togethers with the other wives. What puzzled Jan and many of the other wives was why Louise remained married to Al. It was no secret that he had a string of girlfriends and would flaunt them at golf tournaments and other functions. Jan found it "heartbreaking." Of course, no one could ever discuss it with Louise, who seemed willing to deal with her husband's infidelities by not acknowledging them, at least not in public.

Training for *Apollo 14* began in mid-August 1969. According to NASA records, the obvious emphasis was on the prime crew, but the backup crewmen ensured they got enough time in the simulators to step into any shoes that might be vacated. Inevitably, a friendly rivalry developed between the two groups. Cernan's backup crew began calling itself the First Team. The men joked about buying the prime crew tickets to go skiing or rock climbing or skydiving, but Shepard's crew took it all in good spirits. Nothing spurred them on more than knowing others wanted to take their place.

Like the prime crew, the three backups were all military men, all pilots, and all married. The three couples became very close, enjoying what free time the grinding schedule allowed them. Often they were able to mix business with pleasure while getting together at one of their homes for dinner and drinks. Afterward the wives would talk among themselves, and the astronauts would review the flight plan or take stock of their training progress and any deficiencies that needed attention. Watching Gene, Ron, and Joe working so hard on a mission they almost certainly wouldn't be flying, Barbara Cernan thought they were "perfect together."

Deke Slayton always said he could put any three of his astronauts together, and they would become an effective crew. Ron could see that everyone in the office had a common purpose and a desire to succeed, but not all crews were the same. Mike Collins famously described the *Apollo 11* crew as "amiable strangers" in contrast to the three close friends of *Apollo 12*. Ron Evans called a close-knit team that worked together and played together a "crew-crew." He would look back on his Apollo days and reflect that there were really only two crew-crews—*Apollo 12* and *Apollo 14* backup.

Uniquely for Apollo, Cernan and his fellow understudies even produced their own mission patch as a parody of the official version. The "real McCoy," bearing the mission designation and the names of the prime crew, shows an astronaut's gold pin rocketing from the earth to the moon. By sharp contrast, the parody patch has the names of the backup crew and borrows characters from the popular Warner Bros. *Road Runner* cartoons to drive home the First Team's message. Replacing the gold pin is a version of the hapless Wile E. Coyote with a gray beard and glasses, alluding to Shepard's age; red fur, signifying the redheaded Roosa; and a slight paunch, suggesting Ed Mitchell. The coyote has been beaten to the moon by the Road Runner, which sports a First Team banner and vocalizes the patch's title, "Beep Beep."

Joe Engle is happy to name the culprit: "The idea of the whole thing was Gene Cernan's. It was a joke to kid the prime crew that we, the backup crew, were right behind them and ready to take over. If they stumbled or had any problems at all, we'd take over and fly that mission for them!" He tends to agree with Dee O'Hara that astronaut gotchas provided some relief from the high-pressure training and flight preparation. Ron Evans also gave Cernan credit for the patch, recounting how the idea became infectious with all three men working on its design and figuring out how to make it a reality. The Beep Beep patch was arguably the ultimate gotcha of Project Apollo.

The *Apollo 14* training schedule suffered a number of disruptions and changes in direction. When the crew was selected in August 1969, the mission was expected to fly in February 1970. After the success of *Apollo 11*, the tempo eased, and *Apollo 14* was postponed to October. The repercussions from *Apollo 13* were more drastic. Not only was the mission deferred until early 1971, but the target changed. Shepard would now inherit Lovell's landing site, a hummocky region sharing the name Fra Mauro with a nearby crater. The rocks that formed the site were assumed to have been blasted from deep within the moon by an ancient asteroid impact, which created the Mare Imbrium (Sea of Rains). Conveniently, a more recent impact had punched the four-hundred-yard Cone Crater into the blanket of Imbrium material, ideally exposing the choicest morsels for Shepard and Mitchell to collect from the crater's rim.

The six men ultimately had nearly a year and a half to train, but at times they had to make way for the crews of *Apollo 12* and *13*. Then all training schedules went out the window in mid-April 1970. Nevertheless, all six astronauts accumulated substantial training hours. Stu Roosa clocked just over 953 hours in the command module simulator and an additional 106 hours rendezvous work in the CM procedures simulator.

The prime crew obviously had priority, but Ron was able to amass a very respectable 784 hours in the CM simulator and 83 rendezvous hours. Taking all other forms of mission training into account, Stu Roosa spent 1,868 hours preparing to fly to the moon. As understudies go, Ron did well to achieve 1,480 hours. Not included in those figures were the numerous T-38 and helicopter flights that all of the astronauts made. In Ron's case, this amounted to an additional 342 hours over 212 flights during the training period.

Although the focus of attention during any Apollo flight was on the men who landed on the moon, the third man in lunar orbit—forgotten by the public but not by the geologists—had important tasks to accomplish. During the long months of training, Stu and Ron shared many geology field trips and lunar topography reviews. Far from being dry-as-dust classroom lectures, those sessions placed the two astronauts in the hands of one of the most remarkable individuals to contribute to the scientific success of Project Apollo.

Born in Egypt in 1938, Farouk El-Baz obtained his degree in chemistry and geology in Cairo at the age of twenty. Relocating to the United States, he achieved his doctorate in geology at the Missouri University of Science and Technology. In March 1967 Bellcomm offered him a job as a geologist "to look at the moon." (One of the men who hired him was a geologist named Edward Nixon, the youngest brother of the future president.) Farouk became a self-taught expert in lunar topography as he analyzed and cataloged thousands of spectacular images transmitted by NASA's Lunar Orbiter probes, and he classified all the geological features. As an employee of Bellcomm, not NASA, he soon became involved in training the Apollo astronauts to recognize features and make valuable observations in lunar orbit.

The first astronaut who fell willingly under the brilliant young Egyptian's spell was Ken Mattingly. Many of the earlier Apollo astronauts "didn't want to do a damn thing with geology," but after the first hour, Mattingly was hooked. Farouk saw him as a different kind of astronaut, one who accepted the need to be both a skillful pilot and a keen scientific observer. But their efforts training for *Apollo 13* were laid low by the rubella virus, even though Mattingly himself never fell ill.

Word spread about Farouk. Even before Stuart Roosa was announced as the *Apollo 14* CMP, he introduced himself to Farouk and made a request: "I want you to make me as smart as Ken. Hell no, I want you to make me *smarter* than Ken!"

During his work on *Apollo 14*, Farouk found that he had acquired two keen astronaut pupils for the price of one. The key to the success of his technique was to encourage a sense of curiosity about the moon and instill a sense of friendly competitiveness in understanding lunar features. Some of the training was fairly informal. He would go to dinner with Stu or Ron. Or Stu and Ron. They would talk about the moon and the mission requirements for as long as they wanted (often over many hours of the astronauts' personal time); then they would adjourn for a beer or three.

Farouk quickly recognized that although Stu and Ron got on well together, they had quite different personalities. He enjoyed working with both men and became their lifelong friends, but at times, he found Stu could be rather . . . full-on. Ron was able to apply a useful "cooling effect" and had the knack of dialing Stu down a notch or two. NASA was determined not to have another crew member substitution, but if the unthinkable did happen—if Stu Roosa broke his leg, for example—Farouk was absolutely satisfied the lunar orbit science would be in safe hands with Ron Evans. "He attended all the training sessions, he asked the right questions. . . . He made absolutely certain everything was done by the book, the way it should be done, to the best of his ability, so he had a very good effect on the whole planning of that mission, no question about it."

By contrast, the geologists who were training the men who would collect the rocks around Cone Crater felt that Shepard "did not approach this education seriously," and whatever greater interest Mitchell had "was subordinated to his commander's attitude." In fairness to Shepard, he saw his primary task as getting Project Apollo back on track after the *Apollo 13* derailment. Landing safely at Fra Mauro and returning safely to Earth were what mattered, and any rocks they could pick up to keep the geologists happy would be a bonus. If that was an oversimplification, it was not a huge one.

As launch day approached, NASA was determined to avoid a repeat of the German measles debacle. A strict crew quarantine regime was introduced to identify anyone who might unwittingly threaten the mission. Three weeks before launch, the prime and backup crews entered an isolation period in the crew quarters at the Kennedy Space Center.

On the eve of Ron's departure to the Cape, he and Jan took Jaime and Jon to the Houston Astrodome to see a big trucks event. While they were gone, their good friends and neighbors initiated a civilian gotcha to make Ron's last pre-quarantine evening memorable. When the Evans family returned home, in the front yard they were confronted by a big electric billboard lit up with the words "Jan's Motel." Seeking an early night, Ron and Jan found an ice bucket by the bed, chilling a bottle of champagne. Their friends had short-sheeted the bed. When they were finally able to climb in, they started hearing familiar voices in the bedroom making humorous and suggestive comments. It turned out the culprits had installed loudspeakers in the attic and were crouching at the back fence while talking into a microphone. The "early night" soon became an impromptu party in the kitchen. Ron was unlikely to be flying to the moon this time, but his friends still wanted to give him an unforgettable send-off.

The next day he piloted a T-38 to Patrick AFB with Gene Cernan in the back seat. The quarantine regime did not prevent the astronauts from flying, so Gene, Joe, and Ron all made a number of helicopter flights as the launch approached. Cernan, in particular, seized the opportunity of more vertical landing practice just in case, even at this late hour, Shepard screwed up.

But Shepard wasn't the one who screwed up. On 23 January, only eight days before the launch, Cernan flew a Bell H-13 helicopter along the Indian River tidal lagoon near the space center. By his own admission, he was flying low over the water, showing off to tourists in their boats. Known as flat-hatting, this practice was defined in one World War II safety manual as "a form of flying that discourages longevity." His landing gear caught the water, and the aircraft plunged below the surface, explosively disassembling into hundreds of fragments. Cernan found himself strapped in his seat, upright, on the seabed. Fortunately, those long-ago sessions in the Dilbert Dunker at NAS Pensacola paid off, and he managed to fight his way to the surface. Unfortunately, he popped up in the middle of burning fuel and had to keep diving and resurfacing to find a patch of water that wasn't on fire.

Later that morning, Barbara Cernan answered a telephone call. The voice was familiar but oddly rasping. "Barb, I've had an accident."

"Oh my God, are you okay?"

"Well, barely . . ."

He warned Barbara to stay home and avoid the press. Later he admitted to her it was "a stupid mistake" and groaned, "How foolish could I be to have done this? Deke's going to pull me off the flight!"

Barbara wasn't sure if he meant *Apollo 14* or the distant prospect of *Apollo 17*. And what of Ron? By that stage in Project Apollo, Ron's wagon was hitched pretty firmly to Gene's star, and they would probably fly or flounder together. Barbara doesn't remember Ron saying anything particular about the crash, but she assumes he "probably raked Gene's butt over the coals when he saw him." Tom Stafford, having flown twice with Cernan, had earned the right to speak his mind. Cernan later described the verbal flailing he received by phone from his old commander. "I've never heard words like that in my entire life. It ended up with 'dumb shit!'"

Barbara was consoled by the fact that her husband was alive and surprisingly unscathed apart from singed hair and a seared throat. As for the future and for *Apollo 17* specifically, she could only watch God's plan play out as He saw fit. It was a bit late to pull the backup commander off *Apollo 14*, but it seemed unlikely that any last-minute injury to Shepard would have resulted in Cernan flying to the moon on 31 January.

The eight men sitting around the breakfast table on launch day seemed relaxed and cheerful, particularly Al Shepard. If he had any concerns about his imminent departure from Earth, they weren't visible to the camera filming the men's traditional consumption of steak, eggs, orange, and coffee. In fact, Shepard looked positively eager to get going, exchanging grins and jokes with his supporting cast of Roosa, Mitchell, Cernan, Evans, and Engle, as well as guests Deke Slayton and Tom Stafford. Gene Cernan doesn't look noticeably singed, although his haircut suggests some remedial work. Ron, as always, was smiling and enjoying his food. He had never really expected this flight to be his and was happy to have played a part in the long months of training. It would have been interesting at that moment to compare the innermost thoughts of the three men who were about to fly to the moon with those of the three who weren't.

At some point during the mission preparations, a number of the First Team's Beep Beep patches were smuggled on board *Apollo 14* and hidden in various nooks and crannies among the clothing, the food, and the pages of the flight plans. Whether Ron himself put any patches on board the command module *Kitty Hawk* is unclear. Joe Engle is rather coy about this, suggesting that it could have been NASA technicians or "anybody who was on board, testing or checking out the vehicle while it was on the pad." As for the lunar module *Antares*, Joe recalls hearing about "a patch that went to the surface of the moon with Alan" but modestly claims no credit for that.

As the prime crew made its final preparations, Ron was free to join Jan and his guests for the launch. He had invited a number of his old squadron mates to witness a display of raw power that would make even the Crusader look like a damp squib. Among his guests were John Holm and Lee Hill, both of whom had shared quarters with Ron on the USS *Oriskany* eleven years earlier.

It was not the best day to witness a launch to the moon, as clouds and rain rolled across the Kennedy Space Center and delayed liftoff by forty minutes. The weather improved but was still murky and overcast. John Holm would later see the spectacular night launch of *Apollo 17*, but his memory of *Apollo 14* is that it also launched in darkness. The liftoff actually took place two hours before sunset, but NASA film shows the rocket sitting on the launchpad, almost lost in the gloom cast by the low cloud deck. The ignition of the five first-stage engines dramatically lights up the scene and paints the clouds a lurid, shimmering orange before the Saturn V is swallowed up thirty seconds after liftoff. After that, for Ron, Jan, and their guests, the deep rumbling birth roar of the rocket remained the only lingering evidence that Alan Shepard and his crew were heading into space, leaving the First Team on the ground.

Apollo 14 made it to the moon, but it wasn't pretty. Only a few hours after launch, Stu Roosa's first five attempts to dock with the lunar module failed inexplicably. After much collective head-scratching both in Houston and in space, his sixth attempt was successful. There was some speculation that the pre-launch rain had leaked into the docking probe and frozen in space, thus preventing the capture latches from operating. Had the crew been unable to dock with *Antares*, the best they could have hoped for would have been a couple of days in lunar orbit to salvage some results from Stu Roosa's assignments. Shepard would then have limped home, tail between his legs, and Project Apollo might well have died in a wave of apathy or even derision. ("Two failures in a row? Enough!")

Landing on the moon was never easy, but Shepard and Mitchell had to deal with a floating blob of solder that kept flashing abort signals to the LM computer. After reprogramming the computer to ignore the spurious signals, they then had to cope with a landing radar that wouldn't perform. Houston ordered the simplest of remedies, essentially: "Switch it off, then switch it on again!"

Once *Antares* was safely on the moon, mission control opened a second capcom channel to allow Ron to stay in communication with Stu Roosa. Ron was able to provide the "third man" with a running commentary as first Shepard and then Mitchell stepped onto the surface at Fra Mauro base.

The dialogue between the two CMPs was frequently in rapid-fire technical jargon, a language in which they were both fluent. Much of it involved problems Stu was having with a bulky Hycon lunar topographic camera. The cumbersome instrument was supposed to image scientifically important regions and future landing sites with a resolution better than five feet at the lowest point of the orbit, but the shutter was misbehaving. It shot frames intermittently, making ominous clacking noises. Stu tried everything to fix the Hycon, but finally Houston gave up on it. He had to use a handheld Hasselblad camera to shoot crucial close-up images of the planned *Apollo 16* landing site, simultaneously slewing the spacecraft to compensate for his rapid motion over the target. Fortunately, when the machine failed, the man rose to the occasion. Stu's pictures were sharp and clear.

During their exertions on the moon, Shepard and Mitchell encountered evidence of the Road Runner's having preceded them. As they deployed a special handcart for carrying tools and equipment on their two moonwalks, they found it contained one of the Beep Beep patches, prompting Ed to joke about "visitors."

"Yeah. Hardly worth mentioning," Shepard replied, unwilling to give the First Team too much satisfaction. But the backups arguably had the last laugh as Shepard and Mitchell filmed their activities, raising the U.S. flag and setting up their scientific instruments. Clearly visible, stuck on the rear of Al Shepard's backpack, was a Road Runner patch bearing the names Cernan, Evans, and Engle. Ron had made it to the surface of the moon, in a manner of speaking, but without getting his boots dirty.

Near the end of the second EVA, Ron drew Stu Roosa's attention to his commander's unorthodox use of a contingency sample collecting rod. Shepard had fitted a familiar device to the end of the metal rod. Ron said, "And Stu, for your information, Al and Ed are back at the LM. Al's down there hitting golf balls. Seeing how far he can hit them."

"How's that coming out?" Roosa inquired.

Peering at a color TV image of Shepard taking another one-handed sand-trap shot, Ron observed, "Well, it looks like he had a couple of slices there but then finally got a hold of one and really drove it down the old lunar surface."

Shepard boasted he had hit the ball "miles and miles and miles." The reality was a little less flattering, but it didn't matter.

Ron knew exactly what Big Al was doing. "That was his thing that he was going to do—hit a golf ball on the moon! And it worked! You know, *everybody* remembers that! They don't remember anything else, but they remember Al Shepard hit a golf ball on the moon!"

Some of the scientists griped about Shepard's clowning on the moon instead of paying closer attention to his scientific tasks. Disoriented by the hummocky landscape, he and Mitchell had gotten to within thirty yards of the rim of Cone Crater but never actually found it. That didn't affect their selection of rock samples, which were undeniably ejecta from the crater. They could perhaps have picked up more of them and done a better job of documenting them, but Shepard had achieved his own personal goals. He had returned to spaceflight; he had landed, walked, and even golfed on the moon; and he had certainly put Project Apollo back on track. He figured that would open the door for a new phase of lunar exploration in which the scientists would get all the rocks they wanted. That was also what Gene Cernan, Ron Evans, and Joe Engle were hoping.

18. The Rocky Road to Taurus-Littrow

It is said that the darkest hour of the night
comes just before the dawn.

—Thomas Fuller (1608–61)

Harrison Hagan Schmitt was born on 3 July 1935 in Santa Rita, New Mexico, but grew up and went to school in nearby Silver City. Friends and colleagues came to know him as Jack, and according to Jan Evans, "only his mother called him Harrison."

Schmitt's introduction to geology came as a field assistant to his mining geologist father, who would lower him down mineshafts in ancient ore buckets, having issued warnings about rattlesnakes and other unfriendly critters. He gained a bachelor's degree in geology at Caltech in 1957 before traveling on a Fulbright scholarship to Norway, where he studied glaciated rocks of the coastal fiords as *Sputnik* ushered in the Space Age.

After a spell as a teaching fellow at Harvard University, Schmitt received his doctorate in geology from that hallowed institution in 1964. He took a post at the U.S. Geological Survey's Astrogeology Center in Flagstaff, Arizona, and worked with the renowned planetary scientist Eugene Shoemaker on developing geological field techniques that were later used by Apollo astronauts.

In November 1964 NASA and the National Academy of Sciences invited applications for the first selection of scientist-astronauts. Schmitt thought about it "for about ten seconds" before applying. In June 1965 he became one of six members of NASA's fourth group of astronauts and the only qualified geologist.

Eleven months before naval aviator Ron Evans arrived in Houston, Jack Schmitt was an astronaut who had never flown any kind of aircraft. The remedy was a fifty-three-week training course at Williams Air Force Base, Arizona. He recalls that learning to fly was "more difficult than it would have been had I been ten years younger, particularly instrument-flying," but it was "just something you had to apply yourself to, you had to get through it." Having learned to fly the T-38, he earned his air force jet pilot wings and later received his navy helicopter wings. The successful pilots of Group 4 then joined with those of Group 5 and shared their introductory training.

Schmitt well remembers working with his future crewmate. "Ron and I were together on survival training during that period, in fact we were in the same group of three with Ed Mitchell, so I spent quite a bit of time with Ron during that training cycle.... Ron was a great, friendly individual, very exuberant about everything he was doing, extremely positive in any interaction that he had with anything that was going on. I really found it very remarkable that he could maintain that kind of positive attitude during times of uncertainty, such as prior to crew assignments and things like that. Ron was a unique individual in his view of life as being one great bowl of cherries, I think."

Even as he learned how to be an astronaut, Schmitt established—with the approval of Al Shepard—a geology training program that gave military pilots enough background and experience to perform reasonably well as field geologists on the moon or as useful observers in lunar orbit. Schmitt's dedication and zeal cannot be overestimated. On top of his geology work, and apparently to the virtual exclusion of any social life, he threw himself into mastering the many subsystems of the lunar module. He knew full well that to earn a seat to the moon, his knowledge of rocks would get him nowhere without understanding the hardware.

His efforts finally seemed to have paid off with his selection as the backup LM pilot on *Apollo 15*. The scientific community and certainly his mentor, Gene Shoemaker, had been scathing about NASA's apparent unwillingness to send a scientist to the moon. Now, at last, one of their own stood an excellent chance of landing on the moon on the anticipated *Apollo 18* mission. The announcement of Schmitt's assignment was made in a press release dated 26 March 1970 that dropped a broad hint about future selection criteria. The backup crew of Dick Gordon, Vance Brand, and Schmitt would be "eligible for selection as prime crewmen for any mission subsequent to *Apollo 16*." In case that was insufficiently clear, the next paragraph singled out Schmitt, noting: "It is likely that he will be a prime crewman on *Apollo 17* or *18*."

Richard Nixon was inaugurated as the thirty-seventh president of the United States on 20 January 1969. Having lost the 1960 election by a whisker and won in 1968 by a slender margin, Nixon always feared that the 1972 election would also be tight. Public opinion was therefore very important to him.

Project Apollo had been a Kennedy policy, but shortly after Nixon entered the White House, he was able to bask in the reflected glare of publicity and worldwide adulation as his nation landed two astronauts on the moon. The press castigated him for hijacking *Apollo 11*'s achievement for his own ends, although in fairness to Nixon, what incoming president would not have made the most of the opportunity? He also accepted the risk that he could be blighted by a catastrophic mission failure.

By most accounts, Nixon genuinely admired the Apollo astronauts. He avidly followed the *Apollo 8* mission at Christmas 1968 and referred to the *Apollo 11* landing as the most exciting event of the first year of his presidency. According to his senior adviser John Ehrlichman, Nixon hero-worshipped the astronauts and felt that they represented the best of America. Hugh Sidey of *Life* magazine even suggested that Nixon saw the astronauts as "the sons he never had."

But for the president, there was a big difference between admiring the astronauts and approving the tax dollars for more Apollo missions. Under Nixon, the space program became another domestic policy rather than a great national commitment. Even in the warm afterglow of *Apollo 11*, opinion polls were against further space extravaganzas. In December 1969 presidential assistant Peter Flanigan noted that Nixon "did not see the need to go to the moon six more times." Presidents do not command the tides; they ride them. The tide that had swept the United States to the moon was ebbing fast, and Nixon had no intention of being shipwrecked on the shores of the Sea of Tranquility. For Nixon, *Apollo 11* had been a great national triumph, but now it was maybe time to come down to Earth.

The *Apollo 13* crisis deeply affected Nixon. According to White House chief of staff Bob Haldeman, the successful splashdown left the president "really elated" and ordering celebratory cigars for everyone, but the near disaster reinforced his doubts about whether further lunar missions were worth the risks.

Following a cabinet budget discussion on 13 January 1970, NASA was directed to make extra cuts of $200 million. NASA managers initially concluded that they would have to ax three lunar landings—*Apollo 17, 18,* and *19*—to save money. They initially relented, but during the summer of 1970, hard decisions had to be made to map out future space policy. The announcement in September that "only" two missions were being canceled came as a huge blow to the likely *Apollo 18* crew members of Gordon, Brand, and Schmitt, and would have unfortunate consequences for others in the astronaut office.

The account from eminent space historian John Logsdon of Nixon's space policy makes alarming reading at times. Drawing on presidential papers and tape recordings, Logsdon writes that in December 1970, Nixon was contemplating the cancellations of *Apollo 17*, the *Skylab* space station, and even the preliminary development of the space shuttle. He actually *did* decide to cancel *Apollo 17*.

Intriguingly, Logsdon adds, "The president had somehow gotten the impression that *Apollo 17* was even more risky than the three missions scheduled to precede it." Nixon did not want a space disaster to occur in mid-1972 while he was campaigning for reelection. His concerns for the safety of the astronauts were probably overshadowed by the "possible negative political fallout."

Ron Evans certainly didn't know that the mission he aspired to fly had effectively vanished by presidential command. Fortunately, Nixon's new science adviser Edward David was in favor of seeing Project Apollo through to the last mission. On 31 December 1970 he wrote an eloquent and persuasive memo to the president, recommending restoration of *Apollo 17* and pointing out that cancellation would give the administration "an unfortunate image among opinion-makers in society." David proposed stretching the intervals between missions so that *Apollo 17* would fly *after* the November 1972 election. Astutely, he added that this would "maintain critical employment levels" through to the planned *Skylab* missions in 1973. Nixon agreed to reinstate *Apollo 17* (and retain *Skylab*), but the reprieve was only temporary.

The next threat to the mission came in May 1971, as NASA geared up for *Apollo 15*. On 13 May Nixon suggested to Ehrlichman that five lunar landing attempts were enough and that *Apollo 16* and *17* should be canceled. "Why in the hell would they have to go up there and take a look round the damned thing again?" he asked. Ehrlichman replied, "We're working on it."

As preliminary discussions began on NASA's budget for fiscal year 1973, Deputy Director Caspar "Cap" Weinberger of the Office of Management and Budget sought to clarify the president's intentions. In a memo to Nixon on 12 August 1971, he advised that the "announcement now, or very shortly, that we are canceling Apollo 16 and Apollo 17 . . . would have a very bad effect, coming so soon after Apollo 15's triumph. It would be confirming, in some respects, a belief that I fear is gaining credence at home and abroad: that our best years are behind us."

But Weinberger concluded that if the president insisted on canceling the last two Apollo missions, such action should at least be coupled with a decision to confirm funding for the planned space shuttle. Nixon's cryptic handwritten response was, "I agree with Cap." That did not make the fate of the last Apollo missions any clearer.

On 19 October Bob Haldeman asked the president to clarify his position. In the taped conversation, Haldeman asked, "So you do want to cancel 16 and 17?" Nixon replied, "Yes, I do want to cancel them. . . . I just don't think we should take the risk of a possible goof-off in the damn thing." When Haldeman suggested that the missions could be postponed, Nixon replied, "Postpone and then cancel them, if you could get away with it." He made it clear: there were to be no Apollo moonflights in the election year of 1972.

At an Oval Office meeting on 24 November 1971, space policy was third in a list of eighteen budget items discussed. Nixon asked, "Space, what's the problem here?" Ehrlichman replied, "Well, the problem here is, do we go ahead with the next two [Apollo] shots?" At first, Nixon's emphatic reply was no, but then he confused the issue again by adding, "If we go, no shots before the election." In the following brief discussion about the impact on NASA jobs, Nixon suggested retaining the jobs without actually flying *Apollo 16* and *17*. He again referred to the risks revealed by *Apollo 13* and seemed to conclude that *Apollo 17* could go ahead *after* the election, but *Apollo 16* would be canceled. Finally, as he judged there was a politically "safe" gap between *Apollo 16* and the November election, and given *Apollo 17* would fly *after* the election, the president decided that NASA job security and the interests of the scientific community would take precedence over his fears of a space disaster. The last Apollo missions would go ahead.

None of the astronauts knew about the behind-closed-doors drama in Washington DC. Jan Evans and the former Barbara Cernan only learned of the issue when interviewed for this book.

If at times NASA management seemed strangely reluctant to fight for the final Apollo missions, it is important to bear in mind that the Apollo era was drawing to a close. NASA was hoping to transition to a new era of regular spaceflights via the reusable space shuttle. That meant, however, NASA's eggs were all in one basket. No shuttle meant no more American astronauts.

In a 1986 interview, Ron Evans addressed this issue, pointing out that—individual disappointments aside—he did not recall any strong resistance to the cancellation of *Apollo 18* and *19*. "Most of the astronauts really didn't get involved in the political decisions that were going on." Furthermore, he noted that "the decision had already been made: we're going to build a space shuttle and go along those lines." He also recognized that people both inside and outside NASA were voicing concerns about the impact of an Apollo disaster on future funding, particularly for the shuttle. "If we'd sent *Apollo 18* out there to explore one more site on the moon, it would have cost an extra number of dollars; you're taking a greater risk of losing somebody out there, and if you do lose somebody, your allocations would go down the tubes. You would have a much harder time getting the congressional commitment to go ahead and develop the space shuttle."

Putting it more bluntly, the loss of *Apollo 18* and *19* was the price NASA had to pay for a new generation of space vehicles. Perhaps Ron would have taken a less laid-back approach if he had known that NASA and President Nixon had independently considered applying similar logic to *Apollo 17*.

When the crew of *Apollo 16* was announced on 3 March 1971, clearly Deke Slayton had followed the flight rotation rule of thumb by selecting the *Apollo 13* backup crew (with due allowance for the Swigert-Mattingly exchange). Mattingly would now get his own mission.

That left only *Apollo 17*, and most interested observers would have expected the rotation to post the names of Cernan, Evans, and Engle. Privately, Ron Evans was never so confident. He had seen short-term assumptions about Slayton's rotation system overtaken by circumstances, with the best examples being the sidelining of Gordo Cooper, Donn Eisele, and Walt Cunningham, as well as the unexpected medical issues involving Mike Collins and Al Shepard. Until a crew announcement was made, Ron had every right to be concerned. Rotation into *Apollo 18* was no longer an available option for Dick Gordon's *Apollo 15* backup crew.

Gordon felt he had an ace up his sleeve with the inclusion of geologist Harrison Schmitt. Even Dr. Schmitt himself, interviewed by this writer in 2003, made it clear that he did not give up hope of hammering rocks on the moon just because *Apollo 18* had been axed. "The backup crew of *Apollo 15* decided that we would begin to compete with Gene Cernan and his crew and see if we couldn't do such a good job that they couldn't afford not to fly us. And astronauts are very competitive people, and Dick Gordon and Vance Brand felt like they might have a shot, particularly since they felt like my chances of going were pretty high, since I would be the only scientist."

It must also have occurred to Ron that by August 1971, Gordon, Brand, and Schmitt had completed backup training for an advanced J mission, which was designed for a longer stay on and around the moon. That training was more commensurate with *Apollo 17*'s likely mission profile than the simpler training regime for *Apollo 14*.

With the apparent loss of Schmitt's mission, the space science community was up in arms. NASA seemed to have abandoned any thought of sending a scientist to the moon. Associate Administrator Homer Newell, who had been a driving force in NASA's space science program, traveled to Houston and listened to the grievances of the scientists and scientist-astronauts. They complained to Newell that crew selection criteria had an obvious bias in favor of military pilots. As the only geologist with the necessary skills, Schmitt's name inevitably arose.

Newell reported back to NASA management, including Manned Spacecraft Center director Bob Gilruth, who still favored putting experienced pilots in all three seats for a moonflight. Despite the representations made to Newell, NASA managers did not make a firm commitment to send a scientist to the moon; however, they *did* confirm that NASA should, "if possible," fly a geologist—namely, Harrison Schmitt—on Apollo. Provided Schmitt's training continued to progress satisfactorily, he would be selected for *Apollo 17*. NASA would review that decision and make a final announcement no earlier than August 1971, after Schmitt had completed his *Apollo 15* backup duties.

This not-quite-a-decision was carefully hedged with enough caveats to satisfy the Gilruth camp that no compromises on flight safety would be tolerated. But it was beginning to look as if the writing was on the wall for Joe Engle, if not Cernan's whole team.

Whether Deke Slayton heard the full details of NASA's conclusions is unclear; his later selection recommendation suggests he did not. However, if only on the basis of the March 1970 press release, he must have been generally aware that NASA was *very* keen to see Schmitt walk on the moon.

Slayton undoubtedly felt under pressure as he considered his options. He had often stated that the rotation was not sacrosanct. He could easily apply it for *Apollo 16*, but for *Apollo 17*, the last mission to the moon, he had too many applicants. Six into three doesn't go! He certainly had no shortage of advice from his senior astronauts. Jim McDivitt and Pete Conrad made no secret of their preference for Gordon, while Tom Stafford and Al Shepard, heavy hitters who had both served as chiefs of the astronaut office, were firmly in the Cernan camp.

Stafford acknowledges that either crew would have done a great job, and he praises Dick Gordon as "a super pilot." But one issue, for Stafford, tipped the balance in Cernan's favor: "Gordon was a more mature research pilot. But Cernan had wide experience with the lunar module with me at the moon. Gordon flew the command module."

Stafford recalls "a real push from the scientific community" to have Schmitt selected. The obvious solution for Slayton was to acknowledge that the rotation could be ignored for the final mission and to accept that Schmitt, Dick Gordon's "ace up the sleeve," trumped Cernan's crew. Weighing all the factors, does Stafford believe there was any sufficiently compelling reason—other than the normal rotation—*not* to select the complete team of Gordon, Brand, and Schmitt? In a word, "No."

Perhaps Deke Slayton didn't read the writing on the wall. As he later noted, "I thought it would be nice to send a geologist to the moon, but it wasn't my business to be nice. I had at least one guy I thought was more qualified. So the choice was obvious to me. I submitted Cernan-Evans-Engle as the crew to NASA HQ. And had it rejected." In an earlier era, Henry Ford had told buyers of his Model T automobile they could have one painted any color so long as it was black. NASA's message to Slayton seemed equally stark: "You can select any crew for *Apollo 17* so long as it includes Schmitt."

As the normal rotation was therefore not going to apply to *Apollo 17*, only two options remained. Slayton could either replace Engle with Schmitt in Cernan's crew or bypass Cernan and select Dick Gordon's crew. An obvious logic lay in maintaining crew integrity, thus preserving the almost-intuitive understanding that develops in a close-knit team while working together for a common goal.

Apollo 15 commander Dave Scott remembers receiving a call one evening from Slayton. "He asked my opinion regarding the selection of the crew for *Apollo 17*, either Gordon and Schmitt or Cernan and Schmitt. I recommended Gordon and Schmitt because they had worked so well together as backup to *Apollo 15*. Deke acknowledged, and said the selection would be made soon, with no indication of the direction he would go—a brief but pleasant conversation." No mention was made of Ron Evans or Vance Brand. Their fates seemed tied to those of their respective commanders.

Deke Slayton clearly favored test pilot Joe Engle, the man who had already earned air force astronaut wings. Gene Cernan has praised Engle as "probably the finest stick-and-rudder aviator that I have ever flown with in my entire life." By contrast, he could only rate Jack Schmitt as "a safe and adequate pilot." Of course, the LM pilot didn't actually fly the LM; he monitored the spacecraft systems to provide information to the commander. And Schmitt had toiled long and hard to familiarize himself with every switch, gauge, knob, quirk, and foible of the lunar module. Engle, meanwhile, had seemed reluctant to grasp the nitty-gritty of mastering the computer and other subsystems. According to Cernan, "Jack was not an 'adequate' lunar module pilot; he jumped in with both feet and was an outstanding lunar module pilot." He saw Joe as "only an adequate lunar module pilot."

Though it pained Ron to admit it, he also recognized in retrospect that Joe's being "a great throttle-jockey" was simply not good enough.

Slayton now clearly saw that the key question was no longer which of the two men was better—he had made *his* view on that clear—but whether Schmitt was *good enough* to select as the *Apollo 17* LMP. Cernan was emphatic about this point: "Deke was the kind of guy who could never have been pressured by anybody. . . . Deke would never, ever have put anybody on any mission who he didn't feel could Goddamn do the job and do it well. Deke would not have put Jack—nor would he have put me—on that flight if he didn't think we could handle it."

Cernan could see where it was all heading. But assuming Schmitt got the nod, Deke still had to select the rest of the crew. With memories of that helicopter crash still fresh in his mind, Cernan was not optimistic. He later wrote about it and spoke candidly about it in public: "I said it in my book, and I truly, honestly, believe it to this day: that if I had been Deke Slayton, I would have picked Dick Gordon's crew. There was absolutely no reason to have picked me to command that crew, wherein we would have to change one of the crew members. Keep Dick's crew in place—very capable guys—keep them in place and fly them! Deke had every reason to pick Dick Gordon, and I almost expected him to do that."

Michael Cassutt, who cowrote Slayton's autobiography, is satisfied that Slayton always intended to stick with the rotation for *Apollo 17*. But as the moment of truth approached, Ron Evans didn't know that.

On Sunday, 8 August 1971, Ron, Jan, and the children flew from Texas to Acapulco for a holiday with the Cernans. Perched on a hillside overlooking Acapulco Bay, the Las Brisas Hotel comprised a collection of chalet-style rooms with individual swimming pools. Tension was mounting as the families awaited Deke Slayton's verdict, and they needed a break to unwind.

Barbara Cernan was looking forward to the holiday with Gene and their great friends Ron and Jan. Barbara had liked Ron right from their first meeting. "He had beautiful eyes, twinkly eyes—I'm an eye-person. I always look people in the eye; that's the first thing I notice about a person. And he smiled all the time. And he was so devoted to Jan; they had a deep love for one another and a deep respect for one another. I loved that part of him too—he was devoted to his family—but he was just a fun, fun guy. A very loving man. That's why I think we all had so much fun together: because you could talk to them; you could share anything you wanted to share with either one of them."

Barbara laughs on hearing that Ron had been described by some former squadron mates as "very quiet" and that NASA flight director Gene Kranz had called him "the ultra-quiet Ron Evans." She says, "I'm telling you from a woman's point of view, he was tons of fun. Now whether he was fun in the squadron or not, that I don't know! But socially, he was *so much fun*! And he laughed all the time! When you say he was quiet, I guess he was quieter than some of them, but he had his own persona. He was just—I mean, he was like someone you felt you could sit down with and really talk to. But anyway, I'm just telling you from a woman's point of view. See, a man's going to tell you something different. They're not going to say, 'Oh, I looked in his eyes, and they were sparkling!'"

On Thursday, 12 August, Jaime Evans joined her father and went scuba diving in the sea. After they returned to the hotel, the sequence of events becomes a little hazy, but Jan recalls that all the adults and children were around when

Gene's phone rang. Deke Slayton was on the line. There were no preliminaries. "Congratulations, Geno. *Apollo 17* is yours." The good news included Ron, whose long years of patience had finally paid off. The man from St. Francis was going to the moon.

Jan vividly remembers her husband's reaction. This was no time for Ron to play the "Quiet Man." His face lit up, and "those big old brown eyes were about to pop out of his head! He was really, really, *really* happy!" He swept Jan up in a bear hug as the children Jaime, Jon, and Tracy Cernan joined in the celebrations. Jan remembers her son and daughter jumping up and down, hugging Daddy and Mommy and each other.

The glow from Deke's announcement cast one shadow: he did not mention Joe Engle. The families clearly saw that Joe had been left out of the announcement because he was not in the crew. But human nature being what it is, their celebrations came first. While the adults went off to the bar, the children made a paper banner announcing: "Congratulations Apollo 17 crew." Someone snapped a photo of them holding up the banner. There is a hint in the picture that Jaime, days away from becoming a teenager, thought it looked a bit lame, but if it made the young kids happy, that was okay.

Back in Houston, Gene tackled Deke about the selection and listed all the reasons why it was a bad idea to break up a great team. Unfortunately for Joe Engle, Gene was not an irresistible force, and Deke was very definitely an immovable object. He offered Gene two choices: fly with Schmitt or hand the mission to Dick Gordon's team. Disappointed for Joe though he was, Gene Cernan was not going to throw away the command of *Apollo 17* in a futile gesture of defiance. The *Apollo 17* crew would be Cernan, Evans, and Schmitt.

Whatever decision Slayton made was going to upset someone, but in the end (setting aside the Schmitt issue, which was out of his hands), did he just take the easy course by following the normal rotation? Doing that was simpler to justify than departing from the rotation, but Deke never explained his rationale.

Although the conventional wisdom was "buy Cernan, get Evans," that arguably underplays the weight of Ron's contribution in the delicate balance of the selection equation. Not only did Ron and Gene understand each other and work well together but also one can surmise that Deke Slayton was well disposed toward Ron, admired his combat record, and felt from an early stage that he deserved to be considered for a moonflight. Having no test pilot credentials, Ron was clearly not going to fly an early mission. To prove himself worthy of a later mission, good preparation would have included plenty of support crew assignments and capcom experience, which is exactly what he was given, followed by a good record on a backup crew. By all accounts, Ron had proved himself a safe pair of hands in the command module. On 13 August 1971 the names were officially announced, and Ron became the first Apollo prime crew CMP, and the first Group 5 prime crew astronaut, who had not trained as a test pilot. It does not seem too much of a stretch to conclude that having Ron Evans as Gene Cernan's CMP made it easier for Deke to stand by the rotation. Putting it another way, Ron's presence was one more reason not to adopt the more radical approach of setting aside the rotation.

The selection of Jack Schmitt in place of Joe Engle caused much pain beyond the astronaut office. Dee O'Hara understood NASA's logic, but she could also see the hurt and the bitter disappointment. Jan Evans was as disappointed as anyone for Joe and Mary, but in time she accepted that it wasn't the wives going to the moon or making the decisions. "Nobody asked my opinion! They didn't ask me if it was all right if they did that! I didn't second-guess NASA. There was a lot of camaraderie among the six of us, so of course you were disappointed and surprised and shocked, but that doesn't stop you thinking, 'Hail, hail, the gang's all here!' Let's get down to business!" Jan certainly didn't want to lose sight of the fact that *her* husband was going to achieve his ambition. The last man to steer an Apollo spacecraft to the moon would be Ron Evans.

The real losers in the *Apollo 17* selection controversy—whatever their innermost thoughts—were never less than gracious about it. Dick Gordon would joke that Cernan stole his LM pilot and his mission, but he never moaned about it. He retained his charismatic bonhomie throughout his life. For Vance Brand, the real disappointment ("the biggest disappointment of my life") was the cancellation of *Apollo 18*. As far as he is concerned, that was when he lost the moon. His memory of efforts by the *Apollo 15* backup crew to snare *Apollo 17* are slightly at odds with Schmitt's. Brand wasn't particularly surprised when the mission went to Cernan's crew "because it was the rotation: the backup crew of *14* was sort of expected to go on to be the crew of *17*."

For Joe Engle, the man most entitled to feel bitter, the hardest thing was having to tell his children that their daddy wouldn't be flying to the moon after all. With measured understatement, perhaps made easier by the passage of time, he notes, "Certainly I was disappointed that I was not going to be on the crew. Every crew trains so intensively that you do form a bond in the course of getting ready for a flight." Of the substitution, he is even able to say, "It was not my idea! But I did not disagree with the rationale that Jack Schmitt should go on *Apollo 17* instead of me, because Jack had a doctorate in field geology, and that was the primary purpose of the [J missions], to study the geology of the moon."

Let Dee O'Hara have the final word on the controversy. "I have to hand it to Joe Engle. He took it like a gentleman. He never outwardly complained or said anything in public. He accepted the fact that he wasn't going to fly, and you know good and well he was horribly disappointed. . . . We didn't see any outward resentment from him, so he was very much a gentleman about everything—and still is to this day."

19. Many a Slip

It is adventure of the highest sort. It's one of the few things
that raises us up above the grubbiness that man seems to
be making of much of his life.

—*Time* magazine on *Apollo 17*, 11 December 1972

The people in mission control never got involved in crew selection issues. Flight directors like Gerry Griffin heard rumors and hallway talk about who might be in or out, but usually he only learned the details much later. When Gerry heard who would be flying *Apollo 17*, he thought the choices all made sense. Ron Evans was very much a known quantity, and when Gerry was later appointed lead flight director for the mission, they spent a lot of time working together and became good friends. Gerry felt for Joe Engle but knew that flying Jack Schmitt, the geologist, was the right thing to do. He also knew that Gene Cernan and Evans had already done a great deal of preliminary work on the mission. Selecting them preserved the benefit from their early work.

Gene Kranz, the flight director for the *Apollo 11* landing, wrote that he was "ecstatic" upon learning of Schmitt's selection because "no one deserved a flight more." He added, "The cheering didn't stop there. Ron Evans, the command module pilot, was one of the Capcoms most familiar to the Mission Control team."

Jan Evans had gotten used to aviators transferring out of Ron's navy squadrons. She had seen old friends replaced by strangers and just learned to adapt. In time she realized Jack Schmitt had "one heck of a lot of smarts to be such an accomplished geologist, plus going through flight-training, learning to fly the T-38, learning to fly the helicopter, and everything else." Such work took "a lot of nerve and a lot of smarts."

Although offered grudgingly at first, Jan's respect for the cuckoo in the nest grew. She discovered that although Ron and Jack had very different personalities, they did have some similar characteristics. Unlike so many other astronauts, neither man felt the need to drive a flashy sports car. Practical vehicles worked fine for them. Ron also realized that to become a lunar geologist, having Schmitt on board to guide him on the finer details would be helpful.

The crewmen of *Apollo 17* detested being called "the tail of the dog," but there was no denying that their mission would bring the great adventure to a close. It was inevitably a topic of conversation during their training and influenced their official mission patch, designed by space artist Bob McCall. The theme was "*Apollo 17* is not the end but the beginning," and McCall's artwork shows the Greek god Apollo, superimposed on the American eagle, turning his gaze beyond the moon to the planets and the distant stars.

For all the high-sounding bravado, an unpleasant truth had to be acknowledged: no more moon shots inevitably meant job losses. Workers at any shipyard understood this problem. As soon as the latest gleaming vessel is launched in a heady mix of champagne and razzmatazz, the pink slips are distributed. When the last Apollo lifted off, the same would occur at the Kennedy Space Center. There were concerns that a man who knew he was working himself out of a job might, even unwittingly, do less than his usual best. He might even miss the tiny flaw that could lead to disaster.

As one of the three men with most to lose, Ron didn't think there was much evidence of a morale problem at the Cape, saying it was "a little more hype than reality." But because of the *perception* of a morale problem, the crewmen spent a great deal of time with the Cape workers, cheerleading for *Apollo 17*. They played baseball, turned up at parties, drank a lot of beer, and handed out silver "Snoopy awards" in recognition of work of the highest standard. According to Tex Ward, the mission training coordinator, Ron attended so many gatherings and handed out so many Snoopy awards that Cernan and Schmitt referred to him as the mission's social director.

Ron had no doubt who was the mission's managing director. As the commander of *Apollo 17*, Gene Cernan deployed the organizational skills that his old squadron mate Marlo Holland had witnessed at NAS Miramar in 1958. Very sociable and outgoing, Cernan by 1972 had developed a rapport with most of his managers, so he knew where to go if some problem arose to interrupt the smooth progress toward launch. According to Ron, "He was the commander and acted like a commander. If something wasn't quite right ... man, he would go right to the source, to get it squared away or get things going, or attempt to get his way, to get around some of the bureaucracy that was starting to develop in the program at that time."

But Cernan didn't micromanage. He could and did delegate. If a problem arose with the lunar module, he would resolve it. If it pertained to the command and service module, he left Ron to sort out the problem and report back later. That report also had to include the solution.

As the months passed, all three men clocked long hours in the simulators. While it was only a part of the training, the familiar command module simulator was crucial to Ron's preparations. In all, between September 1971 and early December 1972, Ron spent nearly 623 hours in the CM simulator. As the commander, Gene Cernan had to be familiar with both modules and have an intimate knowledge of the launch phase, so he accumulated nearly 158 hours in the CM simulator. There were no passengers on board. Several functions in the command module were the responsibility of the lunar module pilot, and Jack Schmitt spent 167 hours in the CM simulator learning which knobs and switches to press and which *not* to press.

As the launch approached, Ron was confident he knew the CSM substantially better than Cernan did. "He knew the overview of it, but I knew the nitty-gritty details. So it worked out great because we could both work together; he could keep me honest and basically check me out; but if he wasn't quite sure about something and I *was* sure, he knew I was right!"

Jack Schmitt was as well placed as anyone to offer his opinion on Ron's state of readiness. "He had a variety of responsibilities in piloting the CSM, and where commander Cernan and I had to focus on lunar landing and other activities related to that, Ron spent a great deal of time alone in the training. It was very obvious, right from the very beginning, that he knew more about the CSM than probably anyone who ever flew, and [he] certainly knew how to fly it."

During the first seven months of training, the crew had no landing site. Finally, after much debate, NASA announced on 16 February 1972 that *Apollo 17* would land near the southeastern "shore" of the Mare Serenitatis (Sea of Serenity) at a place to be known as Taurus-Littrow. The site's name derived from the adjacent Taurus Mountains and Littrow Crater. Orbital geometry dictated that the launch—planned for 6 December—would have to take place at night, a first for NASA. The lunar module would touch down in a deep valley surrounded by high mountains. Press releases referred to a valley floor surfaced with dark volcanic ash. An apparently low number of craters suggested that the ash may have been spewed out relatively recently, filling in older craters. On *Apollo 15*, Al Worden had spotted a cluster of dark-rimmed craters in the region that looked enticingly volcanic. At least one of these possible cinder cones lay within range of the landing site.

Or so said the NASA press office. Even before *Apollo 11*, almost every geologist believed that meteorite impacts had created almost all lunar craters. Most, including Schmitt, were skeptical about volcanic activity having occurred in recent times, although it depended on what was meant by "recent." Sampling potential cinder cones in the valley, even if they didn't turn out to be "young," would certainly be fascinating.

For his own geological training, Ron became reacquainted with his instructor from his *Apollo 14* days, Farouk El-Baz. Having worked closely with Al Worden on *Apollo 15* and Ken Mattingly on *Apollo 16*, Farouk was another stalwart of Project Apollo. He saw Ron Evans very much as a known quantity and relished working with him again.

The most basic tool of the observer in lunar orbit was the chunky, Swedish-built Hasselblad camera. Ron Evans was determined to make the most of his visit to the moon and spent a great deal of his own time, often at home, taking test images with the Hasselblad. Ron Ammons remembers sitting with his wife, Audrey, in their friends' living room and "being bored to death while Ron showed picture after picture on the screen." Ron made the camera a part of his life so that using it in lunar orbit would be second nature to him.

Useful though the Hasselblad was, there were several considerably sharper implements in Ron's orbital toolbox. A previously unused bay of the spacecraft's service module had been retrofitted with a suite of photographic and surveillance instruments designed to examine the moon in fine detail and even probe almost a mile beneath its surface. This scientific instrument module (SIM) bay had been introduced on *Apollo 15*. Al Worden had pioneered the imaging techniques allowed by the high-performance cameras and achieved stunning results.

The mapping camera took long strips of images that could be combined into stereoscopic maps. A stellar camera, which simultaneously imaged star fields, acted similar to a mariner's sextant to fix the precise locations of the moon's surface features. Meanwhile, a laser altimeter bounced the light of a ruby laser off the surface features to measure their distance from the spacecraft to an accuracy of around six feet. Combining all of the data allowed precise maps to be constructed.

The panoramic camera, derived from air force aerial surveillance cameras, was similar to an instrument carried on the U-2 spy plane. It could image wide swaths of the lunar surface by rotating the lens assembly during exposure. The massive camera contained more than 6,500 feet of film. At best resolution, the images could—and did—spot lunar modules and even a lunar rover.

Apollo 17's SIM bay also contained instruments to measure tiny variations in lunar gravity and to construct a temperature map of the surface. The lunar sounder, unique to the mission, beamed electromagnetic impulses that could penetrate deep below the moon's surface to map layers of ice, permafrost, or water, assuming such exotic substances were present.

Jan Evans prided herself in having a good understanding of her husband's mission and the tasks he would be conducting, but the workings of the SIM bay instruments were difficult even for her to grasp. "Yeah, he told me about them, but, you know, a lot of it didn't make sense," she laughs. "He did something up there which was measuring clear on down through the lunar surface and telling you what the heat is there, and what water is there, and everything. That will always be beyond me!"

There in a nutshell lies the key to the near invisibility of command module pilots during lunar landing missions. The public would be able to see Cernan and Schmitt exploring the Taurus-Littrow Valley on live color TV. Their work would be visibly adventurous. Almost none of Ron's work in lunar orbit would produce headlines or immediate, tangible results. Even his dramatic photographs of the moon would have to be processed back on Earth.

One drawback of the Taurus-Littrow landing site was that Ron's orbital track would duplicate large areas of *Apollo 15*'s coverage. Farouk El-Baz had already assembled *Apollo 15* mapping camera images into long strips that showed the journey across the sunlit hemisphere from the far side to the near side. He was now able to take advantage of the similar orbital tracks to advance Ron's training. Farouk recalls, "I rigged up a kind of television monitor, a big screen, and put the *Apollo 15* strip-photography on it, and we would move the frame kind of slowly, as if he was looking at the moon from lunar orbit. And he enjoyed that tremendously, and every training session he would want to look at another strip . . . and he would ask, 'Why would that crater have that kind of shape although everything around it is perfectly round?' So we would discuss what makes crater shapes and the color changes on the flat plains—some are a little darker gray than the others. Is that a different kind of material, or is the time of cooling different? He would ask very good questions. That was one of the unique characteristics of Ron: his view from orbit, adding to the whole of the moon, with emphasis on the whole lunar surface."

Apollo 17 provided the last opportunity for lunar scientists to test their pet theories, resulting in more and more experiments being added to the flight plan. As the mission approached, Ron was uncharacteristically visited by pangs of self-doubt. The teacher and pupil had become good friends, and Ron leveled with Farouk about the ever-increasing demands of the scientists. They not only wanted him to observe and report but also to analyze and offer conclusions on issues such as the subtle color changes on the lunar surface.

"He said there were so many scientific requirements he would have to satisfy, that maybe he would not be able to do all of these things," El-Baz recalls. "We loaded him with all kinds of things because we knew it was the last mission—we were not going to have any more observers in lunar orbit—and therefore Ron was kind of our 'last gasp.' Load him up with everything we needed to know! He may have felt a little overwhelmed by all our needs."

Farouk felt Ron's concerns were groundless, although he acknowledges perhaps too many tasks *were* piled up one after the other. Nevertheless, based on their months of training, he was confident Ron had the enthusiasm, the knowledge, and the ability to get it all done, and he told him as much. So no pressure . . .

Farouk was not the only foreign geologist to contribute to Ron's training. In September 1972, while on a visit to the Cape, he arrived a little late for an after-dinner geology briefing on the landing site with a guest geologist, a German named Dr. Lucchitta. Arriving in the dining room, Ron looked around and saw a number of familiar male faces including the backup crew. There was also an unfamiliar, attractive female face. It was perhaps a sign of the times that a puzzled Ron asked, "Where's this Dr. Lucchitta who's supposed to be lecturing us tonight?" The other men all pointed toward the woman.

If Ron was embarrassed, Dr. Baerbel Lucchitta (née Koesters) was not. She remembers being rather amused by Ron's faux pas. After the presentation, Ron redeemed himself when it emerged that Dr. Lucchitta's rental car had suffered damage on her drive to the Cape. He chivalrously offered to look at it for her. When he determined the damage would be a matter for the rental company to handle, he was at least able to drive her back to her hotel.

Being overloaded with tasks was not Ron's only concern as he neared the end of his 1,454 hours of formal training (plus his countless hours of home study). All of the SIM bay's film was contained in the service module, which would be jettisoned to burn up in the atmosphere at the mission's end. That meant he had to retrieve the film cassettes during a spacewalk on the return journey. It would be the highlight of Ron's part of the mission, and he relished the prospect. There was just one problem.

The standard NASA way to introduce astronauts to the microgravity, or zero-g, environment of spaceflight was to take them up in a Boeing KC-135 aircraft, affectionately known to all as the "Vomit Comet." On a typical flight, the pilot would go into a dive, pull out, climb at around forty-five degrees, and push over the top before descending steeply, repeating the exercise thirty to forty times in a few hours. At the top of each parabolic arc, the occupants of the aircraft would experience almost thirty seconds of zero-g. The experience wasn't a happy one for everyone. Often, even the saltiest sailor would lose whatever meal he had been foolish enough to consume before the flight. There was no accounting for it. Some astronauts had cast-iron stomachs and were unaffected. Others, even including highly experienced test pilots, would sometimes wish for the respite of death.

Ron Evans fell into the second category. This was no latter-day secret admission; it was obvious to everybody on a flight who barfed and who didn't.

"The KC-135? I got sick on that stupid thing every time!" Ron recalled. "Oh boy, I think the maximum parabolas I ever got was 14 or 15, and then they had to quit. I mean, once you get sick, you may as well quit!" He may or may not have heard that Ed White had endured similar problems in the Vomit Comet before making spacewalking look easy on the *Gemini 4* mission.

To prepare for the deep-space EVA, Ron suited up and trained in NASA's underwater training facility and in the KC-135. In each case, he worked with a mock-up of the SIM bay and had to retrieve the film cassettes. In the aircraft, the task was hard enough to attempt in half-minute chunks of weightlessness without his body rebelling against the experience. NASA had a particular concern about a space-walking astronaut vomiting inside his suit and literally choking on the mess. A bout of sickness on the *Apollo 9* mission had forced a delay in Russell Schweickart's extravehicular test of the lunar space suit. Ron Evans was only too aware of this issue.

He recalled, "I always got the suit off, or got the helmet off. But I really had a concern about getting sick in space . . . because even in those days some people did, some people didn't. I got sick on that stupid airplane every time I did that. But I told myself it wasn't because of the zero-g, it was because it was hot in those suits, and you're floating, and your stomach goes upside down, then you pull 2 g on the pull-out, and you're lying sideways in the suit and then . . . [graphic sound effect]."

Having found his sea legs on carrier duty, Ron felt that the seasickness that had afflicted him on his midshipman cruises was coming back to bite him at the most inopportune moment of his career. Astronauts had one trick at their disposal, aside from medication, to guard against the risk of space sickness: after entering preflight quarantine, crews engaged in regular aerobatics in the skies over central Florida.

"We'd go and jump in a T-38 and just 'wring it out,'" Ron noted. "There's a certain amount of acclimatization associated with being able to 'uncage the little gyros,' or whatever they are, in your inner ear. And you can! You can get acclimatized by doing aerobatics, that kind of stuff."

Ron's logbook shows seven such flights, the last being on 5 December 1972, the day before the planned launch. All of the flights were solo. As pilots have often discovered, *doing* aerobatics is fun, but having aerobatics *done to you* as a passenger isn't so much fun.

Bodily infirmity nearly impacted the mission in other, more dramatic ways. Gene Cernan has written that he developed a prostate infection a few months before the launch. He also injured a tendon in his right leg. He recovered, but for a time he was convinced he would have to hand over *Apollo 17* to his backup, John Young, who had just returned from commanding *Apollo 16*.

Ron himself later had a scare while tinkering in his garage. A tool he was using slipped and lacerated his left hand. Jan was suddenly confronted by the spectacle of her husband staggering into the house with a blood-soaked towel wrapped around his hand, "cursing like you would not believe." She drove him to seek help at the NASA medical dispensary. Jan later recalled, "I've never heard such words in all my life between home and the dispensary—all through his mind it was going to keep him from going on the flight." It was as if this sudden self-inflicted threat to Ron's cherished goal had unleashed a pent-up torrent of oaths and expletives that had accumulated, largely unused, during his thirty years of exposure to expert navy and NASA cussers. A doctor stitched up the wound. To add insult to injury, it became infected and required another stitching. After a miserable week for Ron, the hand began to heal well, and backup Stu Roosa's heart rate settled back to normal.

Memories of the Munich Olympic Games in September 1972 will forever be overshadowed by the abduction and murder of eleven members of the Israeli Olympic team by the Black September Organization. The massacre generated such headlines that any future threat by the Arab terrorist group was bound to be taken seriously. A few weeks before the launch of *Apollo 17*, U.S. intelligence agencies warned NASA of a credible threat to the mission. The huge Saturn V rocket seemed an obvious target for Black September, so NASA security chief Charles Buckley increased patrols, with regular helicopter flights crisscrossing the area around Pad 39A.

Further alerts from the Pentagon raised the disturbing prospect that Black September might go after the Apollo crew members or even try to take their wives or children hostage. The three astronauts were warned, but a decision was made not to tell Barbara and Jan of the specific threat. Whether a similar decision would be made today is questionable, but at the time NASA, Gene, and Ron felt it was better not to alarm the families.

NASA had always increased security around launch times, but Jan Evans felt at the time that the level of security was much greater for *Apollo 17*. Ron didn't tell her about the credible threat to the families until after his return from the moon. Not once during the mission did she hear any mention of Black September.

In the final buildup to the mission, Ron packed what he would need at the Cape and said his goodbyes to Jan and the children; then he flew to Florida on the afternoon of 13 November with support crew member Bob Parker. Shortly after Ron left, Tom Stafford called Jan and said he wanted to drop by for a visit. That wasn't unusual. The Staffords were friends and El Lago residents. Tom arrived a little later with a man Jan had never met before, and they all sat down. Acting quite casually, they told Jan it was such an exciting time—what, with it being the last flight and everything—that they just thought someone should sit outside the front of her house in a car. All day, all night. They also gave her a phone number, saying to call if she had any questions or concerns. They added that, by the way, someone may inconspicuously follow the children to school and look out for them. In retrospect, Jan isn't sure whether she simply accepted that this was the way NASA did things, but at the time, her only response was, "Okay, anything else?" They said no but cheerily added that most worrying piece of advice: "Just don't worry."

In Jan's experience, Apollo wives visiting the Cape, even for their husbands' launches, had to make their own travel arrangements. NASA never got involved. For *Apollo 17*, a local broadcasting company had offered to fly Jan, Barbara, and their children to Florida, where they would be staying, as always, at the Holiday Inn, Cocoa Beach. Two days before their departure, Charlie Buckley called Jan and told her they would be flown in a NASA Gulfstream aircraft and given guest accommodations at Patrick AFB. Jan protested that they had already made arrangements. If she hadn't gotten to know Charlie so well over the years, she might have taken offense at the way he slowly repeated, as if to a child, what was going to happen. All she could say was, "Well, okay." This change was a big deal. Families didn't normally get to fly in a NASA Gulfstream.

The children were upset by the altered plans. They loved staying at the Holiday Inn and seeing Millie, their favorite waitress. Ron and Jan talked every evening by phone during his quarantine period, and when Jan mentioned Buckley's call and how the family would be cooped up at Patrick AFB, Ron knew all about it. He was reassuring: "Yeah, I heard that. That's all right. You'll still have fun!"

The air base was bristling with armed guards, including one stationed outside each of their rooms. Jan was introduced to "her" security man and to a separate shadow who would accompany Jaime and Jon. Similar arrangements were made for Barbara and Tracy. They didn't know if their guards were military, as the men would only say they came from "headquarters in Washington."

If the children couldn't stay at the Holiday Inn, they could at least visit for a meal and a chat with Millie. On the road to Cocoa Beach, a security car drove in front of their vehicle, and another took up the rear. When the convoy reached the hotel, the "headquarters men" tried to be inconspicuous by occupying three tables. That was too much for Jan, who blurted out, "Oh my God! We're going to be one big happy family and eat together, or I'm going home to Texas!" The senior security man relented, and for the rest of the trip, everybody sat together around one big table.

On the evening of Tuesday, 5 December, President Nixon was at Camp David. It is not known whether he had read the *New York Times* that previous Sunday or if Press Secretary Ron Ziegler had drawn its content to the president's attention during any of their several meetings. We therefore do not know if he was aware of Howard Muson's article, which rehashed comments attributed to Ron Evans in the *Houston Post* ten weeks earlier. Muson had written, "Occasionally one of the astronauts would open his mouth in public and shake the foundations of government buildings in both Houston and Washington—as when Ron Evans of the Apollo 17 crew recently criticized the Nixon administration for supporting the space program with talk but not action, and suggested that senators and congressmen who don't support it are responding to letters 'from kooks who think you ought to be spending the money for welfare.' After the interview, one reporter's sources said Evans was chewed out by Shepard, Slayton and NASA Administrator James Fletcher, in that order."

Even the Kansas press, which revered the state's famous son, felt that the reported comments "made Evans look shallow, out of touch with reality." Joe Engle sprang to Ron's defense, suggesting that the press had made "a goof on interpretation." Ron was quoted as saying he had not intended his previous remarks to be made public, and he had never criticized the president. He did use the word "kooks," but Farouk El-Baz made light of it, referring to "the Day of the Kooks."

When the comments first surfaced in September, Dr. Fletcher lost no time in phoning the astronaut office to have Ron "gagged." Clearly nerves were still raw concerning the troubled history between Nixon and NASA about *Apollo 17*. However, to put matters into perspective, Jan Evans has no recollection of the incident. Nor does Harrison Schmitt or Gerry Griffin. Perhaps it was, after all, a storm in a teacup.

If Nixon was aware of Ron's comments, he gave no obvious indications of displeasure when he telephoned the crew on the evening before the planned launch. In the recorded conversation, the president sounds mostly upbeat and avuncular, recognizing Gene Cernan as a space veteran. He jokes about the two navy men having to cope with a civilian; then he remarks, "I was just thinking, this is the last of the moon landings, but it's not the least, because it's going to be the best, right?" When Ron identifies himself on the line, Nixon responds, "Well, fine, we just want to wish you the best—you're going to be up there circling around while these other fellows are going to be looking around on the moon?"

Ron replies, "Yes sir, but I tell you, that's a mighty important part of the mission."

Nixon agrees: "I'll say it is!"

Ron adds, "I feel honored to be able to do that part of it for my country."

There follows a short dialogue on Ron's role, with Nixon observing, "That's right, you'll be up there . . . and these guys have to depend on you; because if you aren't there, they aren't going to be able to get back!"

"Yes sir!"

"Well, give them a good ride!"

"I sure will!"

A hint of the president's earlier concern seems to emerge when he mentions phoning previous Apollo crews. "It's always a big thrill to talk to you and to know we've got such great fellows in this program, willing to take these great risks for the service of the country and the service of science as well."

Schmitt quickly interjects, "Well, with the people we have backing us up, Mr. President, it's really not that great a risk."

Wrapping up, Nixon says, "Well, we wish you the best, and God speed all the way!"

Instead of simply thanking the president, Gene can't resist mentioning Nixon's recent reelection. "We're looking forward to a super four years with you at the helm."

The launch of *Apollo 17* was scheduled for 9:53 p.m. on Wednesday, 6 December 1972. To prepare themselves for a long night, the crew had been going to bed progressively earlier and rising earlier. They had actually been having breakfast on Tuesday evening when the president phoned. After the call, they all hit the simulators and later enjoyed a full roast beef dinner in the morning light before retiring to bed. As he waited for sleep to claim him, Ron Evans was ready to fly to the moon.

20. Captain America Sets Sail

The long day wanes: the slow moon climbs: the deep
Moans round with many voices. Come, my friends,
'Tis not too late to seek a newer world . . .

—Alfred, Lord Tennyson (1809–92)

Jan Evans was definitely ready for Ron's big day. A month earlier, the *Kansas City Star* had expressed amazement that both seemed so eager to embrace what lay ahead: "It is almost incredible to see the enthusiasm of husband and wife about the prospect of the dangerous mission. Jan, a slender, fun-loving mother of two, says she expects to cry at the launching, because she always does, but that her tears will be tears of happiness. 'I'm just so happy!' she explained. 'Those lucky guys are finally getting to do what they want.'"

Fourteen years later, Jan would explain why she had "no qualms about it whatsoever." "I watched NASA operate: the engineers, the flight controllers, everybody involved. They all knew what they were doing, it was all one common goal." As for Ron, she said, "he was *ready*! Ready in all of his training, and mentally and physically—and rarin' to go!"

For people meeting Jan before the flight, it was only natural to ask whether she was scared for her husband as he prepared to board a rocket to the moon. At that time, and for some time afterward, she couldn't bring herself to say that flying to the moon was safer than flying over North Vietnam. She felt it wouldn't have been fair to the many people she knew with husbands, brothers, or sons still at war.

The spacecraft was also ready. It had been a mostly trouble-free countdown. The gleaming white Saturn V stood on Pad 39A, cocooned by the service structure that would roll away in the early hours of 6 December. The crewmen had visited the vehicle three days before the launch for a final rehearsal of their emergency escape procedures. Wearing orange jackets, hard hats, and broad grins, they were photographed against the colossal bulk of the rocket, its great white exterior studded with vertical rows of rivets. The pictures of the three Lilliputians standing at the foot of the Brobdingnagian Saturn V seemed to beg the question: could something so massive, so solid *really* raise itself off the ground?

Near the top of the waiting monster, hidden behind aerodynamic adaptor panels, crouched the decidedly un-aerodynamic lunar module *Challenger*, named after the pioneering nineteenth-century research vessel HMS *Challenger*. Positioned just below the powerful escape rocket was the command module *America*, and Ron was delighted that his spacecraft would bear the name of their country. Although Gene Cernan was the mission commander, Ron was the pilot of the command ship, and it wasn't long before the others were calling him "Captain America." According to Jack Schmitt, Ron was "ecstatic" about the name, which alludes to the original comic book superhero that first appeared in 1941 and was revived in 1964.

Uniquely, the three crewmen would not be the only warm-blooded creatures to orbit the moon on this mission. Sharing the command module with them would be five thumb-size pocket mice, which would be dissected within hours of splashdown to test for any evidence of brain damage caused by cosmic rays. The mice were never intended to keep the astronauts company. For hygienic reasons, they were locked inside their self-contained Biocore (biological cosmic ray experiment) canister.

Although technical preparations for the last moon mission had progressed relatively smoothly, Apollo's world of the extraordinary found itself in conflict with the down-to-earth realities of labor relations. At the beginning of December, sixty Boeing Aircraft contractors had threatened to go on strike, raising the specter of picket lines. If other workers at the Kennedy Space Center had refused to cross them, a postponement of the mission might have been unavoidable. After intensive negotiations, the strike was averted, but it was not the only industrial dispute casting a shadow over the mission. A squabble between the CBS TV network and its technicians threatened to entangle NASA and the other TV networks, leading to the loss of public coverage of the mission. This problem was also averted, but the idea that a noble voyage to another world could be threatened by mundane earthly disputes over wages somehow seemed inappropriate, like the breaking of a spell or a rude awakening from a dream.

The world's press shared many expressions of melancholic finality as the last Apollo prepared for launch. *Time* magazine observed, "Historians will have a difficult time explaining the decision to abandon the Apollo Program. Having trained the men, perfected the techniques and designed the equipment to explore the Earth's own satellite, having achieved the ability to learn more about man's place in the universe, Americans lost the will and the vision to press on."

The Times of London ventured that space exploration was "an expression of human curiosity which cannot be suppressed without risk of stagnation. Human beings who ceased to wonder about the stars would be less than human."

In a letter published in *Time*, Hugh Carmichael of Jacksonville asked, "What can be the fate of a nation that has the means to further the search for knowledge and understanding through the exploration of space, yet does not do so because of its own seeming loss of spirit?"

But these questions were addressed to a future generation, not to the three men about to climb to the top of the enormous white rocket on Pad 39A.

Many Americans did seem to be aware of what they were about to lose. An estimated seven hundred thousand people, the largest crowd since *Apollo 11*, gathered at the Kennedy Space Center to see the launch. Ron's brother Jay; his wife, Pam; and her parents planned to join Jan, Jaime, and Jon, although neither of Ron's parents would be attending. Jimmy would watch on TV in Bird City, and Marie would in Topeka. Jan's parents would watch from their home in Salina.

Wednesday, 6 December, was a very busy day for Jan Evans. It was all such a whirl that although she has a clear recollection of the events leading up to the launch, their precise timing and sequences have become a little hazy. Early in the morning, before Ron retired to bed, Jan and the children got to talk to him in the crew quarters, separated by a germ-proof glass wall through which their conversation was relayed by an intercom system. Ron confessed to being bored by the quarantine period. He was definitely "rarin' to go."

Prominent among the many festivities surrounding the launch had been the usual lavish party thrown by *Life* magazine, although Jan can't remember if it occurred on the day of or the evening before the launch. In any event, it marked the end of *Life* as she had known it; the magazine's last weekly edition was published on 22 December 1972.

Jan had a very special launch-day opportunity that had never been offered to any other Apollo wife. Her security guy was friendly with most of the helicopter pilots who were constantly buzzing around the Saturn V. Realizing that Jan was no shrinking violet, he asked if she would like to see her husband's rocket from the air. The flight turned out to be one of the most thrilling and emotional experiences of Jan's life. To this day, she still gets "chills and tears" just thinking about it.

The crew, Jan, and her security guard took to the air on an "extremely noisy" military helicopter shortly before dusk, as the rocket was being fueled. Jan had a perfect, unobstructed view out of the side window as the chopper approached Pad 39A from above and spiraled lower toward the glowing Saturn V, bathed in the glare from a ring of spotlights. The beams glinted off a coating of ice enveloping the liquid oxygen and liquid hydrogen tanks, but she could clearly see the giant letters spelling the words "United States" and the prominent American flags on the side of the rocket through clouds of vapor generated by the fueling process. The pilot circled closer and closer, although it was difficult to judge exactly how far they were from the vehicle. By that point, the sun had set, and the combination of spotlights, gleaming ice, and swirling clouds of boiling fuel produced a surreal experience that overwhelmed the senses and created indelible memories. But as with the launch to come, how do you adequately describe it to someone who has had no comparable experience?

For Dee O'Hara, the launch of *Apollo 17* was a bittersweet moment. She was disappointed that the great adventure was ending but pleased for Gene, Ron, and Jack that their day had arrived. Jack Schmitt and Joe Engle had very different personalities, but fifteen months of training with Gene and Ron had "loosened Jack up." Dee was confident *Apollo 17* had a great crew that would work well together as a team.

The process of adjusting the astronauts' body clocks was complete. They were awakened around four in the afternoon, and Dee attended a private Mass with Gene Cernan. All three crew members then attended Dee's little examination room for a final assessment. They had been undergoing full medicals every day for the past five days, and Dee simply checked the basics—such as the men's temperature, heart rate, and blood pressure—to rule out any last-minute problems. She also examined their ears and searched for any skin irritation, since they would have to wear electrodes intermittently during the flight to monitor their life signs. Finally, they climbed on the scales, naked, to measure their preflight weight.

Around the time Jan Evans was circling their launch vehicle, the crewmen attended a very late breakfast to enjoy the usual low-residue offering of orange juice, steak, eggs, and coffee. Photographs of the ritual show three relaxed and smiling astronauts sitting around the table chatting with Deke Slayton, Al Shepard, and backup crew members Stu Roosa and Charlie Duke. Charlie Buckley was also there and no doubt briefed Gene and Ron on their families' security. The crew then—as Dee O'Hara delicately puts it—"went potty" for the last time under gravity for twelve days.

It was time to dress for the moon. Like medieval knights being fitted into cumbersome armor, the three men were pushed and pulled by attendants and technicians. As Ron Evans sat in his bulky white space suit, he did one more thing before his Lexan bubble helmet was attached: he lit and smoked one last cigarette. His crewmates had been trying to persuade him to quit, and Schmitt remembers them advising him, "Take advantage of having to go cold turkey for two weeks, Ron!"

Ron's old friend Marv Miller remembers discussing the smoking issue with Jan Evans at some point before the flight. Laughing, Jan told him, "Marvin, I'll bet he hides a pack of cigarettes in one of his boots!" She may have been alluding to an old sailors' trick of storing cigarettes in the back of a sock for easy access.

How does a long-term smoker cope with twelve days of separation from his nicotine? Al Worden was another regular smoker but claimed not to suffer any withdrawal symptoms on *Apollo 15*. He considered it "a mental thing." A former air force fighter pilot and test pilot, Worden said, "It was always like you turned a switch on or off. When you got in an airplane, you turned the switch off. When you got back on the ground, the switch comes on, and you'd say, 'Gee, I'd really like a cigarette.' In flight you never think about it."

Two hours in a jet aircraft is hardly the same as twelve days in space, but Worden coped. "It didn't make any difference to me. As a matter of fact . . . you're in a very unusual environment when you make a [space] flight, and we were busy enough all day long. I gotta tell you the truth: I never thought about smoking once during the flight. Never once!"

It remained to be seen how Ron would cope. On *Apollo 17*, Schmitt, the scientific observer, would watch for any evidence of Ron's suffering from nicotine withdrawal symptoms. He never did see any physical symptoms, but he couldn't read Ron's thoughts.

Once their helmets were secured, the astronauts began breathing pure oxygen. This would prevent the risk of nitrogen bubbles in their bodies causing the bends as pressure in the spacecraft dropped on the way into orbit. At around 6:45 p.m. on 6 December, Dee O'Hara stood in the corridor and watched as her three "patients" walked from the suit room to an elevator that would take them to the ground floor. They all looked cheerful, and Ron even attempted a little shuffling dance. The astronauts were now cut off from all human contact inside their heavy white carapaces, and all Dee could do was wave them goodbye.

After her exclusive aerial view of Ron's spotlit conveyance, Jan was reunited with her children and the Cernans, who had all been touring around under the watchful eyes of their security team. The families were driven to the crew's training building and lined up outside the doors waiting to bid the crew bon voyage. As the commander, Gene emerged first, waving and blowing kisses to Barbara and Tracy, and catching Jan's outstretched hand. Ron was next. He and Jan attempted to hug each other, a difficult task for a couple when one is encased in a spacesuit. Jan instinctively tried to kiss Ron but only succeeded in striking the bubble helmet with her mouth, leaving a smear of lipstick and gaining a bloody lip for her efforts. As Ron waddled happily to the crew transfer van, it looked as though he tried to click his heels, but neither the restrictions of the suit nor Earth's gravity would permit such acrobatics. Not to be outdone, Jack Schmitt entered the van and pretended to turn back, as if having second thoughts.

The world's press loved the sight of Ron and Jan saying goodbye. In London, a photograph of the embracing couple, with Jan smiling broadly, appeared prominently on the front page of *The Times*. A similar picture later appeared in *National Geographic*.

As the crew traveled to the launchpad, the journey assumed an almost surreal quality as an overhead helicopter illuminated the road ahead with a beam from its spotlight. On arrival the men rode the elevator to level 320, where *America*'s open hatch beckoned. Meanwhile, NASA allowed Jan, Barbara, and the children to relax in the crew quarters for an hour and a half. They were then driven out to a special viewing area for families and their guests, some distance away from the VIP bleachers.

Three and a half miles to the east, Ron was preparing himself for the flight of his life and the culmination of a six-year ambition, to be delivered by a machine with 425 times the thrust of the F-8E Crusader on afterburner. Everything about the Saturn V was a superlative. Standing sixty feet taller than the Statue of Liberty, the fully fueled vehicle weighed just over 6.5 million pounds, with the precise weight depending on how much atmospheric moisture froze on the outside of the fuel tanks. That ice would begin to shake free as the rocket's full thrust was unleashed at launch.

Friends and family gathered near the edge of a man-made lake, a turning basin that allowed barges to bring heavy equipment to the nearby Vehicle Assembly Building. In the darkness of the calm, muggy Florida night, a shimmering reflection of the distant floodlit rocket glinted on the water. Over the public-address system, the booming voice of NASA's Chuck Hollinshead intoned the familiar ritual of the countdown. Jan Evans had heard those numbers dwindling toward zero many times before, but this time it was personal. For her, the flight was also the culmination of a six-year ambition.

Austrian filmmaker Fritz Lang most likely invented the countdown, which he used to crank up the tension in his 1929 science fiction epic *Frau im Mond* (Woman in the Moon). It certainly worked for real launches, but this time, the adrenaline rush was tinged with the inevitable trace of melancholy that accompanies the last of anything worthwhile. That sentiment, as it turns out, was a little premature.

As the moment approached, all eyes turned eastward to gaze at the gleaming white spear pointing skyward. The launch team relayed a message to the crew: "Good luck and God speed!"

With just over three minutes to go, control over remaining events passed to the terminal countdown sequencer, a marvel of 1960s technology that had always worked perfectly. Unknown to anyone, after six years of faithful operation, a diode on a printed circuit within the sequencer had failed.

One by one, the sequencer ordered each of the Saturn V's fuel and oxidizer tanks to be pressurized with helium gas to maintain a constant flow to the engines during launch. At T minus 2 minutes and 47 seconds, the faulty diode blocked the signal to pressurize the third-stage liquid oxygen tank. A diligent human spotted the problem and manually ordered pressurization.

At T minus 50 seconds, the sequencer disconnected the Saturn V from its external power supply and the vehicle's internal batteries came online. "All systems to internal power. We'll be looking for the 'engines start' sequence at the 8.9-second mark in the countdown."

Hundreds of thousands of faces stared at the rocket. Millions more around the world were watching on live TV. And then . . .

"T minus 30 seconds. We have a cutoff. We have a cutoff at T minus 30 seconds. We're standing by at the T minus 30-second mark. We'll bring word to you just as soon as we get it. . . . T minus 30 seconds and holding, this is Kennedy Launch Control."

One of the media observers was James Burke, known to millions of British viewers as the BBC's "voice of Apollo." During the announcement, he audibly muttered, "Cutoff!" Then he advised his audience of a more pressing concern: "I thought I saw a flash at the bottom of the rocket. Something is wrong! They haven't told us what it is yet. . . . There was a flame, a short flame . . ."

During the countdown, Jan Evans also had seen what appeared to be flames around the rocket, but astronaut Al Bean was there to help with such matters. "Oh, that's just venting," he reassured her. "That always happens." The flames were the result of a controlled burning of excess hydrogen gas. In daylight, the hydrogen flames were invisible, but as *Apollo 17* sat rooted to the launchpad shortly before ten o'clock at night, the flames were visible in spite of the spotlights and must have been alarming to anyone who didn't have astronaut advice on tap.

In Topeka, Ron's mother was already apprehensive as the launch approached. She later admitted to Marv Miller that she "almost had a nervous breakdown" when the cutoff occurred.

Jan's initial reaction was alarm, but thanks to Al Bean, her concerns were replaced mainly by frustration. While James Burke was reminding his viewers that the fully fueled rocket had the explosive potential of a small atom bomb, Jan was wondering what would happen if NASA couldn't fix the problem. Would the launch be delayed a month?

Barbara Cernan was initially scared by the cutoff, but her main concern was that the crew might not get to fly. She had been there before with Gene's first flight, *Gemini 9*. The countdown had been halted at T minus 1 minute, 40 seconds, and after two further holds, the launch was postponed two days. Like Jan, she was reassured there wasn't really anything to be frightened about, but the delay was still frustrating.

An explanation soon emerged as to why the sequencer had stopped the launch in spite of the launch team's manual pressurization of the tank. Apollo Program director Rocco Petrone cut through the usual NASA jargon to point out that the launch had fallen foul of simple machine logic. Pressurizing the tank was the sequencer's job. It didn't do it. As the tank obviously wasn't pressurized, the sequencer stopped the launch.

In Houston, lead flight director Gerry Griffin had been sitting at the flight director's console with his colleague Gene Kranz. Kranz was the flight director on duty for the launch and was expecting to take over control of the mission as soon as the Saturn V cleared the launch tower. When the countdown was halted, Gerry's chief concern was seeing a fully fueled rocket sitting there, doing nothing. "With all that volatile fuel, millions of pounds of it, I always liked to get those things off on time—start burning that stuff, so it goes away! But this was a different kind of problem: actually, I thought we were going to have to scrub and probably de-tank all that fuel and then go later. That was my main concern. At least with a problem like that, you're not in a dynamic phase. If you get into a problem after liftoff, like we did on [*Apollo*] *12* . . . now you've got all that energy they're sitting on, and a problem at the same time—and you're flying, you're moving; you're going! [*Apollo 17*] was a case where, if we didn't launch, at least we were stable, so I would put it in a different kind of category."

In terms of levels of concern, Gerry contrasts *Apollo 12*'s post-launch lightning strike with *Apollo 17*'s cutoff: "[*Apollo*] *12* was a 9 or a 10. This was down in the 2 or 3 range because nothing dynamic was going on. . . . While it was a concern, it was more of a frustration. . . . And it kept going on and on. And it got so late, and I said, 'The crew's going to be exhausted, and all of us are going to be exhausted, and the Cape guys are going to be exhausted, and we're probably going to get this thing called off!'"

On board *Apollo 17*, Schmitt recalls that "three pairs of eyes [were] rapidly scanning the instrument panels for any clues as to why the countdown had stopped." After five minutes of "extreme uncertainty," the crewmen were briefed about the terminal countdown sequencer's problem and were reassured nothing was wrong with the spacecraft. While waiting to learn if the launch would be scrubbed or merely delayed, Schmitt briefly dozed off, leaving only Cernan to witness that "the unflappable combat vet Ron Evans" had also fallen asleep, "his relaxed snore a deep undertone to the chatter on the radio net." In future years, Jan Evans often heard Gene teasing Ron about falling asleep while waiting for blastoff.

Time dragged by, and the hold kept being extended. Fortunately, experts at the Kennedy Space Center and at the Marshall Space Flight Center eventually managed to devise a workaround to bypass the faulty circuit and allow a further launch attempt before the launch window closed. But now the space center's clocks showed that it was Thursday, 7 December. It is unlikely that Gene Cernan was the only crew member to remember Pearl Harbor. On Pad 39A, a sleeping giant was about to awaken.

It had been a long night of uncertainty and frustration for the families, but finally the countdown resumed at 12:25 a.m. Jan gathered her children beside her. Once again, the clear voice of Chuck Hollinshead reported the diminishing numbers. Once again, control passed to the terminal countdown sequencer. "Now approaching the half-minute mark. T minus 33 . . . T minus 30 seconds and continuing on now . . ."

If anyone in the Launch Control Center had been crossing his fingers, he might well have uncrossed them at that point. The workaround had worked. This was the final countdown.

"T minus 17, final guidance release. We'll expect engine ignition at 8.9 seconds . . . 10, 9, 8, 7 . . . ignition sequence started. All engines are started. We have ignition . . . 2, 1, zero. We have a liftoff, and it's lighting up the area— it's just like daylight here at Kennedy Space Center as the Saturn V is moving off the pad!"

After a delay of two hours and forty minutes, the reluctant *Apollo 17* slipped its shackles and rose into the night, ponderously at first, dazzling the onlookers and bringing a false dawn to the Florida coast. In the period between the glare of ignition and the arrival of the sound waves, birds were heard singing and taking flight, thinking a new day had arrived. Vibrations that the watchers felt reverberating through the ground were enough to send shoals of fish in the turning basin leaping into the air.

Witnesses to the awesome spectacle have tried over the years, with varying degrees of success, to describe what they saw. The truth of the matter is, those who did not witness it in person can never really know what it was like; and those who did witness it can never adequately convey the sheer majesty of the event.

Tom Klein, one of Ron's former V F-51 squadron mates, was able to attend the spectacle. He had known the deafening roar of a Crusader launch, but the Saturn V was in a different league. "It was amazing! It was an experience none of us had ever had. You hear about the noise, but what you don't realize is the noise causes your entire body to shake, it's so loud. It was just . . . it was something else!"

For John Holm, formerly of V F-142, it was "probably the most spectacular thing I've ever witnessed!" He added, "Once the engines lit, I mean, you just couldn't believe the brilliance of the light, and then everything, the whole area, just turned into broad daylight. . . . And, you know, the birds were singing! All of a sudden, it was like the sun had come up, and the birds were all active. It was really weird!" Then the sound of the engines came rolling across the marshland, "and the thundering and the noise was tremendous. But it's hard to explain because if you haven't seen one, it's almost inexplicable."

Marv Miller was equally impressed. "Next to my marriage, it was the biggest thrill of my life!" Three and a half miles from the ascending rocket, the brutal low-frequency, crackling vibrations were so powerful, Marv was convinced that "something's going to blow up here!" Confessing to an inadequacy of descriptive power, he adds, "You couldn't believe it. It was just a wonderful, wonderful, experience. . . . I would say it was a highlight of our lives, for me and my wife."

For Ron's brother Jay, it was "a real honor" to share this experience. A man of few words, he simply said the launch was "quite a thing to see," and over the years he always enjoyed going back to look at his photographs of the event. He and his family closely followed the flight in the media, and he summed up his thoughts on Ron's participation in very personal terms: "I'm extremely proud of my brother."

Jaime Harner tries to recall all of the memories of her thirteen-year-old self as her father soared into space right in front of her eyes. She doesn't think she was actually scared for her father, but the enormity of the event was difficult to take in. Was this really happening? She certainly remembers the sky lighting up and the waves of noise and the eerie blinding light, but the clearest picture in her head is the bizarre spectacle of all of those fish leaping out of the water. Jan Evans and Barbara Cernan also remember seeing the jumping fish. It may seem odd they weren't all gazing unwaveringly at the ascending rocket, but just how long can anyone stare at a light almost as bright as the sun without averting the gaze downward for momentary relief? And, no, they weren't all seeing spots in front of their eyes. Spots don't flick their tails.

Barbara Cernan had been entranced by the sight of the Saturn V launch that had taken Gene on his first flight to the moon, but that just couldn't compare with what she was now seeing. Like so many others, even those who didn't have a personal stake in the launch, she struggles to find the words to describe the experience. "Of course, the Saturn V was fabulous, but it's just, when you saw it at night, it was just unbelievable. It lit up the whole Cape. I mean, it was just . . . it was . . . gorgeous!"

Rarely lost for words on the subject of *Apollo 17*, Jan Evans also struggles to describe her memories of the launch. She talks of the noise, the vibrations coming up through her feet as the ground shook, the jumping fish, and the brilliance of the fire spewing from the engines, painting everything in the vivid colors of a man-made dawn. But she recognizes the launch as far more than a breathtaking spectacle. Her emotions set this one apart. The launch was a truly emotional experience. "You just can't believe what's going on. It's finally happening! It's . . . it's just the most wonderful thing! You are so happy for the guys up there, because they're finally on their way!"

And you don't need to be a mind reader to understand that Jan isn't just talking about the launch delay.

On board spacecraft *America*, at the tip of the Saturn V, the crew felt the deep rumbling of engine ignition. To Gene Cernan, it felt like "a big old freight train sort of starting to rumble and shake and rattle as she lifted off." Ron Evans "didn't expect the thing to shake quite as much as it did." But it was exhilarating! Five seconds after liftoff, Ron whooped, "Woo hoo!" A short time later, he added, "Thirty seconds, we're going up. Man, oh man!"

If he had a concern at that point, it was that the vibrations were so bad he couldn't actually read the instruments on the panel in front of him. He observed, "Man, this thing shakes like a son of a gun!"

The shaking eased off as the atmosphere outside became thinner, but the rocket was rapidly accelerating, and g-forces were building toward a peak of around four times their normal weight. Although the veteran Cernan warned his rookie crewmates to hold on as they approached the s-ic first-stage sep-aration, the vigorous jolt of staging still caught Ron by surprise. He blurted out, "Jesus Christ!"

Compared to the brute-force s-ic stage, the s-ii second stage was as smooth as silk. Reading the instruments was no longer a problem. Half in jest, Gene asked his CMP, "I forget anything yet, Ron?" This question produced the reas-suring reply, "Nope."

The s-ii served well and dropped away, leaving the s-ivB third stage to conclude the ascent phase. Eleven minutes and forty-two seconds after launch, *Apollo 17* entered low Earth orbit, a little over a hundred miles above the Atlan-tic Ocean and speeding toward a new dawn at 17,450 miles per hour.

All three astronauts began working through their checklists, reconfiguring the spacecraft from launch mode to flight mode. Nine minutes after orbital insertion, Ron spotted huge thunderclouds on the curve of the eastern horizon that were silhouetted against the approaching sunrise. He exclaimed, "Holy mackerel!" A little later: "Here it comes . . . Ka-boom!" The silent explosion of dawn ushered in his first morning in space.

They were not the first crewmen to be distracted by the beauty of planet Earth beneath them, but their schedule was relentless. They had to complete many checks before a second s-ivB engine burn would set them on course to intercept the moon.

As they sped across the rolling map laid out beneath them, Ron Evans must have been reflecting on his new status as a "real" astronaut. His exclamation of delight was addressed as much to himself as his crewmates: "Hot diggity dog! I can't believe we've made it up here!"

Ron Evans definitely wasn't in Kansas anymore.

21. Into the Shadow

That gentle moon, the lesser light,
The lover's lamp, the swain's delight,
A ruined world, a globe burnt out,
A corpse upon the road of night.

—Richard Francis Burton (1821–90)

Once they were settled in orbit, Ron Evans and Gene Cernan exchanged seats. Ron was now the pilot and preparing for translunar injection, or the second burn of the S-IVB. The NASA public affairs officer tried to explain how they would catch up on the flight plan by firing the engine slightly earlier than scheduled and for a little longer. The spacecraft would take a little less time to reach the moon than originally planned, thus canceling out the launch delay and synchronizing the ground elapsed time (GET) with the flight plan. The theory was sound, but the actual execution left many people confused.

Some three hours after the S-IVB had first shut down, its single engine lit again, burning smoothly for nearly six minutes. At cutoff, the spacecraft briefly achieved a velocity of 24,242 miles per hour and rapidly flew through the inner and outer Van Allen radiation belts, like a finger flicking painlessly through a candle flame. Such was the pull of Earth's gravity that within thirty-six minutes, that velocity had dropped to "only" 15,000 miles per hour, but it was enough to place *Apollo 17* on a course that would intersect the moon's orbit after a journey lasting eighty-three hours.

Ron's first major task as the command module pilot was to separate the spacecraft from the spent booster, to turn, and to extract the lunar module. Crucial to the success of any lunar landing mission, the transposition and docking procedure was supposed to be relatively routine. However, as Ron well knew, Murphy's Law had intervened on *Apollo 14* to thwart Stu Roosa's attempt to set a record for fuel economy during the maneuver.

Explosive cord separated the command and service module from the spacecraft adapter, whose four redundant panels spun slowly away into the void. Ron pulsed the CSM's rearward-facing thrusters to achieve separation, and *America* nosed away from the booster amid a shower of particles generated by the explosive separation. Schmitt noted that "excitement permeated Evans" as Ron put into practice the task he had spent hundreds of hours simulating back on Earth. Jack trusted Ron implicitly and got a kick out of his enthusiasm, but he still kept a very close eye on the checklist. He had a personal interest in the fate of that flimsy lunar module, which would soon be sandwiched between the massive booster and the sturdy CSM.

As Ron pitched his vessel around using the thrusters, the dazzling earth passed across his field of vision before disappearing, leaving him with a view of the LM amid the cloud of debris. He couldn't resist exclaiming, "My gosh, look at the junk!"

Gene Cernan chimed in, "We're right in the middle of a snowstorm." The particles consisted of paint fragments and ice particles shed from the S-IVB, but they were only mildly distracting. "Captain America is very intent on getting *Challenger* at the moment," the commander added.

Ron noted that his approach was a little slow, at just over an inch a second, but better safe than sorry. The protruding docking probe entered the LM's drogue, establishing a soft dock. Then came the rapid rattling sound of the capture latches engaging. Ron allowed himself an elated, "Ka-chunk!" It later turned out that only nine of the twelve latches had engaged. Two were operated manually, but one seemed faulty. A decision was made to leave well enough alone. Nobody wanted the latch to refuse to *disengage* prior to the undocking in lunar orbit.

Forty-eight minutes after docking, Ron fired his thrusters again to separate from the spent S-IVB. The united *Apollo 17* spacecraft continued on course for its rendezvous with the moon. The booster would steer a separate collision course to act as a man-made meteorite whose shock waves would excite the seismometers left on the moon by four earlier teams of explorers.

Shortly after the separation, capcom Bob Parker noted, "Looks like you've got a super-conservative CMP up there. . . . You used about forty pounds of RCS [fuel] on the T & D. . . . Beautiful!"

As Gene and Jack purred in appreciation, Ron didn't quite say, "Aw, shucks," but he came close. "Still a little bit too much, but that's not bad."

Jack Schmitt's verdict on the docking: "That was a very important part of the mission, of course, and had to be carried out extremely precisely, which he did. And it was just characteristic of Ron."

The spacecraft's trajectory to the moon gave the astronauts a unique view of their home world, as close to seeing a "full Earth" as would satisfy anyone but a nitpicker. Only minutes after separation from the booster, the rapidly increasing distance from Earth meant that the men could actually see the full disk of the planet clearly out of two of their windows. One of three photographs from the mission showing the planet's complete circumference is now known as the *Blue Marble*. Described as NASA's most requested photograph, it is possibly the most recognizable photograph in history and demonstrates the sheer beauty and fragility of Earth in ways that words cannot match.

Who took the *Blue Marble* picture? Surprisingly, this question has provoked much debate and research. As in the case of the almost-as-famous *Earthrise* photograph from *Apollo 8*, all three astronauts claimed it at one time or another. NASA officially credits the whole *Apollo 17* crew with the *Blue Marble*. But on 8 January 1973, *Time* claimed to have identified the photographer: "Early in the mission, astronaut Ron Evans made his most notable photographic contribution; he took a picture that will rank among the classics of the space program. As Apollo sped toward the moon after blasting into its translunar trajectory, he pointed his camera back toward home and caught a stunning view of the Earth, with the side visible to the astronauts completely illuminated."

It would certainly look good on Ron's record if he had taken one of history's greatest photographs, but it is not that simple. Early transcripts of the crew's dialogue have confused the issue by mixing up the speakers' names. One Hasselblad camera was available at the time, and all three astronauts used it. At four hours and forty-seven mins GET, Ron looked out of a window and gushed, "Wooh! What a beauty! What a beauty! Yes—the earth! Look at that!" He took the camera from Schmitt and captured a sequence of images of Earth, but the telephoto lens could not encompass the whole sphere in one view.

Ron then handed the Hasselblad to Cernan, who took a series of dramatic pictures of the slowly receding S-IVB booster.

Ron had the camera again at five hours and three minutes GET, and he changed to a wider-angle 80mm lens. A few seconds later, he passed the camera to Schmitt, advising him that he had taken one picture exposed at F22. Ron asked, "Here, Jack. Can you see him good? Check the settings there." It is not clear whether, by "him," Ron meant the earth or the S-IVB. A single 80mm picture of the booster and three images framing the whole earth (the first slightly overexposed) do exist. The most likely sequence of events is that Ron changed lenses, took one picture of the S-IVB, and handed the camera to Schmitt, who by then had a better window in the rotating spacecraft. Schmitt then took the three pictures of the whole earth.

The evidence certainly supports that conclusion. Jack was observing and photographing the earth all the way to the moon. About ten minutes after capturing *Blue Marble* on film, Jack reported to Houston: "That view of the earth . . . there was something I was looking forward to, and I was not disappointed." Schmitt himself is certain he took the *Blue Marble*, and it seems inconceivable that he would have missed that unique opportunity to image the full Earth. That NASA played it safe by calling the photograph a team effort, however, is perhaps understandable.

There is one additional piece of evidence. When shown a copy of the *Blue Marble* in 1986, Ron was asked who took it. "Oh, I claim that I did." He then laughs. "Jack claims he did it. I don't know which one did it, but I tell everybody I did it." And that is the point. Ron didn't say, "I took it." He only said that he *claimed* he took it. That is very reminiscent of the jocular way Frank Borman and Jim Lovell both claimed to have taken the *Earthrise* photograph, kidding the real photographer, Bill Anders.

To this day, the astronauts of *Apollo 17* are the only human beings privileged to have seen the full earth. No photograph can truly capture the finest details, the vivid colors, or the subtle contrasts. Ron was neither a poet nor a philosopher. He knew he lacked the descriptive powers to do justice to the vision of the beautiful blue sphere floating in utter blackness beyond his window. But for all his right-stuff self-deprecation, it is obvious from his contemporaneous impassioned reaction—"Wooh, what a beauty!"—that the spectacle penetrated his analytical mind and registered on an emotional level. The experience was a stunning reminder that he really did have the best job in the world.

Jack Schmitt spent so much time reporting on the earth's cloud patterns that capcom Gordon Fullerton told him, "You're a regular human weather satellite." Aside from the fact that he had less to do on the outward flight than Cernan and Evans did, Jack had two good motives for his Earth observations: "If you don't have things occupying your mind, then it will take about three days for most people to feel comfortable in weightlessness. One reason I spent so much time describing weather patterns was because I was very interested in weather as a kid and subsequently. And, number two, it was a way to occupy my mind. That's the way it works."

And it did seem to work. As the crew entered the second day in space, their resident weather forecaster told Houston he had "acclimated." Shortly after arriving in orbit, he had developed a headache, but a day later he was "feeling fine." None of the astronauts had much of an appetite on the first day, and according to Jack, Gene "wasn't feeling too good." Thirty-three hours after launch, they awoke from their second sleep. Ron could be heard singing and laughing in the background. Jack continued to dominate the airwaves with his weather reports. Gene was largely silent. The published medical records show that he took a scopolamine capsule on the second day for "stomach awareness."

None of this surprised the lead flight director, Gerry Griffin. He knew that every crew suffered some degree of queasiness in the early days. It was very unpredictable and was just as likely to affect seasoned pilots as anyone else. Gerry does not recall any concerns on *Apollo 17* and is satisfied that any symptoms cleared up and did not affect the crew's performance.

Not only did Ron's appetite return but he also became notorious on the mission for his food consumption. Later in the flight, giving a daily food report, Jack Schmitt disclosed, "Okay, the CMP, the chowhound of the kennel here, had sausage, grits, fruit cocktail, orange beverage, and coffee. He had ham, cheese spread, peaches, cereal bar, and orange-pineapple drink. Later on, he had tomato soup, half a hamburger, half mustard, vanilla pudding, sugar cookies, grape drink, and tea." Such was Ron's reputation that Jan always thought he gained weight on the mission. While that makes a good story, the medical records show he actually lost a fairly typical two and a half pounds even though his calorie intake was the highest in the crew.

Before the mission, Jan Evans had persuaded NASA to install two squawk boxes, rather than the usual one, in the house. These loudspeakers played the live dialogue between mission control and the crew, allowing the family to follow the mission. One box was placed in the main family room, and the second was near the children's bedrooms. During training, Ron had taken his wife and children through what to expect during the mission, from launch to splashdown, and answered all of their questions. Sleep time in space was synchronized with nighttime in Houston, so Jan expected to be able to use her much-thumbed copy of the *Apollo 17* flight plan to follow all the key mission events on the squawk box. Unfortunately, NASA's campaign to adjust the timings to allow for the late launch completely confused people. Nothing happened when the flight plan said it was supposed to occur. By the third day, having missed several "Ron moments," Jan gave up relying on the flight plan's times. She was not alone. Even official NASA records of the mission leave readers wondering whether the quoted times take into account the catch-up for the launch delay.

Fortunately, Deke Slayton phoned the families every evening and gave them a summary of the day's activities. He was always very positive, telling Jan and Barbara "the guys are doing a great job, and we're just so proud of them." Of course, giving this praise wasn't too difficult as *Apollo 17* was progressing so smoothly, but Jan appreciated the attentiveness of the gravelly voiced Slayton, whom she always affectionately thought of as the "mother hen" of the astronaut corps.

The families were still none the wiser about the terrorist threat, but mission control passed carefully worded messages to the crewmen to assure them that everything was fine on "the home front." Tom Stafford also occasionally dropped by to check on the families for himself.

During the mission, Jan found that she was being treated almost as if she were "NASA royalty." As far as housekeeping was concerned, she got the children to school and did the laundry, but that was mostly it. "The house had people in it all the time. You never had to cook a meal. There was food everywhere. Everyone who came over brought this or brought that. You didn't have to do anything!"

It was essentially an open house, and that was exactly how Jan wanted it. Remembering how the Armstrongs had been so welcoming back in 1966, Jan had written to the latest astronauts who had arrived in 1967 and 1969 and invited them and their wives to "come and share in all the fun." The party atmosphere in the house emanated from the support network of people she knew and trusted to share the experience with her and the children.

Jan and Barbara visited mission control a couple of times during the flight.

The last thing they wanted was to interrupt people doing their sensitive jobs, so they just went and took seats in the viewing gallery. Separated from the flight controllers by a plate-glass window, they could exchange unobtrusive waves.

Although her husband was almost as far away as possible, Jan was elated. If Ron was assigned some future flight, it would be in Earth orbit. Nothing would ever match this experience for either of them. Being a part of *Apollo 17* made her very happy. Even looking back over the gulf of time, she remembers the whole period as "a fantastic family experience."

On board *Apollo 17*, the largely uneventful outward journey allowed "domestic" issues to draw disproportionate public attention, and Ron Evans seemed to generate much of the news. Forty-eight hours into the flight, the crew settled down for a third sleep period. Ron would be the duty officer overnight, prompting Cernan to ask mission control, "Are you going to sing to Ron in the morning to wake him up?"

When the time came to rouse the crew, the task proved unusually difficult. In a nod to Ron's alma mater, his colleagues in Houston played a hearty chorus of the Jayhawk fight song but with no effect. They played the song a second time. Then a third time. They even sounded a crew-alert klaxon, but the doctors could see from Cernan's biosensor readings that he was still fast asleep.

The morning shift in mission control brought a new capcom, Gordon Fullerton, who tried to stir the sleepyheads. "Good morning, *Apollo 17*. It's time to rise and shine." Nothing. Eighteen minutes later and rather more tersely: "It's morning. Time to get up." Still nothing. He then transmitted a high-pitched oscillating tone through the circuit and wheeled out the Jayhawk song one more time.

Finally, after sixty-five minutes of silence from the spacecraft, came a voice. "We're asleep," muttered Schmitt.

"That's the understatement of the year," Fullerton grunted.

"Never let Evans be on watch."

"I think we'll go along with that from here on."

Sufficiently awake to joke, Schmitt added, "That was some party last night, Gordy. Man, that was a humdinger!"

A slightly sheepish Ron Evans later explained, "As much as I hate to admit it, the Power/Audio tone was off in my headset." In other words, the officer of the watch had forgotten to set his alarm. Capcom issued a mock warning that the management would dock them a day's annual leave to make up for the crew's late start. Shortly before the next sleep period, news emerged from the home front that while serving on aircraft carriers, Ron had often placed a Baby Ben alarm clock in a metal dishpan to amplify its ringing. When the news media had approached her about the missed-alarm story, Jan Evans had ratted out her husband and admitted that "he literally just sleeps through anything."

On each journey to the moon, the CMP periodically realigned the space-craft's guidance platform. This was crucial, not least to avoid hitting the moon. The news media, however, seemed less interested in Ron's highly accurate realignments than in another emerging "domestic" story: on the third day, Ron revealed that he had lost his scissors.

Everyone who has misplaced his or her car keys understands the sheer willful cussedness of inanimate objects and their unerring ability to find the most obscure places to hide. In the zero-g environment of an Apollo capsule, anything not nailed down, or at least attached to something else by cord or Velcro, had a whole extra dimension in which to disappear. No nook seemed too small, no cranny too remote to serve as a hiding place for any wayward floating object. Scissors were essential to open NASA's sealed food packages, and each astronaut had a pair. Try as he might, Ron could not find his miss-ing scissors. What started as a funny story actually had a minor operational impact on mission planning, since the moonwalkers would have to leave Ron one of their pairs of scissors and adjust their procedures on the moon. Not that they would tell him for a while . . .

Viewed from the spacecraft, Earth was shrinking rapidly, and night had claimed part of one hemisphere. Ron didn't spend as much time as Jack observing the planet, but it certainly caught his attention. He later recalled, "The earth is a beautiful thing up there, it really is! Catching a glimpse, and watching it change, first of all from something that has very little curvature to it, and makes that transition down to a little ball, and then the ball gets smaller and smaller." But in case that was beginning to sound poetic, he admitted that his reaction to the shrinking earth was more prosaic. "Gee whizz! We're get-ting a long ways away here." He saw nothing philosophical about his home world looking smaller. That was supposed to happen! What it mostly meant to him was they were getting ever closer to their destination. After six and a half years, Ron Evans's long voyage to the moon was nearly over.

Before entry into orbit, Ron had to conduct a simple but vital task to gain access to the battery of cameras and other sensors in the service module's SIM bay. The equipment was protected behind an aluminum panel that had to be explosively jettisoned. Some people called it the world's biggest lens cap. Jettisoning it before lunar orbit meant avoiding any risk of a future collision. As the spacecraft and crew neared their destination, Ron gave Houston a countdown, his voice betraying no concerns about the vitality of the explosive cord that needed to do its job so he could do his.

"Okay, SIM door jet 5 . . . 4 . . . 3 . . . 2 . . . 1 . . . Jet. Oh, I got a good bang!" The panel blew away cleanly. The crew last saw it tumbling off into space.

Two and a half hours later, Gene Cernan reported seeing "the limb of the moon," the tiny part of the rocky world's trailing edge that was sunlit as *Apollo 17* prepared to fly behind the leading edge. Ron could also see the thin curve of light. "Talk about a sliver of the moon! That is a sliver of a sliver."

That was pretty much the last anyone heard from Ron for the next couple of hours as he carried out a final alignment of the guidance platform and began a meticulous set of checks on the health of spaceship *America* to test its readiness for a long sojourn in orbit. Then for the final thirty-four minutes, he dominated the airwaves, reciting a litany of checks on the engine, the fuel, the pressurization, and the orientation of the spacecraft with respect to the moon. It was a good day for acronyms and jargon. But he also found time for a final grouse about his missing scissors: "I'll have a hard time eating if you guys take all the scissors with you. . . . But my teeth are pretty good, though." Cernan and Schmitt said nothing.

Meanwhile, under the influence of the moon, *Apollo 17* accelerated to just over a mile per second and aimed to miss the leading edge of the rugged world by only sixty miles. From Houston, Gordon Fullerton gave them a go for lunar orbit insertion. A minute before the anticipated loss of signal, he wished them "a good burn." As the last lunar explorers of the twentieth century passed out of radio contact with the earth, they were 454 miles high and curving moonward at nearly 5,000 miles per hour.

As it journeys through space, the moon casts two distinct shadows. *Apollo 17* had already passed out of sight of the sun. Now the spacecraft flew tail-first into the lesser shadow, leaving behind the pale bluish glow of earthlight and entering almost complete darkness. As he prepared for the crucial lunar orbit insertion burn, Ron stole a few glances out of his window and saw only a peppering of stars. What really caught his attention, and probably hoisted a few of the little hairs on the back of his neck, was the broad arc of absolute blackness where no stars were visible. The vast bulk of the moon made its presence known by blocking out all light.

The computer, programed with the latest navigational data, ignited the engine of the service propulsion system (SPS) right on schedule without the need for any further human action. The astronauts felt a reassuring, gentle pressure against their couches that lasted for just over six and a half minutes, until the spacecraft had shed two-fifths of its velocity. At the appointed second, the flow of chemicals to the combustion chamber ceased, again without human intervention, and the explorers slipped into orbit around the moon. The return of the sun had nothing of the graceful warm glow of an earthly dawn. One instant it was dark; the next, they were dazzled by the risen glare that cast long, inky shadows across the stark moonscape below. Still out of radio contact, the three men gazed down at the far side of the moon. Gene Cernan had seen it before, but he still found the experience majestic and almost overwhelming. The in-house geologist Jack Schmitt was glued to the windows, mesmerized by features he had previously only known from books and photographs.

For Ron Evans, the feeling was more of bewilderment. Something was very wrong with this world he had studied so intensely: *there were no craters.* "You're behind the moon with no communication with the earth, and you come out into the sunrise, and everybody's looking out the window, and I don't see any craters! All I see is a bunch of bumps. Total optical illusion! The first time I look at the moon and it's humps all over the place! I blinked a couple of times and it was still that way, and I thought, 'My God! I'm going round the moon, and I'm supposed to be a geologist, and observing things, and it looks like it's got mounds on it instead of holes!'"

He had finally reached his destination, but the view wasn't quite what he had expected. This optical illusion was not uncommon. In his autobiography, *Carrying the Fire*, Michael Collins would illustrate it with a photograph that showed several extraordinarily mound-like features. The picture is playfully captioned, "Do you see what I see?" If the book is turned upside down, the mounds become craters again.

After thirty-three minutes out of contact with Earth, the spacecraft reappeared around the eastern limb of the moon. Gene Cernan, only the third earthling to visit the moon twice, turned a greeting into a mission statement: "Houston, this is *America*. You can breathe easier. *America* has arrived on station for the challenge ahead."

Near the high point of their elliptical orbit, the astronauts looked down on the landing site. Although they spotted nearby large craters, including Littrow, the touchdown point was still lost in shadow while awaiting its own dawn. As their orbital track took them back into darkness, the astronauts could see this part of the moon's nightside was lit by a pale radiance.

"Hey," Ron remarked, "you can even see the horizon in the earthshine out there."

The time had come to swap the role of tourist for that of geological observer. After *Apollo 17* disappeared behind the moon for a second time, Ron began to power up his SIM bay instruments. As they flew back into sunlight over the craggy lunar far side, first the mapping camera with the laser altimeter and then the panoramic camera began taking pictures. With long rolls of film spooling through the cameras' mechanisms, the spacecraft's sharply observant glass eyes peered down at craters and basins bearing the names of Korolev, Apollo, Icarus, Daedalus, and Gagarin. Captain America was hard at work.

22. The Big Picture

We *must* get to the Pole; but we shall get more too. . . .
We want the scientific work to make the bagging of
the Pole merely an item in the results.

—Dr. Edward A. Wilson (1872–1912)

Beardmore Glacier, Antarctica, 8 February 1912

When Capt. Robert Falcon Scott turned his back on the South Pole on 16 January 1912, it would be forty-four years before anyone would return to leave footprints there. Scott had been bitterly disappointed on reaching the pole to find that the Norwegian Roald Amundsen had arrived first. Scott had to console himself with the knowledge that his was primarily a scientific expedition, continuing the groundbreaking work of his first expedition a decade earlier.

As Scott and his polar party—Edward Wilson, Henry Bowers, Lawrence Oates, and Edgar Evans—trudged back toward their base, 880 statute miles on foot from the pole, they stopped along the edge of the Beardmore Glacier. This broad, fractured river of ice not only provided a route down from the polar plateau but also allowed Wilson, the expedition's chief scientist, to collect what would prove to be crucial geological samples at the foot of an exposed vertical sandstone cliff.

Wilson noted bands of limestone and coal in the cliff face, and Scott enthusiastically recorded the finding of "beautifully traced" fossilized plants in a coal sample. Wilson gathered fallen rocks containing fernlike fossil imprints, which were later identified as leaves of the long-extinct *Glossopteris* tree. As Scott mused in his journal, where on Earth might these samples have come from? Clearly no trees or ferns could have possibly survived in the frozen Antarctic wastes of 1912.

Scott agreed to Wilson's request to add thirty-five pounds of carefully documented geological samples to their sledge. They resumed their arduous northward slog, but over the next five weeks, first Evans and then Oates succumbed to the unseasonably awful conditions, the likes of which were later seen only once in weather data recorded between 1984 and 1999. Scott, Wilson, and Bowers struggled on but became trapped in their tent by a howling blizzard just eleven miles from a large supply dump. On 29 March 1912 Scott made his last journal entry: "We shall stick it out to the end but we are getting weaker, of course, and the end cannot be far. It seems a pity but I do not think I can write more. For God's sake, look after our people."

On 12 November a relief expedition following the line of supply cairns found the half-buried tent containing the frozen bodies of Scott, Wilson, and Bowers, together with their journals and photographic plates. Outside the tent, still resting on their snow-covered sledge, lay the bag of rock samples from the Beardmore Glacier.

Frankfurt, Germany, 6 January 1912

A few weeks before Dr. Wilson unearthed his plant fossils where no plants could possibly grow, a German meteorologist named Alfred Wegener delivered a public lecture at the Senckenberg Natural History Museum in Frankfurt. Wegener was not the first person to notice that the eastern bulge of South America could fit fairly neatly against the west coast of Africa, but his lecture was a first attempt to expound his hypothesis of continental drift, which suggested that all of Earth's continents had once been joined in a single landmass and had then drifted apart.

Wegener never managed to explain to his critics how the massive continents were supposed to have wandered across the surface of the planet and across the different climate zones. He died in Greenland in 1930, and for many years, his brainchild was largely unloved and unsupported. However, evidence was accumulating. Fossils of *Glossopteris* had been identified in South America, Africa, India, and Australia, and the samples Edward Wilson obtained in Antarctica provided a missing piece of the jigsaw puzzle of the once-united continents. In 1952 Dr. Frank Debenham, one of Wilson's fellow scientists on the Scott expedition, described the fossils as perhaps the most important of all geological findings from Antarctica. They were "of the character best suited to settle a long-standing controversy between geologists as to the nature of the former union between Antarctica and Australasia." He was right. Continental drift is now an accepted fact.

The Moon, December 1972

Alone in lunar orbit, Ron Evans was finding spaceship *America* remarkably roomy for one man. Not only was he now the captain of the ship, he was its navigator, engineer, chief cook, and bottle washer as well. And zookeeper, if he counted the five Biocore mice, although he never even heard them squeaking from inside their canister.

His previous twenty hours had been a whirl of activity, starting with the third-orbit engine burn that dropped *Apollo 17*'s perilune, or low point, to seventeen miles above the moon. Firing the service propulsion system's engine for twenty-two seconds was the plan, but an overrun of only two more seconds would have put the perilune *below* the lunar surface. In the event, the trusty computer shut the engine off exactly on time, but fingers had been poised for a manual override, just in case.

As they began to drop lower, Ron was still having a problem with seeing the craters as humps. "Gosh, dang it! I wish those craters would turn around the other way!"

At least they had resolved the scissors standoff. The imp within Ron Evans began to talk about dismantling part of the instrument panel to access the hidden recesses where the scissors might be lurking. Cernan and Schmitt knew he was joking, but the commander finally relented—as everyone knew he had to—and agreed to leave Ron one of the two pairs from the lunar module.

"He's not taking the spacecraft apart to find his scissors, and I'll not let him go hungry," Cernan growled.

Ron said his goodbyes to his companions as they boarded *Challenger*. Nobody mentioned or even consciously thought that it just might be the last time they would see each other. They wouldn't be flying the mission if they didn't trust the vehicles and themselves, as well as each other.

On his final pass over Taurus-Littrow before the landing attempt, Ron used his onboard sextant to make a vital contribution. Marking the precise locations of several small craters on the valley floor, he passed the data to Houston, where it was processed and transferred up to the lunar module computer. Ron hit all his targets, earning praise from Schmitt for making *Challenger*'s updated guidance information "as good as it could get."

Landing on the moon had never been easy or routine, even if most of the Apollo commanders had made it look that way. Supplied with just the right amount of flight data by his LM pilot, Gene Cernan rode *Challenger*'s translucent plume of fire across the mountains and down into the valley of Taurus-Littrow. Overhead, having returned to a higher orbit, Ron listened to the dialogue between the LM and Houston, and knew not to interrupt. This was Gene's moment. With two minutes of fuel left in the tanks, Cernan cut his engine, and the lunar module touched down safely. Almost echoing Neil Armstrong, he excitedly reported, "Okay, Houston, the *Challenger* has landed." A few moments later, and more personally, he added, "Epic moment of my life!"

Cernan didn't forget the third man, asking Houston to pass on the good news, but Ron quickly interjected, "Hey, *Challenger*, this is *America*. I heard you all the way down."

Cernan had become the fifth carrier-trained naval aviator to land a lunar module safely. His next comment, from one navy man to another, probably meant nothing to most of the civilian audience: "Ron, I had the meatball all the way!"

Ron understood. "That's great! Beautiful!"

And now Ron had the interior of *America* all to himself. One thing that became increasingly clear in the greater freedom of a capsule built for three was just how much he was enjoying weightlessness. After varying degrees of stomach awareness in the early days, all three men had acclimated, as Schmitt had put it.

Ron now appreciated that nothing anyone had ever told him about the experience could have properly prepared him. Scuba diving? Hell, scuba diving didn't simulate it at all! And as for that stupid KC-135 . . . In the same way he had told Jan that no ground pounder could understand high-performance jet flight, he now realized you just could not understand or even relate to weightlessness unless you had actually experienced it *in space!* He could turn lazy somersaults. He could arch over on his back. He could maneuver across the capsule with the gentle push of a fingertip. It was absolutely delightful!

He found that the simulators had done an excellent job of mimicking most aspects of flying the spacecraft, although one substantial difference was the sound from firing the reaction control system's thrusters. In the simulator, a firing produced a very limp sound: Plink! Plink! In space, the "plink" of the

valves opening was followed by a sound that vibrated across the capsule—
BAROOOM!—and by a flash of light through the windows.

Plowing his one-man furrow around the moon, the chowhound of *Apollo 17* continued to enjoy his food. The meals NASA had supplied for him were now accessible courtesy of the LM scissors and were appetizing, if not actually haute cuisine. However, fourteen hours after the lunar landing, hunger began to gnaw at Ron as he continued his photographic tasks.

"Did I miss lunch, or was I supposed to get any lunch today?" he inquired. "Long time since breakfast, I think."

The word from Houston was not helpful: "You've got about four more hours until scheduled eat time." Usually so meticulous, the people who had drawn up the flight plan had indeed forgotten to include a lunch break for Ron that day. There were wry smiles all round, but the problem was, with such a packed schedule, Ron could not stop working to eat simply because he was hungry. Nor could he take a sneaky unscheduled meal break on the far side of the moon. Just because nobody could hear him didn't mean he could magically insert an extra half hour into the flight plan.

Eventually, the capcom suggested he could have a meal during a relative lull while the SIM bay's lunar sounder was probing the upper layers of the moon's surface. News of the missed meal reached El Lago, probably via the squawk boxes, because after Ron eventually had his turkey and gravy, he was told that Jan had been in touch with mission control. She had pointed out that her husband was "the last person she'd ever think would miss a meal up there."

Messages from El Lago were important to the lone orbiter. Ron also received word from the home front that all of his neighbors had come to his house to put up the outside Christmas decorations. When the capcom suggested that flying to the moon was a pretty extreme way of escaping the task, Ron agreed it was "pretty darn nice of the neighbors."

The capcom also used the official NASA airwaves to tell Ron that Jaime was "real tickled" to have scored straight As on her school tests and knew her father would be happy to hear it. Jon passed along a message for his father: "[I'm] so glad you're there because you worked so hard all these years to be there." Jan simply sent her love. Ron was also reminded that his family was listening to the squawk boxes "and hanging on every word."

Undoubtedly Ron was enjoying himself as a lunar explorer. Jack Schmitt recalls Ron's onboard humming, whistling, laughing, singing, and talking to himself—all features of his personality that his family knew well. Schmitt considered it was all "very typical of Ron" and notes that he had "a great midwestern laconic and humorous view of life, and he just enjoyed himself very much." During Ron's three days alone in orbit, mission control got used to his laid-back and folksy references. One of his navigational signposts, the star Canopus, was "the old blue star." Another was "good old Sirius." When praised for a perfect session with the lunar sounder, he enthused, "Oh, that's dandy!"

The schedule was relentless, but his enthusiasm gave him an edge. "Your adrenaline is up all the time," he later reflected. "The effect is, you can get by on four hours' sleep a night. Really! Which I normally can't do. I couldn't do that down here on Earth for fourteen days!" At one point, the capcom warned him he was going to be "busier than the proverbial one-armed paper-hanger."

But he paid a price for living and working in a shower-less trash can for days on end. As Ron pointed out to nobody but himself (and the voice recorder) on the far side of the moon, "Man, I stink! Whew!" The inadequate remedy was a sponge bath. Jon had wanted his dad to grow a beard on the flight, but Ron couldn't get past the "itching stage." He preferred to shave, even though it was a slow process in the dry, pure oxygen atmosphere.

Farouk El-Baz and the other orbital scientists were delighted with Ron's work and noted that he hadn't missed any mapping camera, panoramic camera, or lunar sounder operations. They were "all on schedule and right on the flight plan." Ron admitted there was "a little more hustling up here than I thought there would be."

The reply from Houston made him laugh. "Keeping you busy, huh? Just don't want you to get lonely up there."

A decade later, Ron and Jan had a dinner guest at their home in Scottsdale, Arizona—Ron's old friend and fellow Crusader instructor John Allen. It was just like old times, but inevitably the conversation turned to *Apollo 17*.

"We were talking about his moon mission," John recalls, "and he said the loneliest he had ever been in his life was when he was in the capsule going 'round the back side of the moon while the other two were down on the surface. . . . He said that was a very lonely time."

That would be in stark contrast to the reported remarks of others. Al Worden always said he was delighted to get three days of peace and quiet to do his own work while his crewmates were on the moon.

The "lonely" remark also puzzles Jack Schmitt. "I can't imagine Ron focusing on being lonely, because he had so many other things on his mind, and he enjoyed looking out of the windows so much. . . . I suspect not having Houston interrupting your thoughts and what you were trying to get done was a great relief for all of those crewmen who were alone on the far side of the moon." But, of course, Jack never had that particular experience.

No one knew Ron better than his wife, and Jan never heard him say he was lonely while he was out of touch with Earth. She recalls he *did* tell her something along these lines: "When you're on the back side of the moon it was unlike anything you had ever experienced, or ever would again, because you were all alone, and it was the blackest black you could ever imagine."

Something may have been lost in translation somewhere. On the one hand, there is a difference between being alone and feeling lonely, and it seems unlikely that a former combat pilot would feel lonesome simply because he was temporarily out of touch with others. Anyone flying behind the moon knew one thing for certain: Earth would soon rise again.

On the other hand, Ron was a family man who was friendly and gregarious, and enjoyed the company of other people. Based on what he told Jan, even an unphilosophical man must surely have been affected by the view out into the depths of infinity on the far side of the moon. Everyone he knew and loved—indeed, every other member of the human race—was separated from him by more than two thousand miles of rock and a quarter million miles of space. Does a brief period of total and absolute isolation generate a feeling of loneliness or just a temporary sense of aloneness? On the *Apollo 17* mission, only Ron Evans knew.

Just how "unphilosophical" was Ron anyway? He is on record as saying he couldn't really process philosophical thoughts or profound descriptions, but that didn't seem to apply to his thoughts about Earth. As with every other Apollo astronaut, he would always steal glimpses of the home world and watch out for earthrise and earthset. Perhaps his emotion came primarily from knowing who was riding that blue marble through the darkness.

He was certainly able to look back on the experience in emotional terms: "As you look out on that vast, vast infinity of space, up there is this beautiful Earth. And you watch the continents rotate into the daylight . . . and then your position changes with respect to the stars in the background, so you look past the earth and you see different sets of stars as the moon is rotating around the earth and you are rotating around the moon. So it looks like the earth is moving with respect to the stars in the background. You don't see any strings holding it up. It still rotates with a precision. And it's my belief that no one could have participated in such an adventure without reinforcing whatever belief you had before you left. In my case, it confirmed that there is a God."

But Ron had not traveled a quarter of a million miles to study Earth. When the need arose he was perfectly capable of summoning up suitable memories to wax lyrical about the rugged, cratered world that had filled most of his view: "It's a beautiful place, you bet it is! It has its own beauty, not like the earth at all. Beauty associated with changes in the albedo, changes in the color as you go around the moon, and changes in the sun-angle as the sun reflects off the moon. And beauty is purely from the standpoint that it's not the earth, it's another heavenly body."

He didn't quite say "beauty is in the eye of the beholder," but that might be a reasonable interpretation. During the mission, the moon presented him with a constantly changing face as the westward march of the sunrise termi-nator opened up new vistas on the near side. Meanwhile, on the far side, the sunset terminator was swallowing features he had previously studied. The ever-changing geometry of the sun-moon-Earth relationship also gradually reduced the portion of the near side that was bathed in the glow of earthshine. In spite of the nearly full Earth's reduction to a crescent over the course of the mission, Ron didn't see much difference in the level of its illumination. Picking out certain craters, valleys, and mountains, even when most of the big blue marble had disappeared, was still easy.

Ron was sorry to have missed one particular experience by several days. "I would like to have been able to go around the moon when the earth was full, and be able to see how bright that earthshine really is!"

He quickly learned that even the rocky moon could be incredibly bright, particularly around local noon. On one of the early orbits, Gene Cernan had found the glare from below dazzling him so much he couldn't find a switch on the control panel in front of him. When the sun was nearly overhead, Ron noted that the surface was "bright—*bright!* Really, really bright!" He had a remedy, if he could find it. "You look around—where are my sunglasses? And the damned things have floated off somewhere!"

The moon was a world of mysteries to be solved. Shortly after arriving in orbit, Jack Schmitt had seen the flash of a possible meteorite striking the sur-face. Ron saw two flashes: the first was close to the western limb of the moon, and the second, intriguingly, appeared very close to Schmitt's sighting. The most interesting explanations would have been eruptions of incandescent volcanic gas—so-called transient lunar phenomena—or actual meteorite strikes. However, Houston suggested a more prosaic third explanation: cosmic rays passing through the brain or retina generated flashes, a phenomenon the three astronauts had actually studied on the outward journey.

Ron's work in lunar orbit amounted to much more than just operating the SIM bay's instruments and taking pictures out of the windows. Every Apollo flight had a specific set of preplanned mission requirements. Earlier CMPs like Al Worden and Ken Mattingly had certainly been trained to report anything unusual, such as patches of color, but during preparations for *Apollo 17*, Farouk El-Baz asked Ron to make a particular point of looking for color changes. They talked about the "extensive dynamic range and color sensitivities" of the human eye. In other words, the eyes of Ron Evans were expected to be much better at detecting small, subtle color variations than the film in the Hasselblad camera, let alone the mapping and panoramic cameras that were limited to black-and-white photography.

Ron had at his disposal a booklet of annotated maps of his target areas, a pair of ten-power binoculars, a reference color wheel, and a voice recorder. He used the recorder to preserve his observations and any other sound effects while out of touch with Earth.

To Farouk El-Baz, this flight was the last chance, possibly for many years, to unlock the moon's secrets. He was particularly wedded to the notion of "the big picture." Cernan and Schmitt would make meticulous observations at one small location, a valley near the shore of one of the moon's many frozen lava seas. When compared with the moon's total surface area of over fourteen million square miles (greater than Africa), Farouk wondered how representative their findings would be.

This was where the third man played his part. As Ron Evans orbited the moon in his flying laboratory, he could study large swaths of the rocky world over many days. By comparing the detailed observations on the ground with the global observations from orbit, El-Baz and his colleagues could later determine whether Taurus-Littrow had unique characteristics or whether its features matched other places around the moon. The young Egyptian geologist found in Ron Evans a kindred spirit who appreciated the significance of the big picture approach. In fact, he found Ron the most receptive of all the CMPs on the subject.

The issue of color variations on the moon had been a real problem for the scientists working with the Apollo astronauts. The traditional idea of the moon's color as being nothing more than shades of gray was reinforced when the crew of *Apollo 8* arrived in orbit. Jim Lovell observed, "The moon is essentially gray, no color. Looks like plaster of Paris or sort of a grayish beach-sand." He later referred to "a vastness of black-and-white—absolutely no color." In fairness to Lovell, Frank Borman, and Bill Anders, they were pioneers first and foremost; the science came second. But the photographs they brought back did hint at subtle color variations.

Al Worden, another of Farouk's star pupils, noted in his autobiography that "changes in color and shading fascinated me as I circled the moon. Looking toward the sun, the lunar surface appeared light brown. Away from the sun it looked gray." But Worden's brief was essentially to observe and report on geological structures. Aside from a few mentions of brownish areas, he summed up the moon as "overwhelmingly majestic, yet stark and mostly devoid of color." Ironically, his *Apollo 15* crewmates, Dave Scott and Jim Irwin, had brought back samples of green volcanic glass beads, but their color was too subtle to be spotted from orbit.

On *Apollo 17*, Ron Evans had been tasked with concentrating on a list of target regions, including specific areas that seemed similar to the dark floor of Taurus-Littrow. The valley that Cernan and Schmitt explored was only a small part of a noticeably dark patch on the southeastern edge of the Sea of Serenity, the patch where Worden had seen those mysterious dark-haloed craters. But a dark ring extended around much of the rim of Serenity and, less obviously, around parts of the rim of the near-circular Mare Crisium (Sea of Crises). Small, dark-haloed craters had also been detected in those regions, raising the intriguing possibility of volcanic activity around the dark shorelines of two nearly adjacent seas.

Shorty Crater, Taurus-Littrow Valley, 12 December 1972

In the latter stage of their second spacewalk, Cernan and Schmitt disembarked from their lunar rover close to the rim of Shorty Crater. It had been a tiring day. Schmitt was flagging, but he was eager to uncover whatever secrets Shorty might be hiding. He was about to receive a real adrenaline boost.

THE BIG PICTURE | 271

The crater was an impressive hole in the ground at 360 feet across and nearly 50 feet deep. Shorty clearly was darkly rimmed and had a blocky central mound. A large, fractured boulder sat on the crater's rim, close to the rover, and Schmitt loped over to examine it. Turning back to take a set of photographs, he noticed something odd about his own footprints and what he had scuffed up in the dust.

"Oh, hey! Wait a minute . . ." What Jack was about to report would create near pandemonium among his fellow geologists in a back room in mission control.

In lunar orbit, Ron Evans was passing out of radio contact. Shortly after he reappeared around the eastern edge of the moon, capcom Bob Parker had some intriguing news for him: "You may be interested that at Shorty, the surface crew found some very, very orange soil—a great deal of it. Indicates strong oxidation and probably indicates water and/or volcanics in the area, and they're really— Jack's kinda like a boy at Christmastime, I tell you, a little kid at Christmastime."

Ron laughed. "I'll bet he would be! Yeah, that's a great find, by gosh!"

Scientist-astronaut Parker clearly shared Schmitt's enthusiasm: "It's orange, boy, you can see it in the television. It's just bright orange soil, no question about it."

"I'll be darned!" said the man who had been sent to the moon to look for subtle color.

The initial conclusion, on the moon and in Houston, was that Shorty was a volcanic vent and that whatever had belched out of the vent had contained enough water to oxidize the content of the soil and turn it orange. Cernan and Schmitt could even see orange streaks on the inner slopes of the crater. They raced against the clock to collect samples and make a photographic record. They needed more time, but that was the commodity in shortest supply. They had to conserve enough oxygen to walk back to *Challenger* in case the rover broke down on them.

Although Schmitt's excitement at the find initially allowed him to give voice to the most startling possibility, he was aware of the more humdrum prospect that Shorty had been caused by an impact that had blasted out a spray of underlying dark and orange material. Even in the midst of the volcano euphoria, Schmitt's old mentor, Gene Shoemaker, was saying Shorty was more likely an impact crater.

But the media sank their teeth into the story like a dog with a bone, speculating that Shorty had been volcanically active within the past three hundred thousand years, contrary to the prevailing orthodoxy that the moon had been largely dormant for three billion years. Suddenly geology was sexy, and Apollo was in the news again.

Once Cernan and Schmitt had concluded their sampling at Shorty and driven away, scientific attention transferred to lunar orbit. The notion of the big picture had snapped into sharp focus. The big *question* now was whether Ron's observations could place the Shorty findings into a more general context.

It would take approximately twenty-one hours before the laws of orbital mechanics brought *America* over the landing site. While Cernan and Schmitt were making their third and final lunar excursion, Ron passed high overhead. He picked out Shorty and wondered whether the orange soil had been found on its north rim. No, replied Bob Parker, Station 4 was on the south rim. "Looks like they barely got into the stuff," Ron replied. But if the north rim was better, that didn't detract from what Cernan and Schmitt had actually found on the south rim.

Parker coyly asked Ron what color he saw. "Is it one of the different tans?"

Ron probably realized the man on the ground was trying not to put words in his mouth. He laughed, "Yeah, the color—yeah, it's kind of a different. . . . Would you believe kind of an orangish-tan through these binoculars?" Later, he added that the color was "more orangish closer to the crater than as you got away from it."

Parker passed on congratulations from teacher to student. "Good show, Ron. Farouk just came out and said a real good showing. He's really charged about what you saw there and real pleased with it."

But perhaps Shorty was anomalous, merely a lunar curiosity? The really big question was whether there was evidence of similar orange material elsewhere, whatever its causes. Passing over one of his targets on the eastern edge of the Sea of Crises, Ron had spotted brownish tints. Around the western rim, he had spotted numerous small, dark-haloed craters that resembled Shorty, but he did not report any orange material. However, he *did* report seeing large boulders around the craters with "a greenish cast to them."

More significant, as his orbital track carried him from the southeastern edge of Serenity around the sea's dark border to its southwestern edge, he passed over the vicinity of the crater Sulpicius Gallus. This roughly textured area had been added to his target list to check for similarities with Taurus-Littrow. On orbit 24, *before* the discovery at Shorty and with long shadows beneath him, he reported, "Sulpicius Gallus region looks kind of brownish to me. I will have to check that when the sun gets a little bit higher."

A day later, *after* news from Shorty, and with the sun higher in the sky, Ron reported, "As soon as you cross that area, we have a dark-tan material that essentially covers the highlands. It is a hummocky material." Referring to a long channel that branched into three, he added, "Those rilles, again, have the dark-tan material on them." Ron didn't actually use the magic word "orange" over Sulpicius Gallus, but the color photographs that he took of the region show numerous distinctly orange-brown patches.

Two days later, with the three astronauts reunited in lunar orbit, the professional geologist wanted to check out his colleague's earlier observations over Sulpicius Gallus. "Yes, we are seeing an orange moon now," Schmitt reported. "In this whole dark mantle in here around Sulpicius Gallus there are scattered craters with a variety of orange to red-brown hues. And they all . . . seem to be small impacts that apparently are penetrating just far enough into the dark mantle material to tap this zone of orange to red-brown material."

But the mission commander was worried about these proliferating reports of orange material. Cernan later noted, "When I heard that Ron had seen some orange patches from orbit, I was concerned that maybe the power of suggestion had taken over." Some members of the press had similar concerns and made the point during a televised press conference on the journey back to Earth. Ron acknowledged that everyone's perception of color is slightly different, and he even conceded that you sometimes see what you want to see.

However, Jack Schmitt brushed aside this apparent veil of modesty and publicly acknowledged Ron's efforts in confirming the wider presence of orange material around Shorty and elsewhere. He pointed out that Ron had obviously looked carefully for orbital corroboration of what his colleagues on the ground had discovered, "and that leads you to see things. Now, that's not seeing things that *aren't* there! It makes you look for things that *are* there! And that's extremely important, and that's where the kind of training that all three of us have had, I think, has made it possible for us to find a lot of things that might not otherwise have been found."

For Cernan, any concerns about "wishful thinking" affecting Ron's observations were swept aside when he personally observed the Sulpicius Gallus region and drew attention to one of the small craters that Schmitt had also noted. "Its entire ray pattern was this color material, with a definite contact between it and the dark material around it, and it had that orange-brown hue to it, without any question at all."

The orbiting observer's angle of view and the location of the sun always affected the interpretation of subtle colors. When the orange soil was sent to the Lunar Receiving Laboratory in Houston, the lab's bright artificial lighting made it look disappointingly yellow-brown. But just as the material that the *Apollo 15* crew collected was genuinely green, the *Apollo 17* orange soil sample was definitely orange. It also contained black, brown, and reddish components.

It turned out that the material had not been oxidized by geologically recent volcanism. Far from being young, the material was actually very old—at least 3.5 *billion* years old—but it *was* volcanic. The orange "soil" actually consisted of tiny glass beads that had been explosively propelled from the lunar interior by fire fountains, similar to the dramatic eruptions of Hawaii's Kīlauea volcano. Over the eons, the layer of glass beads was covered by a veneer of rock and soil. Somewhere between twenty million and thirty million years ago, a meteor punched through the valley floor, creating Shorty Crater and exposing layers of the volcanic glass material.

While the man who first spotted the orange soil might have been understandably disappointed to learn that it was not actually the youngest sample from the moon, Schmitt was unperturbed. "No, as a scientist and geologist you take what you get, and you go on from there. . . . But the orange glass *is* volcanic, and it did come from deep within the moon, and it tells us more than any other sample about the deep interior."

As for the sightings around Sulpicius Gallus, Don Wilhelms, one of the investigating geologists, later credited Ron Evans with detecting and photographing outcrops that "out-Shorty-ed Shorty in orangeness." At the Fifth Lunar Science Conference in March 1974, Dr. Baerbel Lucchitta and Dr. Harrison Schmitt presented a paper that referred to the orange material observed and photographed in the Sea of Serenity's dark ring adjacent to Sulpicius Gallus. They stressed the similarities with the orange soil directly sampled hundreds of miles farther around the dark ring in the valley of Taurus-Littrow. The general similarity of the geologic settings was "remarkable," and they concluded the common causal factor was "multiple fire-fountain eruptions."

Back in Kansas, the *Saint Francis Herald* could perhaps be forgiven for singling out an appraisal of Ron's orbital observations by Farouk El-Baz. In a postflight interview, Farouk said, "I would give him an A plus. Honest to God, his performance was superior. During the mission I had a TV camera where I would feed questions to the Capcom. More than a dozen times Ron was giving the answer before he even got the question—that's how logical his thinking was."

Apollo 17's lead flight director, Gerry Griffin, has a similar appraisal of Ron's scientific work in lunar orbit. "He did it in a superb and calm fashion. He was very efficient. I've looked, from time to time, for different reasons, at the transcripts of the air-to-ground conversations, and he anticipated things. Of course, he spent a lot of time in mission control, so he almost knew before mission control asked for something, he knew it was coming. He was efficient, calm, unflappable, an easy guy to deal with. He was special."

Like Edward Wilson's Antarctic plant fossils, the orange glass beads from *Apollo 17* and the similar green glass beads from *Apollo 15* lay in storage for decades before their true value to science could be measured.

If the objective of individual Apollo missions was to place the surface crew's observations in the wider context of the orbital observations, then the bigger picture was understanding the actual formation of the moon. Even after six landings, it was still frustratingly far from clear how our blue planet had acquired its single, and singular, satellite. Before the Apollo Program, astronomers and planetary scientists argued about whether the moon was Earth's breakaway "daughter," or co-formed "sister," or captured "spouse." After scientists spent twelve years analyzing the Apollo record, a conference in Hawaii in October 1984, sponsored by the American Astronomical Society's Division of Planetary Sciences, produced a remarkable degree of consensus, and a new model emerged to dominate the debate—the Giant Impact Hypothesis.

Today, most people with a passing interest in astronomy and space will have seen dramatic animations showing a Mars-size object sideswiping the recently formed Earth. As portrayed, this cataclysm blasted a huge mass of incandescent debris into orbit, where gravity drew it together into a body we now call the moon. The impactor has even been given a name, Theia.

One of the most compelling arguments supporting the hypothesis was summed up by geologist Don Wilhelms in 1993: "The enormous energy of the collision vaporized and ejected part of Earth's mantle.... The heat of the collision drove off the water and volatile elements that are conspicuous by their absence or rarity on the moon." All researchers who had examined the Apollo samples agreed that the moon was bone-dry, essentially waterless. Even non-scientists could readily understand that a moon from which all moisture had boiled away tied in very neatly with the image of hellish temperatures generated by the giant impact.

There the matter rested for a quarter of a century. Just to underline the point, in 2006 a prestigious scientific study in the United States concluded that the bulk water content of the moon was "less than one part per *billion*."

The hypothesis did have a few detractors. A certain Dr. Harrison Schmitt admitted in 2003 to being "one of a small minority" who considered the Giant Impact Hypothesis to be "highly unlikely." But what would *he* know about the moon?

In July 2008 a team led by Dr. Alberto Saal of Brown University published a paper reporting on a fresh analysis of the volcanic glass beads from *Apollo 15* and *Apollo 17*. Using techniques that were unavailable in the 1970s and 1980s, Saal's team detected traces of primordial, indigenous lunar water trapped inside the tiny beads. One sample contained forty-six parts per million of water. Allowing for evaporation into the lunar vacuum before the molten glass hardened, the scientists deduced that the original water content of the lunar glass beads was not much less than that of the magma in Earth's mantle today.

Another team from Brown University then carried out an even more detailed analysis of some of the orange glass beads from Shorty Crater and tested tiny fragments of olivine trapped inside individual beads. The glass coating had protected the olivine from the lunar vacuum during the fire fountain eruption. This time the eons-old water samples were comparable with Earth's magmas.

Dr. Erik Hauri, who led the second study, summed up the problem for the prevailing hypothesis: "It's hard to imagine a scenario in which a giant impact completely melts the moon and at the same time allows it to hold onto its water. That's a really, really difficult knot to untie."

Since 2011, the fortunes of the Giant Impact Hypothesis have ebbed and flowed with a series of confusing and sometimes-contradictory scientific studies that mostly center on comparing the chemical signatures of oxygen and other elements found on Earth and in the Apollo samples. One team of researchers has concluded that the giant impact must actually have been a head-on collision resulting in a thorough mixing of Theia and Earth before a mass of debris was ejected. But that doesn't explain the moon's water content either.

More recent studies have suggested that most of the water content of both the earth and the moon arrived in asteroids from the outer solar system. Perhaps not one but multiple impacts occurred, and the resulting moon accumulated from sizable chunks of debris that had not been vaporized by the impacts and therefore had retained their original water. Whatever the true answer may be, we now know that the moon has substantial internal reserves of water. Indeed, the accumulations of water ice now believed to exist in permanently shaded polar craters may have migrated to the craters' interiors from deep within the moon. In October 2020 NASA announced that an infrared telescope on board a high-flying aerial observatory had detected the specific wavelength unique to water, widely distributed across sunlit regions of the moon. Unprotected, the water would evaporate but not if trapped, for instance, inside volcanic glass beads. Future explorers will have no shortage of water to slake their thirst or to release oxygen and hydrogen for rocket fuel.

A scientist today would be foolish to claim to *know* how the moon was formed. The argument continues to rage politely. Those who were convinced they had found an elegant solution did not expect tiny volcanic glass beads in storage for decades in Houston to fuel a challenge to their scientific orthodoxy. The work of the Apollo astronauts was not "of its day." It was not a mere historic curiosity to be read about in dusty books. When Ron Evans and his companions detected widespread evidence of ancient fire fountains around the rim of the Sea of Serenity, thus inferring substantial supplies of water within the moon, they simply proved that scientific knowledge advances best by challenging prevailing hypotheses. Not a bad few days' work by two flyboys and a rock hound.

23. Spaceman

The Moon on the one hand, the dawn on the other;
The Moon is my sister, the dawn is my brother;
The Moon on my left and the dawn on my right,
My brother, good morning; my sister, good night.

—Hilaire Belloc (1870–1953)

A week or so before Ron flew to Florida to enter preflight quarantine, Jan asked him a question that had been gnawing at her: "What if they can't get off the moon?" With that particular genie out of the bottle, Jan wanted to know if Ron could do anything to help Gene and Jack, or would he have to come home on his own. She remembers how her husband "kind of cocked his head a little bit" and looked at her before replying, "That isn't something I plan on or think about."

Jan doesn't know if the three crewmembers themselves ever had a cozy chat about it before the flight. Probably not. After the mission, Gene Cernan always said the issue never even crossed his mind "until the time came to leave." Schmitt, when asked, always stressed how many redundancies were built into the lunar module and how many different ways there were to fire the engine. But when pressed, all Cernan could say was, "We didn't have any special plans to stay there, quite frankly, and neither of us planned to be martyrs. I'm not sure what we've have done. I don't know the answer to that question."

Ron certainly told his family that if the landing crew managed to coax a misbehaving LM ascent stage into any kind of an orbit, he would be ready to swoop down and pick them up. It wasn't quite that simple, but the versatile command and service module's capabilities gave Ron multiple options to carry out an orbital—but *only* an orbital—rescue. In a 1983 radio interview, armed with memories of a successful mission, he felt able to say, "If they could not have gotten off the moon, I would have been forced to make a decision: 'Sayonara, guys, I'm heading back to the earth.' Basically, that's what it boiled down to."

On *Apollo 11*, with Ron Evans sitting at the capcom desk in Houston, Michael Collins in lunar orbit had been "like a nervous bride." Even after seventeen years of aviation and spaceflight, he later wrote of the experience: "I have never sweated out any flight like I am sweating out the LM now." His secret terror since his selection for *Apollo 11* was possibly being forced to leave his companions—either dead or doomed—on the moon and to return alone to Earth. If that had happened, he knew what it would mean. "I am coming home, forthwith, but I will be a marked man for life, and I know it. Almost better not to have the option I enjoy."

On *Apollo 15*, Al Worden trained extensively to mount a rescue if his companions had an engine misfire and ended up in "some pretty crazy orbit," perhaps grazing the mountaintops. "The idea was, as long as they were alive, and as long as there was a possibility that I could reach them, then I'd go after them. If we got to a point where there was no way I could reach them, or get them, then I would have to go home by myself." Inevitably, that would have been "a very, very tough decision." Perhaps NASA would have tried to ease the burden on a devastated command module pilot by asking his commander in chief, the president, to issue the recall order. Worden wasn't convinced it would have gone that high up the chain of command, but *someone* in authority would have said, "Okay, Al, you've gotta come home. There's nothing more you can do."

As for flying the spacecraft home alone and carrying out the reentry and splashdown procedures, he was satisfied that the CMP could do all the essential tasks solo. And, of course, he would have had the full resources of mission control to guide his somber journey.

Worden could even explain in convincing detail how he might have mounted a solo spacewalk to recover the SIM bay's film cassettes. While plausible, perhaps as an exercise in exploring the limits of the possible, it is very hard to imagine NASA's allowing the sole survivor of an Apollo disaster to take *any* unnecessary risks, even for science. That is where the operationally possible meets practical reality. Michael Collins wrote that coming home alone would have made him "a marked man for life." In the cases of Worden, Ken Mattingly, or Ron Evans, NASA would not have allowed a situation to occur where it later might be said: "He couldn't save his friends, but at least he saved his pictures."

On the evening of 14 December 1972, it was the turn of Ron Evans to play the "nervous bride," although he most probably would not have thought of himself in such terms.

On the moon, Gene Cernan and Jack Schmitt had completed three days of exploration. They had planted a sixth U.S. flag in the valley of Taurus-Littrow and had bagged Apollo records for the longest distance covered and the largest haul of rocks, including the intriguing orange soil. For Cernan, this mission would be a triumphant conclusion to his astronaut career and a vindication of his decision to hold out for his own command. For Schmitt, becoming the first geologist to make a field trip to another celestial body was a greater achievement than he could possibly have imagined when he followed in his geologist father's footsteps. But if they hadn't thought much about it before, they must have realized as liftoff approached that everything depended on one little rocket engine. Not that they were concerned. Just aware.

It was time to go. President Nixon would later congratulate the crew and—to Schmitt's intense displeasure—refer to theirs as the last manned flight to the moon in the twentieth century. As liftoff approached, the two astronauts would never have guessed that they were about to cut humanity's link to the moon for at least two generations.

Schmitt said, "Proceeded. Three, two, one, ignition."

Cernan reported, "We're on our way, Houston!"

On live TV, the LM's ascent stage darted upward through a shower of insulation debris blasted out in all directions. In Houston, a NASA wizard named Ed Fendell worked his electronic magic. The TV camera on the parked lunar rover tilted under his remote control and broadcast the rapidly rising spacecraft until it disappeared into the blackness.

The longer the ascent engine burned the more rescue scenarios Ron could strike off his mental checklist. Finally, the engine shut down automatically, and *Challenger* settled into a near-perfect elliptical orbit with its low point a safe ten miles above the moon.

On his fifty-second orbit, Ron was nearing the end of his time as the captain and crew of spaceship *America*. Gene Cernan had gradually nudged *Challenger* toward a rendezvous some seventy-one miles above the landing site.

As the vehicles converged, Schmitt took a series of pictures, one of which shows Captain America's ship floating gracefully across the lunar horizon.

With the moon below and the heavens above, the photograph magnificently captures the spirit of adventure embodied in Apollo. But it was not a time for philosophical musing; the temporarily divided team was looking forward to a reunion.

"Good to see you," Gene enthused.

"Good to have you all back up here," replied Ron.

"It's been a good trip!"

"That *Challenger*'s a beautiful vehicle!"

"You betcha."

The linkup itself was not without incident. The docking procedure with the complete LM on the way to the moon was rather different from the more crucial maneuver in lunar orbit. Attached to the huge S-IVB booster, the two-stage LM had been as solid as a rock. Now, minus the descent stage and most of its own fuel, the little ascent stage was almost skittish; Neil Armstrong had called it "a very light, dancing vehicle." The reaction control system (RCS) thrusters mounted on the spacecraft had been designed to control the fully fueled two-stage LM. Al Worden thought they were "way oversized" for the depleted vehicle. He explained, "It was very difficult for the commander to know all the motions in the ascent stage. Very difficult! Every time he touched an attitude-controller he got too big a push in that direction. Then he had to counter it the other way, so the ascent stage basically would sit there in a dither, kind of back-and-forth a little bit."

In an effort not to disturb the ascent stage too much on contact, Ron's approach was not quite fast enough. The docking probe cleanly entered the LM's drogue, but its three capture latches failed to engage.

"Didn't get it," Ron muttered. "Okay, might have been a little bit slow."

After backing off, he fired his rearward-facing thrusters to approach a little faster. This time the latches clicked into place. "Okay, that's a good one," he noted, more cheerfully.

Now came phase 2. On *Apollo 11*, when Collins tried for a hard dock, "all hell broke loose." He later wrote: "Instead of a docile little LM, suddenly I find myself attached to a wildly veering critter that seems to be trying to escape." The problems that Armstrong and Collins encountered—and resolved— helped train future crews in what to do and what not to do.

Ron Evans now had a passive ascent stage hooked on his docking probe, but it was gently swinging back and forth. There was no rush to dock, but the oscillations seemed to take an age to dampen. Finally, after nearly six minutes, everything was calm. Ron retracted the probe. The crew heard the reassuring sound of the main docking latches snapping into place.

"Houston, we're hard-docked," reported Commander Cernan. *America* and *Challenger* were once more united as *Apollo 17*.

At their home in El Lago, Jan and the children were still hanging on every word from mission control as they waited to hear the crew emerge from behind the moon for the last time. Jan still trusted NASA. Recalling how she was feeling at that crucial moment, she says, "I just can't use the word 'apprehension.'" Then she heard a familiar voice.

"Houston, do you read *America*?" asked Gene Cernan.

"That's affirmative, *America*."

"Roger, Houston," reported Cernan, the navy man. "*America* has found some fair winds and following seas, and we're on our way home." It was the eighth time an Apollo crew had entrusted their lives to the big service propulsion system's engine at the back end of the spacecraft. As always, the SPS did its job.

Earth was rising above the lunar horizon for the last time, at least in the sight of twentieth-century man. As *America* began to climb away from the moon, the crew transmitted spectacular live TV pictures featuring parts of the lunar far side, including the huge crater Tsiolkovsky. The TV camera would prove useful again, but as far as NASA and the moon were concerned, this was the last picture show. It was a good one. By the time it ended, the spacecraft had climbed so far that even Jack Schmitt was having difficulty spotting the Taurus-Littrow Valley where he had so recently left footprints in the dust.

Ron took the opportunity to declare for the record that he was "honored and proud to have been a part of this Apollo Program." He was already using the past tense to describe Apollo, but he ended on an optimistic note: "We will continue to explore, and I hope that someday we may all have the opportunity to see mankind enjoy the benefits of the exploration of the Apollo Program."

On any moon landing mission, the crew inevitably felt a sense of anticlimax on the return journey. Jack Schmitt had little more to do until the rock and soil samples were safely in the Lunar Receiving Laboratory. Gene Cernan was nominally in charge on the return, but Ron Evans was navigating and had to conduct regular realignments of the guidance platform. He also had to maintain a backup capability for navigating safely back to Earth and through the atmosphere in the unlikely event that communications with Houston were permanently lost. Broadly speaking, Cernan and Schmitt had done what they had trained for and accomplished their part of the mission. Ron's work would only be finished when *America* was bobbing safely in the Pacific.

But first came the deep-space EVA, Ron's moment of glory in retrieving the film cassettes from the SIM bay. His would be only the ninth "spacewalk" by an American astronaut.

Before the mission, Ron was concerned—hardly helped by his experiences in the KC-135—that he might turn out to be one of those crackerjack pilots who just couldn't cope with weightlessness. Fortunately, that worry was behind him, but another medical issue would threaten his EVA preparations. Early in the mission all three men had experienced some degree of "stomach awareness," but Cernan continued to suffer from stomach and intestinal gas. This condition was attributed to hydrogen bubbles in their water supply, and the treatment was simethicone, an anti-flatulence medication.

Fifteen hours before he was due to step outside the spacecraft, Ron made a rather oblique request over the open radio link. "Okay, Gordo, as a result of 'number five' today and a little bit of a feeling of a little bit of gas right now, with the possibility of a 'desire,' I feel that it may be worthwhile for me to take a Lomotil, and I'd like to get your concurrence on that."

NASA's doctors did not concur. Simethicone was the treatment for gas. Lomotil was the treatment for, as Houston delicately put it, "slowing down intestinal activity." Houston wasn't recommending a treatment for a condition that hadn't developed, but Ron had been trying to tell them without getting too graphic that the condition had *already* developed. Houston offered, and Ron accepted, a discussion over a private communication channel with the flight surgeon, Dr. John Zieglschmid. NASA later provided a summary of the discussion, reporting that the doctors recommended Ron take two Lomotil tablets before going to sleep and another two after breakfast.

The crew on any Apollo mission had very little privacy. With NASA and the media always watching, the experience was akin to spending two weeks in a big goldfish bowl. Ron's private consultation was about a matter nobody would want aired in public, but he was going to be inside a pressurized spacesuit for a couple of hours, and there are certain things an astronaut just didn't want to happen in a spacesuit. Fortunately, the Lomotil did its job, and any concerns were averted.

During the sleep period before Ron's big day, he and Jack had an interesting conversation. It was reminiscent of two boys in a boarding school talking in the dormitory after lights out.

Unlike Al Worden, Ron did not seem to have an "off switch" to banish all thoughts of smoking during the flight. As picked up by the command module's voice recorder, Ron joked about lighting up a cigarette that night. He admitted that he would just have to get through the invading thoughts by himself.

Jack asked him, "Well, are you going to start smoking again or are you going to . . ."

"No, I don't think so, Jack," Ron replied. "I'm going to quit."

"You going to quit? It'll take a lot of courage and fortitude."

"Yes!"

"People around you smoking theirs . . ." Jack was clearly thinking about Ron's having to walk downwind of other keen smokers in the astronaut office.

Their chat was beginning to resemble a confession. "I was up to—oh, two packs a day," Ron revealed. "I—I never went beyond two packs a day, but every once in a while, I hit two packs. It got so it was very seldom just one."

"That's a lot of cigarettes!" Jack exclaimed.

Later, Jack wondered how Ron was feeling, no doubt with the spacewalk in mind. "I feel great, I really do," came his reply. A few hours later, Ron had a cheery greeting for mission control: "Good morning, Houston. This is the command module pilot of the spaceship *America*. And we're up and ready to participate in another day's activities."

A little later, Gene Cernan happily reported to Houston that his crew was "capable of carrying out everything that is required today." He specifically mentioned that "the health of the crew is excellent." In Houston, Dr. Zieglschmid didn't need a translation.

Houston had some news that was good for Ron and even better for his family. His spacewalk would receive the same treatment as the lunar surface explorations. Channel 8 TV in Houston would show the whole event live.

"I think that's outstanding. Thank you!" came the enthusiastic reply.

It was no ordinary Sunday afternoon in El Lago. Family, friends, and neighbors crammed into 1310 Woodland Drive to share Ron's big day. Just to be sure nobody missed the live coverage, one of the NASA contractors had installed several big color TV sets around the house. They had done it for the Cernans as well. For an event like this one, NASA always arranged for "two or three of the guys" to be present in the house to explain things. Jan remembers Jack Swigert and Jerry Carr sitting in to answer questions, and many other astronauts popped in and out as their duties permitted.

The media had congregated in the cul-de-sac, but it was a reduced presence compared with the three-ring circus outside the Armstrongs' house forty-one months earlier. Jan couldn't help but be disappointed by the flagging media and public interest in what her husband and his colleagues were doing. She knew the next year would bring great missions to the *Skylab* space station and that a space shuttle was on the horizon, but the response wouldn't be the same. Her nation's big goal had been sending men like Ron to the moon—to the moon!—and now the program was coming to an end. But not before her husband had his hour in the sun.

Jan had established a good relationship with the press. From the outset, she had told the reporters she knew they had a job to do, and she was willing to work with them if they did right by her and her family. They couldn't have been nicer. For most of the mission, they were content to stay outside, but for the spacewalk (and later the splashdown), they asked to film inside the house. The reporters had gotten to know young Jon Evans and asked Jan if they could put a microphone on him to record his reaction to his father's spacewalk. Jan told them it was okay with her if it was okay with Jon, and it was. As his father prepared to step out into the vacuum of space, Jon sat on the floor, wired for sound and staring at the TV screen.

Ron probably wasn't actually thinking, "Now I get *my* turn in the spotlight!" Jan knew her husband just didn't see things that way. But he was certainly looking forward to doing his job and enjoying the experience. He had trained for it repeatedly but knew that nothing would match the real thing.

The first step was to "get the dirty old suits back on again." His had once been clean and white, but after being stored with two suits stained gray with moondust, it was smeared by a little of that most exclusive of contaminants. During the spacewalk, Ron would wear Gene Cernan's lunar visor assembly, complete with red stripe, and the top part of Cernan's lunar backpack with the "oxygen purge system," which contained a thirty-minute emergency supply of oxygen. His primary oxygen source was through a long, flexible tube that also doubled as a tether.

Donning the suits was a major undertaking for three men in a small, cramped spacecraft, even in zero-g, but at least closing the airtight zip fasteners was easier with someone to help. When all three were satisfied their suits would maintain pressure against the vacuum of space, Ron rotated a pressure valve on the hatch to release the cabin's oxygen supply. He kept his hand on the valve, just in case someone's suit actually did leak, and a close eye on his own wrist-mounted pressure gauge in case that suit was his own.

Perhaps surprisingly, the three men who conducted the deep-space EVAs had exchanged little information. Ken Mattingly has no specific recollections of discussing his "especially memorable" experience with Ron in advance of *Apollo 17*. He assumes they did talk about it, possibly over a beer or a coffee, but he did not give Ron a formal briefing on the conduct of the spacewalk. Of course, Ron's mission commander had conducted only the second American EVA, a perilous episode Cernan described as "the spacewalk from hell." It had been man's first attempt to do proper work in space, but the training program and the spacesuit had proved woefully inadequate. Many lessons were learned from his experience, but on *Apollo 17*, Cernan was particularly attentive as his old friend Ron prepared to step into space.

"Okay, babe," he cautioned, "when you get out there, just take it nice and slow and easy. You got all day long." As Ron began to open the hatch, Cernan added, "Just don't let your body start moving too fast down there."

Jack was more upbeat. "Nice day for an EVA, Ron. Go out and have a good time."

Before leaving the spacecraft, Ron pushed "a humungous trash bag" out into space, an unscheduled housekeeping action that didn't appear in the flight plan. He then set up the TV camera and the movie camera on a boom to record the spacewalk. On *Apollo 15*, the movie camera had jammed. Although NASA released several good *Apollo 16* movie frames of Mattingly recovering his film cassettes, there were no proper photographs of a deep-space EVA. Ron and Jack conspired to remedy that situation. The main obstacle they had to overcome was the attitude of Deke Slayton.

"Deke was never a real believer in EVA, and he wanted everybody to know that!" Ron later explained. He himself acknowledged that an EVA was "inherently dangerous" and added, "It was awful tough to get Deke to 'okay' anything other than the specific mission that had to be accomplished. Any deviation, to extend it or to do something kind of 'non-nominal' was not allowed."

Slayton had stopped Al Worden from taking a camera out on his deep-space EVA, believing the task was complex enough without his astronaut's having to fiddle around with a camera. But if it was verboten for Ron to carry a camera, why not let Jack, who was simply leaning out of the hatch in a supporting role?

Schmitt recalls, "That was a very interesting experience for me, popping in and out of the hatch and taking the cassettes from Ron as he brought them back from the SIM bay. I had worried that on previous missions there had been no pictures taken on those EVAS . . . so I sort of 'bootlegged' one of our lunar surface cameras into the command module. We brought it up with the samples, and I had one of the support crewmen make sure there was a bungee cord on board that I could attach to the camera to keep it from floating out into space. And that's why we have pictures of Ron. So Ron gets to be known as 'the EVA guy' on the return to Earth because we had that camera! That was not an official NASA plan. That was a Ron Evans and Jack Schmitt plan!"

It was time for Ron to step out into the void. *Apollo 17* was some 183,000 miles from home, flying at the relatively modest speed of just over 2,200 miles per hour. He could see the crescent earth ahead of him, and off to one side was the blinding sphere of the sun, provoking the cry, "Man, that sun is bright! Whooo!"

This exclamation, in turn, provoked Cernan to remind Ron to pull down the reflective gold visor.

Unhooking a foot from some hidden obstruction within the cabin, Ron floated out of the hatch, asking, "Forward? Go now? Am I clear?"

"You're clear, babe. Go!" said the commander.

Ron was flying free. He announced it with his now-familiar exclamation, "Hot diggity dog!"

To most viewers of live television and to those watching later on news broadcasts, his expression came as a surprise. Astronauts didn't usually say such things! But it was just the start. Now gripping the nearest metal hand-hold, this most-undemonstrative man proceeded to wave at the TV camera; then he moved with apparent ease along the metal rail.

None of this behavior surprised the people who knew Ron best and certainly not Jan, who was sitting in front of her TV. "Well, he said that a lot! He said it at home and around the kids. That was just one of Ron's 'things.'" Jan was elated as she watched her husband make the spacewalk look easy, but one thing puzzled her. He was floating along sideways and waving downward.

She had somehow expected him to be in an upright position. "I think that I was just surprised that he floated out on his tummy!"

With his gold visor down, apart from the sun, only two objects were visible to Ron in the blackness beyond the spacecraft. "I can see the moon back behind me. Beautiful! The moon is down there to the right—full moon—and off to the left, just outside the hatch down here, is a crescent Earth." But the refraction of sunlight in Earth's atmosphere meant that the skinny sliver of a planet didn't resemble the crescent moon that is so familiar to earthbound observers. "It's got kind of like horns, and the horns go all the way round, and it makes almost three-quarters of a circle."

Now humming happily to himself, Ron deftly turned his body using his wrist muscles until he was facing in the opposite direction. The moon was now to his left, Earth to the right. He was hovering over the SIM bay, gradually inserting his feet into a set of gold-colored foot restraints. Now able to work with both hands, he waved again and laughed with obvious delight. "Hey, this is great! Talk about being a spaceman! This is it!"

Ron had already earned his gold astronaut pin on this mission, but there was more to being an astronaut than floating inside a spacecraft. As he described it years later, his experience sounded like a rebirth. "You're not really a 'spaceman' when you're in the confines of your spaceship. Now you make that transition, you go outside your spaceship, you're hanging on. Whenever you get out there, from the safety and security of your mother ship—if you ever want to be a spaceman, that's the way to do it! Because you're going out there and the suit's the only thing between you and the vacuum of space." For the benefit of any amateur Freudians, the usual NASA term for the long hose supplying oxygen from the mother ship to the spacewalking astronaut is the "umbilical."

But the spaceman knew he had a job to do. Audibly telling himself, "Okay, back to work!" he removed the insulation around the Lunar Sounder Experiment, hooked a safety tether to the vital data cassette, and drew it free before floating hand over hand back to the open hatch. Schmitt took possession of the cassette, then brought it back into the cabin, where Cernan packed it safely away.

Next, Ron had to retrieve the big panoramic camera cassette. Laughing joyfully, he maneuvered himself back over the SIM bay. "Oh, this is great, I tell you!" He stole another glance at the full moon; then he looked to the right, beyond the far edge of the service module. "I can see the—the engine bell sitting back there. That's a pretty good-sized thing too!" His handholds didn't extend far enough to allow him to make an inspection of the SPS engine. Furthermore, he knew there might be dangerous rough edges where the service module had separated from the LM adaptor panels.

Having removed the protective covers from around the panoramic camera, Ron extracted the bulky cassette. Despite its being weightless, the cassette still retained its original mass. "And out she comes! Nice and easy. This is a heavy son of a gun! Not 'heavy' up here, it just has a lot of—a lot of momentum to it. Once she starts pulling in one direction, it just takes a lot of force to stop it."

Safely tethered to the cassette, Ron worked his way back to the hatch. "Ahhh, there it is! Delivered it right to you," he laughed, transferring the burden to Jack Schmitt. For his third retrieval, Ron had developed sufficient confidence to maneuver backward from the hatch. "That's an unorthodox way to enter the SIM bay, but it works!" he reported, slotting his feet into the foot restraints.

He was just hitting his stride. Soon TV viewers were treated to a succession of his chuckling, whistling, and "dum-de-dum-de-dums." Space fans had heard nothing like this since Pete Conrad had cackled and giggled his way through the *Apollo 12* moonwalks.

Resting for a moment before tackling the mapping camera's cassette, Ron waved again and sent some interplanetary greetings back to Earth. "Hello, Mom! Hi, Jon! How are you doing? Hi, Jaime!"

Back home, the TV reporters were hoping for some reaction from Jon Evans, but he never said a word. "He was dumbstruck!" Jan recalls. "Because his daddy said hi to him and waved his big old gloved hand to him."

Jaime might not have been the world's biggest space fan, but that was *her* father, live on TV, floating in space. "Well, I think it was kind of neat!" she laughs. "I'm not sure if I realized how real it was, that he was out *doing* that. . . . I don't think I could really picture it, but I do know that when he said, 'Hi, Jaime,' that was, like, '*Wow!* That's pretty cool!' So I felt kind of special then, I think."

But did Ron overlook anyone? He didn't actually say Jan's name. She is quick to explain that she was always "Mom" at home, and Ron always called her "Mom" in the presence of the children, so there were no grounds for a post-mission scolding. And perhaps Ron was covering all his bases by including his own mother in the greeting.

Before retrieving the final cassette, Ron looked back at his assistant. "There's

Jack!" he laughed. "Hey, how are you doing? You're looking right into the sun, and I'm looking right at you. I should have a camera, and I could take your picture. And there—see the moon back over there?"

It is tempting to think Jack Schmitt had a cushy job, just standing there in the open hatch and staring at infinity, but it wasn't the sky that drew his attention the most. "Well, what was most impressive was Ron and his excitement, his interest, his 'wave to mom,' and his getting those canisters out of the SIM bay. I was focused primarily on supporting him. My job, my primary job, was to make sure he didn't get tangled up, or his hose did not get tangled up with the hatch, or anything else. Ron was having the great view! He could look either way and see both the earth and the moon, whereas I had to really strain to see them both, because the hatch was open and blocking part of my view."

Having extracted the mapping camera cassette, Ron made his way back along the railings, hand over hand, maneuvering his whole body with deft wrist movements. He may have caused a flutter of concern when he exclaimed, "Whoops, come back here, little cassette!" But, of course, the prize was securely attached to his suit with a tether, whose whole purpose was to leave his hands free.

Two things emerged clearly from the welter of whistling, humming, and laughter: Spaceman Evans made the spacewalk look easy, and he made it sound like *fun*. Here was a man who was enjoying his work and having a ball. He himself later described the experience as "delightful." Jan could see he was enjoying every second of it, and if his sound effects reminded anyone of Pete Conrad, well, they were both naval aviators!

With the three cassettes safely on board *America*, Ron's tasks were officially complete, but he was keen to inspect the condition of his spacecraft. Although Project Apollo was ending, four more missions would fly Apollo hardware in Earth orbit, and some of those would last months. He might find useful lessons to be learned. As he waited for Schmitt to retreat into the cabin, leaving the hatch clear, Ron examined areas of damage where the RCS thruster firings had bubbled the silver paint. He wished aloud that the handrails extended around to the far side of the spacecraft to allow a more complete examination, but the capcom was sticking rigidly to the flight plan. "Okay, Ron, we don't need any more spacecraft commentary. We'd like you to go ahead and terminate the EVA. Everybody's real pleased, and we'd just like to go ahead and terminate." One can easily imagine Deke Slayton miming a finger across his throat to the capcom.

The spaceman had finished everything he was supposed to do. By his own admission, he was now "just diddling out there," but he was in his element and would happily have stayed out longer. "I wanted to do more! I wanted to go around the other side of the spacecraft, although it was black over there, I tell you! . . . Really, what I was trying to do was make a visual inspection of what was out there. See, nobody else had done that. I'm up here, I'm an engineer, and I'm out doing my thing! It seemed like Houston ought to have wanted me to go out and report . . . if I see a meteorite hole, or anything. So what did I get a look at? I got a look at the urine-dump [valve]. 'Yeah! There's some yellow ice, a little bit around the corner, here!' Houston comes back: 'Yeah, big deal, Ron! Get back in there!'"

Orders are orders. But as he prepared to float in through the hatch, Ron spotted something. He was already experiencing the warm glow of satisfaction that comes with a job well done, but here was a bonus. Undamaged by the RCS thruster firings, the Stars and Stripes and the words "United States" still brightly adorned the side of spaceship *America*. That made him feel proud.

After briefly showing TV views of both the moon and the crescent earth, Captain America climbed back through the hatch to face the hassle of three men in bloated, air-filled suits trying to wrestle the writhing coils of the umbilical. The EVA had lasted one hour and seven minutes—exactly per the flight plan—and Ron had been outside the spacecraft for forty-eight minutes. He would fondly remember that time for the rest of his life.

Afterward, a question lingered in the minds of some people. Was that joking, laughing, whistling, humming spacewalker really the same guy Gene Kranz had called "the ultra-quiet Ron Evans"? But nothing that Ron had said or done (or even "dum-de-dummed") during the spacewalk particularly surprised his family. His neighbor Ron Ammons wasn't in the least surprised that Ron got the job done, on schedule, and clearly had enormous fun doing it.

The two men who flew to the moon with Ron were not surprised, either, having spent well over a year training with him and experiencing his sound effects. Schmitt had seen how diligently Ron had applied himself to every aspect of the mission, and when he was in charge of a particular part of it, he took the responsibility very much to heart. The man who had once expected to fly to the moon with Ron, fellow Kansan Joe Engle, was not at all surprised by Ron's joyous spacewalk. As he would later confirm for himself, "Spaceflight is exciting; it produces exhilaration."

Even though Al Worden saw Ron as one of the quieter astronauts, he knew him well enough not to be surprised by his expression "hot diggity dog." That was "a typical kind of Ron Evans thing to say." He thought Ron was clearly delighted to get his chance to be "a real spaceman." Al had been happy during his experience too. Going outside the spacecraft was "a wonderful thing to do," although as the first to conduct the procedure, he had been very preoccupied with the mechanics of going out and recovering his two film cassettes. "I was probably overtrained," Al suggested, "because I did the whole thing in about thirty-eight minutes, which was a lot quicker than I thought it would be." Having demonstrated the techniques, Al felt that Ken Mattingly and Ron Evans could relax a little. "So I think Ron's mind was clear on the details of what had to be done, and I think he could enjoy it. And Ron was the kind of guy who *would* enjoy that!"

Farouk El-Baz had seen all the personality types the astronaut office had to offer. Between *Apollo 14* and *Apollo 17*, he saw an interesting role reversal, with Ron Evans now prime crew and Stu Roosa his backup. Farouk knew that Stu was the type who would say, "Man, I'm going to do this job better than anyone else has done it!" Ron never did that. He was measured in his evaluation of himself. His moment of self-doubt shortly before the mission was fairly typical of gifted people who occasionally wonder if they really can live up to other people's high expectations. But when Ron was actually in space and gazing at the moon, any lingering doubts dissipated, and he demonstrated a whole new confidence. That was how Farouk saw it. He heard it in Ron's voice and judged it by the descriptions that came from the command module. He felt that whether Ron was aware of it or not, he was giving himself a "stamp of approval." In Farouk's assessment, Ron was one very happy and contented astronaut when he realized, "Man, it's all in my hands, and I will do it very well."

As lead flight director, Gerry Griffin was responsible for keeping the show on the road, and on *Apollo 17*, the "road" was free of potholes, and the "show" was a real humdinger. When it came to the spacewalk, Ron's obvious happiness wasn't much different from what Gerry had expected. He recalls, "You know, I think one thing that happened to astronauts who did the 'big stuff'— that EVA, and like the guys on the moon—I think once they got in the suit and got comfortable, particularly now they were living on a different kind of life-support system, they got comfortable with it, and they had fun. It was fun for them! And I think they liked the freedom and the view: the view of the earth, the view of the darkness. They all talked about how dark the sky was, you couldn't see anything, it was as black as black could be. So I think it was a kind of euphoric setting for all of them, and I suspect that brought Ron out a little bit. He was now outside the spacecraft with this task to do, and I think he probably figured out, 'Hey, I'm going to be able to do this, and it's not going to be that difficult. And it's going to be fun!' And I think that was basically Ron Evans. He liked the fun stuff, and I think that's why he was humming and whistling: because he was having fun!"

If the opinions of an Egyptian geologist and a Texan engineer seem to coincide on this matter, perhaps that is no coincidence.

After Ron closed and locked the hatch, the crew began to repressurize the capsule. As the oxygen gushed in, Gene Cernan remarked to Houston, "Not a bad performance by my CMP, was it?" Ron just laughed.

America continued its almost imperceptible acceleration toward the distant blue and white crescent.

24. Return to *Ticonderoga*

We shall not cease from exploration
And the end of all our exploring
Will be to arrive where we started
And know the place for the first time.

—T. S. Eliot (1888–1965)

Ron Evans had seen his home transition from the full blue marble to a three-quarters earth, then a half earth, and finally a crescent. He had watched the diminished planet rise above the craggy lunar horizon, then sink behind it. On the homeward journey, the glowing crescent seemed to swap girth for height.

Just after crossing the halfway point, the crew of *Apollo 17* held a live TV press conference. This gave Ron's family the first and only opportunity to see his face during the mission, but the networks didn't carry the transmission live, and they probably only saw a few clips on the evening news. In those fleeting moments, Ron was no longer the disembodied voice from the squawk box. There he was on the screen, bobbing alongside the others and clean-shaven in spite of Jon's hopes of seeing him wearing a beard. The session was inevitably dominated by questions about surface geology in general and the orange soil in particular. As would become apparent, it was too early to write the *Apollo 17* guidebook to Taurus-Littrow, but Jack Schmitt spoke enthusiastically and gave Ron unstinting credit for his orbital observations.

More reflectively, the crew members were asked what each would remember most about the flight. For Ron, the unforgettable Saturn V launch stood out in a mission full of wonders. He admitted that the experience had been more profound than he could grasp in purely engineering terms. He agreed with Gene Cernan that they all had been gifted an obligation to share their experiences with everybody back on Earth.

When asked about their future plans, the ever-practical Evans pointed out that his most immediate short-term goal was "to make a real good reentry." He added, "Beyond that, I also have a strong desire to continue participating in manned spaceflight in whatever capacity that I might be able to."

. For his final comments, Ron thanked the army of workers who had toiled so diligently on the equipment that had made the voyage possible. He would never know whether his morale-boosting pep talks had made a significant difference, but he did know the people who made the machines and the spacesuits, and was not in the least surprised they had done such a great job. Earlier, Gene Cernan had reiterated his belief that *Apollo 17* should be seen not as the end but as the beginning of man's journey into deep space. Acknowledging that a hiatus was coming, he asserted that restricting the quest for knowledge and the desire to explore would be "an abnormal restraint of man's intellect at this point in time." His firm belief was that "the whole Apollo Program is really the true beginning of what's to come in the future."

Gene reserved his final thoughts and prayers for American servicemen still held prisoner in Southeast Asia or missing in action "who may not have the opportunity to get home and enjoy the Christmas that we're looking forward to." He finished by speaking on behalf of the whole crew: "And with that, from *Apollo 17* spacecraft *America* on December 18, 1972, we all wish you a very, very Merry Christmas and a happy holiday season. Godspeed and God bless you all."

The broadcast was very nicely done, but most of it was only seen by NASA personnel and the press corps in mission control.

The saga of the missing scissors rumbled on, with press reports beginning to reveal some degree of disquiet. The heavy surgical scissors could do no harm in free fall, but any loose object in the command module threatened to become a dangerous projectile under seven times normal gravity. When *Apollo 12* splashed down, the sudden deceleration caused a camera left on a bracket to break free and strike Alan Bean a glancing blow to the forehead. A few inches to the left could have been fatal.

During preparations for reentry, Jack Schmitt spotted the stray scissors in the lower equipment bay. He adds, "I was able to alert Cernan to the fact that I had found them, but we kept it from Ron and, of course in a good-natured way, kept warning him that there was going to be a lot of interest in the fact that he supposedly had lost his scissors in space." With the scissors secretly and safely stashed away, Gene and Jack ribbed their companion about all the forms to be completed—in triplicate—arising from the unexplained loss of U.S. government property. Ron himself had checked the vital areas carefully; so even if he didn't know where they were, at least he knew the scissors *weren't* lurking above the three couches in which he and his fellow crew members would ride out the reentry.

Not until weeks later at a delayed splashdown party with the NASA flight controllers in a Houston restaurant did the truth finally emerge. Gene and Jack presented Ron with the scissors in a big ceremony that brought the house down.

Was it a gotcha?

"It was very much a gotcha!" Schmitt confirms with a laugh. It was also the last gotcha of Project Apollo. While opportunistic rather than planned, it was still one of the best.

America plunged toward the thin, bright limb of the enormous horned crescent that dominated the view ahead. Cernan explosively jettisoned the faithful service module containing the SIM bay instruments. "That thing really bangs!" Ron observed, as the now-redundant vehicle took a separate path to its fiery destruction.

The command module continued to accelerate with almost perfect precision at an angle of 6.49 degrees below the horizon until it reached a point that NASA arbitrarily identified as its "entry interface" at four hundred thousand feet (almost seventy-six statute miles) above the earth. For the last time in the twentieth century, humans attained a velocity of over 24,000 miles per hour, although *Apollo 17*, at 24,607 miles per hour, was the slowest spacecraft to return from the moon. *Apollo 10* was 152 miles per hour faster, not that Cernan noticed the difference.

Just before reentry began, Ron had caught a final glimpse of the moon, which was now back to pre-mission normal size. That it had been filling his windows three days earlier almost defied belief.

The spacecraft began to decelerate as the blunt heat shield plowed through the wispy traces of gas at the outer edge of the atmosphere. *America* exchanged velocity for frictional heat, graphically illustrated by the movie footage shot through one of the windows. A pink shimmer gradually turned mauve and then orange with green streamers billowing back along the flight path. Seventeen seconds after the entry interface, the extreme heat of reentry literally tore the atoms of air apart, enveloping *America* in a sheath of plasma that would block radio communications for around three and a half minutes—the so-called blackout.

In El Lago, the Evans family, together with the usual suspects and the press, watched as the last minutes of the mission, and Project Apollo, played out. A tragedy at this final hurdle was unthinkable. Quite apart from the impact on those most immediately involved, it would forever taint the historic achievements of the Apollo Program. Yet the harnessing of forces necessary to reduce the spacecraft's velocity all the way down to a safe splashdown left very little margin for error.

It was probably better not to know the precise details of reentry if there was someone on board the spacecraft you loved. Jaime Evans was reassured by her mother's calm confidence, but she couldn't sit through the radio blackout without some feelings of apprehension. "Everyone's sitting there, waiting, and then when they come through, it's like—'Whew! What a relief!' But I don't remember being '*scared*-scared' or anything."

For Ron himself, reentry was an eerie sensation. "You're being slowed down, decelerated, and you have a lot of g's pushing you back down into the seat, and you can look out the windows behind you, and it looks like you're flying through a tunnel of fire. You can see the black-blue sky out through the center of that tunnel, it's just a big round tunnel of fire—it's absolutely fantastic!"

The crew had been warned to expect a maximum of around 7 g's at peak deceleration. Although they had trained in the centrifuge, they recognized that conducting multiple practice runs in the command module simulator under normal gravity had an element of unreality. They knew what peak g-forces would be like, but none of the simulations could truly prepare them for deceleration sustained over several minutes.

"You've got all the g-forces," Ron explained. "First seven, then it slacks off to four, but all those g-forces coming back onto you after you've been weightless! The adrenaline's pumping at that time, but you really notice the *length* of time you're under those g-forces, as you're decelerating and slowing down.... That's the significant part."

Neither Ron nor the others felt anywhere close to their physical limit, but only by enduring a real reentry could they truly understand what was missing from the simulations.

After emerging safely from the blackout but still experiencing three times their normal weight, the crew monitored the systems as *America* carried out a series of rolling maneuvers. As the pilot, Ron occupied the left seat as he provided Houston with a running commentary, checking each step in the intricate process. "Okay, about 3.1 g's. We're about 4,500 feet a second . . . roll . . . 22 miles . . . 4,000 feet a second . . . zero plus 88 degrees. Okay, that's good." But he was still capable of injecting some humor into the dialogue. "We're stable. Looking good. Coming down like a son-of-a-buck. Man, oh man!" he laughed, apparently enjoying being squashed into his seat.

The heat shield had done its job, charring and bubbling and dissipating the hellish temperatures of reentry. As *America* began a long plunge toward the ocean, the internal temperature was a pleasant 80 degrees.

It was time to arrest the plunge. "There go the drogues!" Ron reported, as two small parachutes were fired from the nose of the spacecraft by pyrotechnic mortars. "Man, oh man! It really vibrates!" The drogues lowered their vertical speed to around 180 miles per hour. Then, 10,500 feet above the ocean and almost five minutes from splashdown, three more mortars fired, leading to what John Young always called "the most beautiful sight in spaceflight": three huge main parachutes bloomed above the command module.

For Jan Evans, the TV picture of her husband's spacecraft descending majestically toward the Pacific Ocean produced understandable feelings of great pride and joy. "It was a gorgeous sight, a proud moment. It's . . . it's what our country had accomplished and what the guys did. . . . To me, that was a beautiful, beautiful sight!" She knew the white parachutes actually had bright orange stripes, but against the color of the sky and the ocean, everything looked red, white, and blue.

For the crew, the climactic twenty-two-mile-per-hour splashdown that concluded their mission came a little earlier than expected. As Ron later explained, "Contact with the water's a pretty good smash! As a matter of fact, the altimeter was wrong! I was calling off the altitude to Gene and Jack. You get down to five hundred feet . . . four hundred feet . . . three hundred feet . . . *Boom!* Hit the water! So I wasn't prepared, and neither were they! You can't really see, as you get close to the water, looking out the side!"

For Jaime Evans, the splashdown was "really neat," but it was "even neater" when she later saw her father emerging safely from the capsule. Even in a distant TV view, there was no mistaking his balding head flanked by the heads of command gray and science black. As a bright thirteen-year-old, Jaime knew it was important that the mission had been successful, but it was far more important to her that she was getting her dad back safe and sound.

By contrast, many TV viewers around the world who were smiling with relief and delight at the triumphal splashdown also felt a palpable sense of loss, now that the greatest adventure of the twentieth century had ended. Rocco Petrone, the Apollo Program director, spoke for many when he said, "*Apollo 17* goes into the record books as the most perfect mission." This writer's 1972 diary records a surge of pride at having "lived through the Age of Apollo."

CM *America* made a perfect splashdown 1.3 miles from the target point. Statistics can be confusing, but after a journey of 240,000 miles, that represents an accuracy of better than 99.999 percent. Ron was not alone among the command module pilots in joking that he was right on target and that it was just as well the recovery ship was off course, or they might have parachuted onto the deck.

Never in the history of the U.S. space program had there been a more suitable recovery ship for a returning crew. As Ron Evans sat bobbing in a life raft in the Pacific, the aircraft carrier steaming in his direction was the very same USS *Ticonderoga* where he had learned of his selection as an astronaut nearly seven years earlier. Jan doesn't remember exactly when Ron found out the ship was picking up the crew, but he had been delighted when he heard. Of course, he realized everybody he had known on *Tico* had moved on.

When the Sea King recovery helicopter carrying the *Apollo 17* crew landed on deck, the band struck up "Anchors Aweigh," and the astronauts stepped out onto a carefully positioned red carpet. Ron had come full circle, and he savored the moment. Watching on TV, Jan had experienced relief seeing those huge parachutes and the thrill of the splashdown. Now she could actually see Ron's smiling face as he rubbed his hands in delight while drawing in lungfuls of fresh sea air spiced with hints of navy kerosene.

Capt. Norman Green introduced the crew and two leading figures from the spaceflight recovery services. Rear Adm. John L. Butts expressed the navy's great pride in the mission's accomplishments. Then Maj. Gen. David Jones (USAF) made a rather unfortunate introduction, stating to the gathered sailors and to everyone watching on TV, "I am proud to present the crew of *Apollo 17*: Captain Gene Cernan, Doctor Jack Schmitt, and Commander Ron Cernan." Pardon?

Jan says she doesn't remember the mistake. Jaime didn't notice it at the time but believes her mother made some comment. It's hard to tell from the TV coverage, but if the astronauts reacted, they gave the merest twitch of surprise.

Gene Cernan confirmed that even after such a wonderful and rewarding journey, it was always good to get back. On a more philosophical note, he added, "It's a fundamental law of nature that either you must grow or you must die. . . . I thank God our country has chosen to grow."

Gene then correctly introduced his friend and first officer, Cdr. Ron Evans, who stepped forward with his trademark grin. He addressed the crowd of mostly young sailors who reminded him very much of the sailors with whom he had once served. "I think it's quite fitting that I should happen to have the opportunity to be picked up by the *Tico* after our journey to the moon. To me, that's really something! The fact that it was the *Ticonderoga*, part of the United States Navy, the fact that I'm a commander in the United States Navy, and I flew the United States spacecraft *America* to the moon and back— . I'm really honored and I'm proud. That's about the best way I can express it."

Ron then introduced Jack Schmitt, who made probably the shortest speech of his life, expressing his thanks to the navy, the greatest team he would never be a part of.

After nearly thirteen days of weightlessness—and without three days to stretch his legs on the moon—Ron realized he was going to have to get used to walking again. "You're supposed to be on the right side of the carpet, walking in a straight line, but you flat can't walk in a straight line. You've got people helping you around. It takes a while. I felt the ship was in heavy seas, and it was straight and level, no wind. But it was rolling for me!" It took about five or six hours before he could walk along a passageway without banging into the walls.

To the disappointment of both Gene and Jack, one of the first things Ron did on *Ticonderoga* was to ask one of the crew for a cigarette. He had gone through the flight without any observable cold turkey symptoms, but he seemed content to pick up where he had left off on launch day.

It probably came as no great surprise to the crew of *Apollo 17* when they received a congratulatory telephone call from the president. .While Nixon had turned up in person when the USS *Hornet* recovered Armstrong's crew, the least he could do now was to phone the men who had safely and successfully brought down the curtain on Apollo. There was a certain irony in that the man who had decided several times to cancel the flight had now lifted the White House phone to congratulate its crew on a "magnificent job."

According to the president's diary for 19 December, he placed a three-minute long-distance conference call to *Ticonderoga* beginning at 4:35 p.m. eastern time. Gene Cernan answered, and Nixon welcomed the crew back to Earth. Another speaker came on the line, thanked the president, and identified himself as Commander Evans. Nixon responded with a noticeably flat "Right," then asked whether the "good doctor" was there. Dr. Schmitt confirmed both his presence and his good health.

Nixon praised the success of *Apollo 17*, which had "capped the other Apollo flights." He acknowledged some sadness on the conclusion of Apollo but added that this "opens the way for new adventures and the exploration of space in the future."

The president then mentioned "the usual drill" of promotions arising from a successful mission. In Schmitt's case, he noted that "you can't promote a doctor to anything." As for Cernan, he had already had two space promotions, "and that's the maximum, right?" That left one crew member. "But, on Ron Evans, who had that lonely task . . . while the rest of you were down on the moon, we are able to promote him, and from now on, instead of commander, it's captain." He then addressed Ron directly and in rather odd terms: "If that's all right, or do you want to not accept it?" It is tempting to wonder if his convoluted syntax alluded to Ron's pre-mission press comments, as if Nixon were asking if Ron really wanted to accept a promotion from an administration that only paid lip service to space exploration. It should be acknowledged that Jack Schmitt recalls nothing untoward about the conversation, but, of course, he didn't remember "the Day of the Kooks" either.

In any event, Ron certainly didn't have to think twice. "Mr. President, I accept it very definitely. Yes sir, thank you very much!"

Nixon then tried a little small talk: "Tell your wife she can buy more for Christmas this way. I think a captain gets a bit more than a commander, as I recall."

"That's great. Yes, sir." What else could Ron say?

Nixon finished with an invitation to the crew to visit the White House. Then he added, "We're all proud of you, and wish you a very merry Christmas and a happy new year." The whole conversation lasted three minutes and twenty seconds.

There followed a round of medical tests, during which Ron was the subject of several sketches (felt tip pen on paper) by NASA artist Chester Jezierski. One shows the newly promoted Captain Evans (unflatteringly looking more like his own father) sitting on a bicycle ergometer with a breathing tube in his mouth and a series of cables wired to his body. The sketch is headed: "Ron Evans does the bicycle thing to measure respiration, heart rate, blood pressure among other things to include the astronaut's humor." It does not mention the heady draught of adrenaline that must have been surging through the veins of the man who had returned safely to Earth after fulfilling his long-held ambition.

25. End of an Era

Success is getting what you want.
Happiness is wanting what you get.

—Dale Carnegie (1888–1955)

Gene Cernan had already seen the adulation that accompanied a successful moonflight. Now Ron Evans and Jack Schmitt, newly scraped clean of the grunge of thirteen days in space, would learn what it meant to be called "hero." The Apollo currency may have devalued in the eyes of some Americans, but plenty of taxpayers, patriots, and admirers were still ready and willing to surround the astronauts and express their undying admiration.

The crewmen were flown from *Ticonderoga* to Pago Pago, capital of American Samoa. Adorned with traditional Polynesian lei, they were photographed on a platform waving to a crowd of enthusiastic islanders and U.S. service personnel. The gathering was rivaled in size only by the visit of Pope Paul VI in 1970.

The Apollo crew then flew via Honolulu to Ellington A FB, Houston, where the warmth of their welcome sharply contrasted with the weather. They had a respite from the rain that had fallen through much of the month, but an icy wind strained the flags and the astronauts, who looked decidedly chilly and underdressed in their NASA jumpsuits and VIP caps.

Ron had made it back for their wedding anniversary the next day, and Jan vividly remembers being reunited with her husband. "It was *cold*! And *windy*! We really weren't expecting it!" A car took the crew from the military transport aircraft to an impromptu platform, and finally Jan, Jaime, and Jon could give Ron a big hug.

Jaime hadn't been able to embrace her father since he had left to go into quarantine. "I remember just that I was so glad to see him, mainly because it seemed like he had been gone so long. But it was kind of tough, because there were all those people there too, and it was hard to break through all the people to get to him and hug him!"

Despite the cold and the cheering crowd of well-wishers, at that moment time seemed to stand still. "Yep! It's, like, 'officially back!' and maybe things will be back to normal," Jaime recalled. "You know, I'm sure I was proud of him, but I think I was mostly just happy to see him!"

With his teeth chattering from the cold and all too aware that nobody wanted to stand through long speeches, Gene Cernan robbed Ron of the line, "There's no place like home." Recalling that President Nixon had just quoted Sir Isaac Newton, saying that *Apollo 17* had seen farther by "standing on the shoulders of giants," Cernan told the crowd, "There are no small people in this program, there are no small people on *Apollo 17*." Adding that Apollo had "taken mankind into a new era," he praised his crew of "Captain America and Doctor Rock," and introduced the onlookers to "the newest captain in the navy."

Ron Evans looked like the cat that got the cream as he addressed the cheering crowd. "You know, they ask you, 'What's the most exciting thing about your flight?' I think we had let you know our enthusiasm while we were up there on the flight, but the best part of it is coming home and being here with the people in Houston, it really is. . . . One serious little thought that I would like to leave, I think, is that we—and by 'we' I don't mean the three of us, or even the people who worked on the spacecraft . . . I mean the United States of America—can be proud that we culminated the beginning of man's exploration with *Apollo 17*. I think that's very important."

But there was time for some humor. Ron bent the facts just a little by reminding everyone how Jack Schmitt had "sworn his allegiance to the U.S. Navy" on the deck of *Ticonderoga*. Introduced by Ron as "Seaman Apprentice Schmitt," Jack played along but concluded the ceremony on a suitable note: "We have evolved into space, and let's keep going there!"

It was time to go home, the last few miles after a very long voyage. Having provided most of the transport on the mission so far, NASA fittingly offered to drive the crew members to their respective front doors. As the car bearing the Evans family entered El Lago, they saw big American flags lining both sides of the street. "That was just a bit overwhelming," Jan admits. They even saw people on horseback carrying flags and local children riding bicycles decorated with red, white, and blue streamers. What seemed like hundreds of well-wishers filled their cul-de-sac, milling around beneath a large billboard that the neighbors had erected. Spelled out in lights, the board's message brought a broad grin of delight to Ron's face: "Hot Diggity Dog! Dum-de-Dum."

After much handshaking and backslapping, the crowd finally dispersed, and Ron could close his own front door behind him. At last he could exchange hugs and kisses with his family without a cheering audience. He was genuinely delighted to be back. Over the previous six years, Ron simply hadn't had time to think much about Christmas, but this year they would have a relaxed family celebration. Thinking about the flight of *Apollo 17*, he made a comparison with a wonderful holiday. While on the flight, he was so totally enthused and absorbed in what he was doing, he didn't want it to end. But once it was time to fire up the service propulsion system, nothing was more important or desirable than getting home. Except maybe the spacewalk . . .

After the successful completion of the mission, the threat from Black September seemed to evaporate. But the world was still a dangerous place, and NASA continued to keep a discrete eye on all of its astronauts and their families. If Jan and Barbara had anything to say about being kept in the dark about the threat, they kept it within the families.

At some point after his return, when he had a quiet moment to himself, Ron retrieved his aviation logbook and made the final entry for December 1972: "7 thru 19: Apollo 17—301.9 hours."

Following his promotion in the field by President Nixon, Ron received paperwork in mid-January from the Department of the Navy formalizing the upgrade. The icing on the cake was a handwritten note on the covering letter: "Congratulations, Ron! Al Shepard." Both men had traveled a long way since their telephone conversation in 1963.

That was only the beginning. A whirl of parties, receptions, and award ceremonies culminated in an *Apollo 17* tour around the United States that covered some sixty centers in twenty-five states. At the Super Bowl VII game in Los Angeles, the fans cheered for the astronauts as they were driven around the field in an open-topped car. Such was the breakneck pace of events that Jan—renowned for her amazing powers of recall—has no memory of the pin party that must have taken place to honor both Ron and Jack.

She does remember traveling with Ron to Washington, where he received his U.S. Navy gold astronaut wings from Secretary of the Navy John Warner (a future five-term U.S. senator and the sixth husband of Elizabeth Taylor). The winged decoration incorporated a version of NASA's gold "shooting star" emblem. Warner also presented a miniature version of the insignia, which Ron proceeded to pin on Jan's lapel. It was a fitting reversal of the roles when Jan pinned gold aviator wings on Ron's uniform in Memphis in 1956 and left her bursting with pride.

The *Apollo 17* crewmen very willingly and gratefully visited the builders of their spacecraft to thank the people whose diligence had allowed them to do their jobs and return safely. One visit was particularly memorable. North American Rockwell, as it was then called, had built the spacecraft *America*, and that made Ron Evans "their guy." They knew Ron from many visits and were familiar with the factory lights reflecting off his balding head. When Ron, Gene, Jack, Jan, and Barbara made their goodwill visit, the function was held in a hangar full of historic aircraft.

There was something different about Ron as he mounted a stepladder to address the assembled workers. He seemed to be sporting a full head of hair. Where light had once glinted, the workers now saw the luxuriant brown waves of a hair piece that Ron's barber had given him just after the mission. Ron offered some ridiculous explanation about how his exposure to space radiation had triggered rapid hair growth, leaving him as trichologically well-endowed as Cernan and Schmitt were. Nobody was buying it, and the expressions on the faces of his fellow astronauts were priceless.

Jan remembers Ron getting "the biggest kick" out of people's reactions. On a later visit to Salina, her parents were treated to the same spectacle, "and I thought my mother was going to die! She just hooted and hollered!" Ron Ammons remembers having one of the photographs from the Rockwell event blown up to life size and mounted on cardboard to tease his friend and neighbor. "He has this damn hair piece, and he's got this grin on his face, and he had a cigarette in his hand, come to think of it. And it was Ron Evans to a 'T.'"

When Ron returned from the moon, Jan didn't notice any changes in his personality or his demeanor. Just as she had gotten the same Ron back from Vietnam, the same man came back to her from the moon. But Ron did acknowledge one change, not in his basic personality, but in his interaction with members of the public. In his early years at NASA, he had been one of the few astronauts who came to enjoy the obligatory "week in the barrel" when the public got to meet an astronaut and hear how their billions were being spent. Spacewalking apart, Ron readily acknowledged that he was "basically sort of a quiet person." After returning from the moon, he recognized that the success of the mission allowed him to be more expressive, more outgoing, and more confident in his dealings with the public. The rounds of presentations and meetings gave him the opportunity to expand his public speaking, and the more he did it, the more he liked it. He recognized that to millions of Americans and non-Americans, he and the other Apollo astronauts were heroes. One way he could see this was in the eyes of the people who flocked to the many gatherings after the mission. They were no longer meeting the silver-pinned, space-virgin Evans. Now they were meeting a man who had flown to the moon and back, and that clearly impressed them. Even before they opened their mouths to speak to him, Ron could see it in their eyes.

Heroes come in all shapes and sizes, and Ron came to realize that heroes also have their own heroes. When the crew members visited Las Vegas, they were told that Elvis Presley was very keen to meet them. Elvis was apparently a big fan of the space program and admired the astronauts' achievements. For Jan, actually getting to meet Elvis was exciting. Sadly, he was no longer the slim, hip-swiveling young star she had seen performing in Topeka in 1956, but he still had the voice and sang the songs she remembered. Ron was delighted to have a personal meeting with such a big star, and it was difficult to know who was more impressed. Jan could see that meeting the celebrities on their tour was something of a two-way street. "They want to meet you, and you want to meet them."

But it wasn't always like that. There was one discordant note during an otherwise harmonious tour around the States. On another visit to Las Vegas, two guides met the Cernan and Evans families at McCarran Airport and told them of an addition to the arrangements: "You guys are getting to meet Muhammad Ali."

Gene and Ron looked at each other and, without having to debate it, knew they were of one mind. Gene spoke for both of them: "We're not meeting Muhammad Ali."

The guides were incredulous. "What do you mean? He's there waiting to meet you."

Gene explained very simply: "We're military men, we'll have nothing to do with him." Years before Ali became universally known and admired as "The Greatest," the biggest sporting icon of the twentieth century, he was a controversial figure—particularly in the eyes of members of the armed forces. In 1967 he had refused to serve in the U.S. Army. For rejecting the draft, he was tried, convicted, and sentenced to five years in prison. Although the Supreme Court unanimously overturned the conviction in 1971 on the grounds that Ali was a conscientious objector, Vietnam was still an open wound. It was just too soon for the two military astronauts to contemplate meeting such a controversial figure.

Undoubtedly flying to the moon opened doors and sprinkled stardust around. Soon after their return, the crew was invited to dinner with Vice President Agnew. Among the guests was the singing superstar Frank Sinatra. It might just have come up in conversation that the *Apollo 10* crew had played Sinatra's "Fly Me to the Moon" to accompany live color TV of the faraway earth. You just can't buy publicity like that! Sinatra obviously enjoyed meeting the astronauts and their wives because he invited them to dinner in New York the next day. Gene, Barbara, Ron, and Jan also received an invitation to Sinatra's home in Palm Springs, where a golf tournament was taking place. Jan was a little apprehensive at first, knowing that Sinatra didn't have a squeaky-clean reputation, but she judged the man as she found him: a softly spoken, wonderfully friendly, and attentive host who clearly admired what the astronauts had done for the country.

Sinatra hosted a big party in a room that had three huge TV sets. That day North Vietnam released a group of POWs, and their arrival in the Philippines was shown live. Suddenly Sinatra was shushing everybody and making it very clear that nobody was to move or talk. The hubbub of voices fell silent during the news broadcast as a line of American servicemen, mainly air force pilots and naval aviators, stepped down from their aircraft and onto the airfield. Jan was one of many who watched through tears of joy, occasionally spotting the face of someone she and Ron knew and sometimes the face of a man they had thought had been killed in action. As for Sinatra, he was visibly moved by the sight of his fellow countrymen coming home. Jan recalls, "That grown man was sitting there crying like a baby."

Jan always said, "Nothing fazes Ron." But that day, as he sat in Frank Sinatra's home and watched the latest stage of Operation Homecoming, Ron Evans was moved. "Here we are, watching friends of mine walking down the ladder after landing at Clark AFB. They'd been POWs—some of them were in for six years, from before the time I was over there. They didn't know that I'd gone to the moon, although I found out later some of them had an inkling somebody had landed on the moon. But for the grace of God, I could just as well have been in the 'Hanoi Hilton.' Instead, I got selected in the Program, went to the moon and returned a month before they did. I tell you, that gets to you. It really got to me."

The razzmatazz of the meetings with big stars provided a certain degree of pleasure and enjoyment, but for Ron, some of the most viscerally satisfying plaudits came from his own home state. In 1971 the Native Sons and Daughters of Kansas had (a little prematurely) made him the Distinguished Kansan of the Year. The Kansas State legislature voted him Kansan of the Year for 1972. In 1973 his alma mater, the University of Kansas, awarded Ron its Citation for Distinguished Service, the university's equivalent of an honorary doctorate and its greatest accolade.

But you can't get closer to the grass roots than where you were born. The little town of St. Francis announced that Thursday, 1 February 1973, would be Ron Evans Day and made plans to honor its famous son. Giant billboards were positioned at the eastern and western edges of the town proclaiming "St. Francis, Kansas, Birthplace of Astronaut Ronald E. Evans, Command Module Pilot, Apollo 17." A seven-foot replica of the *Apollo 17* patch was set up at Ron's old school. For the first time in its history, the *Saint Francis Herald* printed a special eight-page color section carrying the headline "Welcome Home, Ron Evans" above a full-page color photograph of the man himself. His old friend Marv Miller provided background features about Ron's boyhood in the town. Among the many Apollo-themed commercials, the message from the Tri-State Lumber & Supply Company probably best summed up the mood: "Ron, when you went to the moon, you helped put St. Francis on the map. We're proud to be known as your Birthplace and Boyhood Home!"

After being honored in Topeka, where he addressed Governor Robert Docking and the state legislature, Ron flew with Jan, Jaime, and Jon to Goodland Airport. A caravan of vehicles then drove the thirty-five miles to St. Francis. NASA had argued against an official parade through the town, possibly influenced by severe snowstorms five days earlier, but that didn't stop "crowds of friends and relatives and school children" cheering the vehicles along River Street and Main Street. It was indeed a cold day but one of bright sunshine out of a crisp, blue Kansas sky. Ron's father had greeted him at the airport, and press reports indicate that Marie Evans was also at the celebrations to greet her son. A press conference was held at St. Francis High School, where the corridors were lined with posters, poems, and drawings produced by the children. Ron answered numerous questions about the mission and about spaceflight in general, hinting that within ten years NASA would fly women astronauts. On the events of the day, he noted: "This homecoming and reception day has been an outstanding experience. It's amazing that no matter where you go the fondest memories are those you have when you were growing up." He added that it was "something very special to be able to come back to St. Francis." The festivities culminated at a ticket-only banquet held at the school, for which all six hundred tickets had already been snapped up.

Ron would be feted at other banquets, attended by better-known people, but nothing ever beats the taste of home.

President Nixon made good on his promise to the astronauts. First, he and Pat Nixon hosted the families on a private visit to Camp David. Then on Saturday, 3 March 1973, the crew was invited to the White House for "an evening of entertainment." The president had an eleven-minute audience with the three astronauts, Barbara, Jan, and Jack Schmitt's date, Adair Atwell, a receptionist for Senator Jesse Helms. A White House photographer pictured the smiling group, and if the spacemen were feeling nervous, they certainly didn't look it.

The entertainment was provided by Sammy Davis Jr., a genuine star. Of the other 250 or so guests drawn from politics, commerce, and industry, it could hardly be said that any one of them had ever set this world on fire, let alone visited another one.

Until the White House visit, there seemed little chance that the crew would be sent on a goodwill world tour, as had been the case with many of the preceding crews. Ron later recalled that a cash-strapped NASA had not been keen on footing the bill for any tour outside the United States. Then, on the night of the grand party, the three astronauts found themselves in conversation with Charles Gregory "Bebe" Rebozo, a wealthy Florida banker and businessman who was also a close friend and confidant of Richard Nixon. Some concerns may have been expressed that the last Apollo crew had pulled off a stunning success but was not going to be sent on an international goodwill tour. Rebozo seemed receptive and sympathetic. Perhaps it was just a coincidence, but a few days later, the astronauts got word that NASA and the State

Department had determined there *would* be a world tour. Their destinations weren't quite what the crew had anticipated—six African countries, India, Pakistan, and several Pacific island nations—but the view was that previous astronaut tours had overlooked those countries. So be it. The crew saw it as another adventure.

On 18 May the White House announced that the astronauts and their wives would conduct a presidential goodwill mission in what Nixon called "the spirit of Apollo." The Apollo findings would be shared with scientific communities, and the astronauts would also address university audiences "to share space knowledge with other nations and to promote peaceful space exploration."

They flew in a presidential aircraft that was considered Air Force Two when the vice president was on board. A doctor always accompanied an Apollo goodwill tour, but the astronauts and wives all agreed they wanted Dee O'Hara for their medical support. Gene Cernan did his "commander thing" and made it happen. Dee, who had never accompanied an astronaut tour before, fondly remembers the "wonderful trip." For her, it was "a once-in-a-lifetime thrill."

The group had an eventful month of startling contrasts and colorful experiences. In Senegal, Jan, Barbara, and Dee stayed in the U.S. ambassador's residence while the crew was flown into the interior to address various meetings. On their return, Ron told them he was relieved they hadn't come along as they "would have been bounced to holy heck not only in the air but on the ground."

Moving on to another nation, open jeeps took the whole party into the desert to visit outlying villages. As they passed through these communities, the roads were lined with people who bowed and shouted in French, "Apollo Dix-sept! Apollo Dix-sept!" They knew who these visitors were! In the middle of nowhere and with no obvious sign of T V aerials, the locals knew about the U.S. space program and wanted to honor the astronauts. For Jan Evans, it was one of the most amazing experiences on the tour.

In Niger, President Hamani Diori provided a state dinner at his presidential residence. After a day of meetings and speeches, Ron and Gene were surprised to find that Jan and Barbara were missing, as was Madame Aissa Diori, their hostess at the dinner. When the three women turned up late, the men learned that Madame Diori had insisted on taking the wives on a drive to see a region of spectacular baobab trees. They never did find the baobab trees, but in an effort to return in time for the dinner, the little convoy left the road and cut across a patch of desert. All of a sudden, they encountered a herd of giraffes. As she stood in the African desert within arm's reach of an unperturbed baby giraffe, the former Janet Pollom from Topeka, Kansas, marveled at the journey that had taken her to that place.

But the tour was no holiday—certainly not for the astronauts. One day in particular tested their stamina. The three men were transported from meeting to meeting, where they addressed press conferences, scientific groups, and college student gatherings. Everywhere they went, they were impressed by the audience's enthusiasm, warmth, and level of knowledge. But there is a limit to anyone's endurance. After the men had given eight addresses at eight different venues, Jan recalls that "the guys were literally babbling idiots, they could barely remember their own names. That was the hardest day for them."

In Agra, India, the pace was a little slower, and two members of the group took the opportunity to sample a more sedate form of transport. Dee O'Hara is relieved that photographs of her on a camel seem to have disappeared, but the Evans family has a splendid view of Captain America steering a "ship of the desert" in the arid outskirts of a city better known for the Taj Mahal.

Looking back on her experiences before, during, and after *Apollo 17*, Barbara Butler reminisces, "We were all so young! And you think, God! Where did all the years go? What a fascinating life we had! What a challenge and what a wonderful time we had sharing it with so many great people. It's hard to believe that God gave us the opportunity!"

26. Apollo-Soyuz and Beyond

Man, I tell you, this is worth waiting sixteen years for!

—Astronaut Donald K. Slayton (1924–93)

When Richard Nixon campaigned for the presidency in 1968, he had a reputation as an ardent anti-communist and foreign policy hawk. Many were therefore surprised when President Nixon later reached out the hand of friendship first to China and then to the Soviet Union.

Nixon's policy of détente with the Soviets resulted not only in treaties aimed at limiting nuclear missiles but also in the signing of the Agreement Concerning Cooperation in the Exploration and Use of Outer Space for Peaceful Purposes. The signing ceremony on 24 May 1972 included Nixon and Soviet premier Alexei Kosygin in the opulent splendor of the Kremlin. The British *Daily Mail*, never a friend of communism or the Soviet Union, enthusiastically headlined its report of the event as "A Joint Step for Mankind."

The agreement envisaged a joint docking mission between an Apollo spacecraft and a Soviet Soyuz vehicle, with hopes for future missions and "space rescue" capabilities. At a press conference later that day, NASA administrator James Fletcher noted, "I do not know of another more visible way that two countries who are the leaders in the space program can be shown to cooperate on a very complex endeavor like this docking mission."

The *leaders* in the space program? That and similar remarks ruffled a few red, white, and blue feathers back home. Hadn't the United States won the space race? This plan treated the two space programs as equals. *Apollo 7* astronaut Walt Cunningham later wrote: "It was an odd kind of progress, achieved by reaching down and lifting the also-ran up to our level." He was not alone. Ron Evans's friend and neighbor Ron Ammons was candid: "I didn't like the idea, quite frankly."

A joint flight with the Americans certainly lent prestige to the Soviet space program, but it also provided something of a lifeline to NASA, which was facing a long drought between the *Skylab* missions and the most optimistic predictions of a first space shuttle flight in 1978. The Apollo-Soyuz Test Project, as it was called, would keep a lot of people employed and the launch facilities operational. The joint operation would also mean three more seats in space. Pilots exist to fly, and if that meant flying with Russian pilots, so be it. Ron Ammons recognized that his astronaut friend was perfectly capable of being patriotic without being overtly political, and he does not recall Ron expressing any political views indicating that he would have a problem flying with Russian cosmonauts. Of course, when the joint project was announced in April 1972, Ron already had a mission and was too immersed in his *Apollo 17* training to pay it much attention.

Not so for the former Mercury astronaut Deke Slayton, whose heart fibrillation had cleared up. Deke was restored to flying status in March 1972, but by then there were no vacancies in the Apollo or Skylab Programs. For Deke, it was Apollo-Soyuz or bust.

In May 1972 Slayton saw himself as a candidate for the joint mission and felt obliged to pass responsibility for crew selection to Chris Kraft, who had succeeded Bob Gilruth as the director of the Manned Spacecraft Center (soon to be renamed the Johnson Spaceflight Center). Kraft asked Slayton who he thought should fly, and Slayton was honest: he suggested himself as commander with a crew of Vance Brand and Jack Swigert. For backups, he had "fifteen guys to choose from" and proposed moonwalker Alan Bean and rookies Ron Evans and Jack Lousma. At that point, Ron and Jack were still training for their forthcoming missions. Meanwhile, veteran Tom Stafford made it known that he was interested in commanding the American half of the joint mission, and he began to learn Russian. It helped also that he was involved with the working groups discussing the nuts and bolts of how to conduct a docking mission involving the two incompatible spacecraft. Those discussions took Stafford to Moscow in October 1972. On his return, Kraft told him he would be the American commander, with Slayton on his crew. Slayton's account slightly differs: he records that shortly after *Apollo 17*'s return, Kraft called him into his office, where Stafford and Vance Brand were already present, and told the three men they were the prime crew. Nearly fifty years later, Brand cannot adjudicate between the two recollections.

It is not entirely clear when Kraft settled on the backup crew of Bean, Evans, and Lousma. He may simply have accepted Slayton's original suggestion. Tom Stafford confirms that he also recommended Ron Evans for the backup CMP slot. He considered Ron to be "a very talented individual [who] always did a great job." In any event, NASA officially announced the crew selections, including the backup and four-man support crews, on 30 January 1973.

Jan Evans doesn't recall any talk about Apollo-Soyuz backup duty *before* Ron flew to the moon, but she acknowledges that "there were a lot of things the guys weren't supposed to tell anybody." That included their wives. It seems unlikely Ron could have been the favored candidate for over six months before *Apollo 17* without someone mentioning it to him. Just to be sure he wouldn't turn it down . . .

Not that Ron Evans would have ever turned down an assignment. Such an offer went with the territory, and the astronaut office was still territory Ron enjoyed inhabiting. Not every Apollo astronaut in training for a moonflight would have wanted to pitch straight into backup duties on an Earth-orbital mission, but anyone who might have thought Ron wouldn't do it just didn't understand the man. Jan knew Ron wouldn't see it as an anticlimax or a dead-end job. He would see it as a new challenge and a new adventure, and that was fine by her. There was always a chance he might get to fly again, an opportunity she knew he would relish.

Vance Brand, whose health and bodily integrity stood between Ron and a second flight, had no doubts that his backup was fully enthused about his new assignment. "Even though it was hard to beat *Apollo 17*, I think he was pretty fired up to be a good backup crewman. He was a real good pilot. I remember him as a quiet, pleasant man, a very experienced naval aviator. I don't remember him being combative with his fellow workers or anything like that. He was good to work with. And he was a good backup in the sense that we went through the same training, and he was ready to step into my shoes if I had broken a leg a few days before the flight or something!"

An interesting feature of the crew assignments was that by the time of launch, the backup crew (with a total of 141 days in space) would be significantly more experienced than the prime crew. Deke Slayton, NASA's oldest rookie, would be fifty-one years old at the time of launch. Jan Evans was delighted that "the skipper" finally had a flight. "Everybody was thrilled that Deke was going to be flying. Nobody deserved it or earned it more than Deke.

And he was thrilled that he was going to go. Oh yeah! Definitely! But he was still the same old laid-back Deke!"

Even the backup crew realized that training for Apollo-Soyuz would be a unique experience, not least because of the need to learn Russian. Here Ron had a head-start on many of the others, having attended that summer course in Russian at the Naval Postgraduate School, but Jack Lousma had also planned ahead. Concerned at one point by rumors that Nixon might cancel the Skylab Program, he had studied in his own time to obtain a diploma in Russian. Just in case Apollo-Soyuz became the only game in town.

Many outsiders must have wondered how these Cold War warriors could forget the past and work with communists. They must have forgotten that several astronauts had already visited the Soviet Union during more frigid periods of the Cold War. Even through interpreters, they had found common ground, talking pilot to pilot. Jack Lousma sums up his training experiences this way: "None of us thought of them as enemies. At the time, we just thought of them as people we were going to fly with in space. The cosmonauts felt the same way. The camaraderie we had was that we were all pilots, we were all fliers, we were going to fly in space, so that's what we talked about."

Tom Stafford, who was promoted to brigadier general in 1972, never had any trouble working with the cosmonauts. "We got on great. We never discussed politics. *Anything* but politics! We were there on a professional mission."

Evans and Stafford were the first Apollo-Soyuz astronauts officially to meet the Soyuz crew. The Soviet crew announcements were made on 24 May 1973 to coincide with the first day of the Paris Air Show. In all, there were two duplicate crews and two backup crews, but the men who expected to fly were veterans Alexei Leonov and Valeri Kubasov.

NASA sent the *Apollo 17* crew, accompanied by Tom Stafford, to Paris to represent the nation's space program. One of the focal points of the air show was a full-scale display of Apollo and Soyuz docked. Circulating among the numerous visitors, the four astronauts rubbed shoulders with Leonov, Kubasov, and two other veteran cosmonauts. Early indications were that they could all work well together.

Barbara Cernan and Jan Evans accompanied the astronauts to Paris. Jan recalls that they had the use of two black Ford cars with drivers who picked them up every morning to drive them from their hotel to the air show. She couldn't help noticing that the cosmonauts had limousines.

Alexei Leonov's birthday was on 30 May, and the Americans surprised him with a birthday cake. They received a bigger surprise when the other cosmonauts suddenly grabbed Leonov by his ankles, turned him upside down, and shook him until coins fell from his pockets. Despite the language barrier, the Americans quickly realized that Leonov had fallen victim to an old Russian custom.

That same day Leonov arranged for Ron, Jan, and the others to visit the star Soviet exhibit at the air show—the Tupolev Tu-144 supersonic airliner. As Jan puts it, "The cosmonauts were there, and they could do anything." The aircraft designer, Alexei Tupolev himself, actually approved the tour. Jan remembers sitting with Ron on the flight deck, sharing vodka shots with the cosmonauts and admiring the array of instruments.

By an interesting coincidence, Jimmie Taylor, Ron's old shipmate from the USS *Oriskany*, was at Le Bourget Airport to showcase Grumman's new F-14 Tomcat supersonic fighter. The Russians must have held Taylor in high esteem, for on 3 June, the last day of the air show, they also gave him a tour of the Tu-144. Later that day, as he prepared to take off on another demonstration flight, he witnessed the Tu-144 breaking apart in mid-air and crashing to the ground. Expecting his flight to be called off, he was surprised when he was cleared for takeoff, and he actually launched through the rising cloud of black smoke.

Ron and Jan were home in Houston when they heard the news of the terrible crash, which claimed fourteen lives, including the members of the Tu-144 flight crew who had proudly shown them their aircraft. It was a somber start to their *Apollo 17* world tour.

The prime and backup Apollo crews went through the familiar grind of classroom studies, simulator sessions, and visits to the contractors that were preparing the spacecraft and the new docking module. The mission would include a program of extensive Earth observations, and Ron was delighted to be working again with his old friend Farouk El-Baz. The Egyptian geologist had a particular interest in subtle color variations, both in the desert sands and the oceans. He produced a special color wheel with multiple sand colors on one side and shades of blue and green on the other. Being familiar with Ron's work, El-Baz was confident that if Vance Brand were to suffer that broken leg, Ron could take over seamlessly.

Inevitably, the exchange visits set the Apollo-Soyuz training apart from that of earlier missions. On 9 July 1973 a thirty-four-member Soviet delegation arrived in Houston for a three-week getting-to-know-you session. The cosmonauts attended classroom lectures on the flight plan and the basic elements of the Apollo hardware, but they also had time to sample Houstonian hospitality. In fact, on the day of their arrival, the cosmonauts disappeared off the radar along with their astronaut hosts. Jan and the other wives all called each other to find out where their husbands had gone. When the men surfaced the next afternoon, the wives learned that the two groups, with their interpreters, had gotten together for dinner and a party to cement the new relationship.

Ron's first visit to the Soviet Union in November 1973 was essentially a mirror image of the cosmonauts' Houston visit. The prime and backup Apollo crews stayed at the Intourist Hotel in downtown Moscow, and every morning a bus, complete with police escort, took them to Star City, the Soviet Union's training center for cosmonauts. The six men discussed the mission's flight plan and familiarized themselves with the Soyuz spacecraft and its systems.

In those early days, Ron still only had basic Russian. Learning the language proved "difficult but not *very* difficult." Jan knew Ron had always been a good student, and he persevered with his Russian lessons. They were even inspired to call one of their dogs Chetyre (Four).

Most nights, the astronauts joined the cosmonauts for dinner, and they took turns offering a toast. When Ron realized it would be his turn the next night, he somehow found time, accompanied by his Russian interpreter and guide, to visit a Moscow bookstore. He found inspiration in a little book of children's nursery rhymes, especially "Humpty Dumpty," which he memorized. That evening after dinner when he was called upon to deliver his toast, he recited: "Humpty Dumpty sat on a wall / Humpty Dumpty had a great fall . . ." in broken but apparently intelligible Russian. His effort brought the house down, inspiring much consumption of vodka, and the cosmonauts would always remind him of it whenever they got together.

Jack Lousma recalls that Ron was a good traveler who enjoyed the new experience of visiting the Soviet Union. The astronauts got along very well with the cosmonauts, touring the country and sharing weekend sightseeing trips and cultural events. "I liked those guys! Especially, I liked Alexei. . . . He spoke English better than all the others. He was very savvy politically and also public relations-wise. He made a good impression wherever he went, he was hospitable, he had a nice family. We went to have banquets or dinners in his apartment at Star City occasionally when we were training down there. Kubasov was the same. Like other nonmilitary cosmonauts, Valeri lived in Moscow, rather than Star City, and had us over for dinner as well. We had good camaraderie together, we enjoyed a lot of the same things, we talked about hunting and sports. We didn't get into politics much! The Russians were really good hosts for us, probably better than we were for them when they came over because they had government money to host us and take us around places. We didn't get government funds to take them places on weekends during their visits to America!"

When the cosmonauts returned to Houston in April 1974, the training stepped up a gear, with sessions in the command module simulator. When sharing the simulator with Ron on 15 April, Alexei Leonov was well aware the astronaut sitting beside him had been the last man to steer the real thing to the moon and back.

An unexpected distraction arose for Ron in the middle of his training. On 2 March 1974 Robert Docking, the popular Democratic governor of Kansas, announced his imminent retirement. Republican Party representatives in Kansas approached Ron and attempted to persuade him to seek election as the next governor of his home state. It didn't take Ron long to turn them down. With the exception of Docking, who had treated Ron and his family with great courtesy and respect in Topeka the previous year, not too many politicians had impressed him. Moreover, he was very happy with the job he had.

Jan remembers the approach and confirms that running for governor, or any other elected office, did not interest her husband. Ron and Jan always voted Republican, but Ron had no strongly held political views. He was a proud Kansan and a patriotic American, but he was definitely not a political animal.

In late June the astronauts and their entourage returned to Star City for the next phase of their training. This time the prime and backup crews took turns in a Soyuz simulator, learning firsthand that the vehicle lacked the sophistication of Apollo.

The Americans were staying in a brand-new hotel in Star City. After attending a grand reception hosted by the American ambassador on the evening of the Fourth of July, Ron Evans revealed his own contribution to the celebrations. He had brought "half a suitcase of fireworks" with him, an unthinkable action in today's post-9/11 world. Tom Stafford remembers, "We had a ball shooting those off!" The astronauts lit a string of firecrackers in front of the hotel. Lights went on in buildings all around, and soon a police patrol arrived but stayed on the other side of a small lake. Stafford "ordered Evans to fire a bottle-rocket in their general direction." It burst high over the heads of the police, prompting the senior officer to approach the astronauts, apparently more in bafflement than concern. In his best Russian, Stafford exclaimed: "Good evening, Comrade! We are celebrating the birthday of our revolution!" The officer politely declined the offer of a drink and left the crazy Americans to their revolutionary activities.

In Houston that September, the cosmonauts were involved in more intense simulator work, with further training in docking procedures. But as always, they had time to relax. Ron and Jan invited them to El Lago for a Texas-style party on the last night of their visit. Jan was delighted to welcome the cosmonauts, particularly Leonov and Kubasov, to their home. "Personally, I don't think they could have picked two better fellows," she recalls. "They were a lot of fun, they were sociable, and they were very much on the job and knew what they were doing."

Some of the space travelers had been out on a quail hunt during the day, and Ron had arranged for a caterer to cook the quail to supplement the rest of the food. He had also taken down part of their back fence, providing access to a big open hummocky field, and borrowed several trail bikes from the neighbors (including the Ammons family) to add to his family's four. The cosmonauts thoroughly enjoyed letting off steam by charging around the field, and clearly they only understood one setting on the powerful little machines— full throttle. The NASA wives exchanged glances, wondering whether it was such a good idea, but they consoled themselves with the knowledge that the cosmonaut team included plenty of backups. Jan saw it didn't seem to matter which side of the Iron Curtain the men were from. "Those guys were just as much a bunch of idiots as our guys!"

The Apollo-Soyuz Test Project brought down many barriers. At the end of April 1975, astronaut Ron Evans climbed inside the actual Soyuz capsule that would fly in July. Tom Stafford had insisted on inspecting the interior of the capsule to which he would briefly be entrusting his life, and he made sure all of the astronauts—prime, backup, and support crews—had their turns to do the same. They were visiting the Baikonur Cosmodrome and were only the second group of Westerners ever to do so after French president Charles de Gaulle. Ron didn't say much about the visit, but Jan remembers his telling her with a smile that they flew to the huge launch complex in darkness and flew out again in darkness. As if the Americans had never seen spy satellite views of Baikonur. Better than that, in 1973 Bean and Lousma had made a point of observing the complex many times from orbit, on board the *Skylab* space station.

The quarantine period at the Kennedy Space Center for this flight was not as rigid as for *Apollo 17*. As final preparations for the joint mission were being made, Ron and Jack Lousma would occasionally make a break for it and go fishing in the channels and lagoons near the launchpad. It was some of the best fishing in Florida, and what they caught they gutted, cooked, and ate in the astronaut quarters, to the apparent displeasure of the NASA cook.

Even astronauts couldn't roam around near the launchpad without clearance, but Charlie Buckley happily assigned a security man to prevent potentially germ-ridden members of the public from approaching and disturbing the men's fishing. The guard's sidearm might also have been a reassurance when fishing in the ominously named "Gator Hole" between Pads 39A and B. Jack remembers NASA tour buses driving by, "and they'd be looking at the launch sites, and they'd see these two guys out there fishing, and they didn't know who we were."

On 15 July 1975, the day of the launch, Ron and Jack joined the prime crew for breakfast. Later, Lousma recalls the backup crew being sequestered at the crew quarters "until it was clear we would not be needed." This begs an interesting question: what if Vance Brand actually had tripped getting up from the breakfast table and broken a leg? Leonov and Kubasov had successfully launched into orbit at 8:20 a.m. Florida time and would return to Earth in fewer than six days, so NASA was under pressure, both operationally and politically, to launch on time. A delay would have been embarrassing. In theory, Ron was trained and ready to take over. In practice, there was no question of a simple substitution to allow Apollo to be launched on schedule. Tom Stafford confirms that the incapacity of a crew member on launch morning would have resulted in the launch's being scrubbed at least for that day. As with Jack Swigert in 1970, Ron would have had to demonstrate to his commander in the simulators that he was ready to fly. "I'd say we might have been able to do it in three days," Stafford suggests. A launch window was available on 18 July, although the joint mission would have been shortened.

Vance Brand never did break his leg, so there was no second Apollo flight for Ron. Once the prime crew suited up, the backups were able to leave the crew quarters to watch the launch. Jan had traveled to the Cape, courtesy of one of the contractors, but because of Ron's backup duties, the couple could not get together. They separately watched their three friends bore a hole in the sky in pursuit of their new Russian friends. Meanwhile, about a mile to the north, a young law student from Bangor, Northern Ireland, watched in awe as the Saturn 1B roared into life. But that is another story.

After Apollo-Soyuz, Ron found himself in an unusual situation. Previously he had experienced only a few periods in his astronaut career without a mission to work toward, in whatever crew capacity. It was no consolation that everyone else at NASA was in the same boat. Along with the rest of the astronaut corps, Ron was assigned to work on the development of the space shuttle, the multibillion-dollar basket into which NASA had placed all its eggs. The shuttle would fly at some point, but NASA's hopes for 1978 had evaporated. Technical problems and further budget cuts made even 1979 look unlikely.

Some astronauts had been working on the shuttle since President Nixon gave the go-ahead, so Ron was already playing catch-up. NASA documents refer to his being "responsible for the operational aspects of the ascent phase of the space shuttle orbital flight tests." That phrase pops up in any internet search about Ron but without any elaboration.

Other astronauts were busy with their own specific parts of the program and can't remember anyone else's assignments. Vance Brand spent a lot of time in the simulators planning the shuttle's ascent to orbit, reentry, and landing, but he didn't compare notes or cross paths with Ron. Jack Lousma's main responsibilities were the cockpit controls and displays, including a new head-up display, as well as the hydraulic power units and controls. He doesn't recall working with Ron but points out that the available astronauts were assigned to do "thousands of things."

Ron certainly didn't spend his days cooped up in his office. His logbook entries for the period August 1975 until January 1977 show visits to ten states, including multiple visits to Los Angeles, where Rockwell was building the shuttle orbiter. Flying was always a pleasure, but the lack of a truly inspirational goal driven by a relentless timetable made him start to question his future path.

Charlie Duke felt the same. In the days of Apollo when a technical problem arose with the spacecraft, it was examined from all directions. Someone took responsibility for troubleshooting the problem, a solution was found, and the fix was made. Not so with the shuttle. Now a committee would consider the problem, and maybe a few subcommittees would be involved. Somebody would draft a report, the committee would consider it, a recommendation would be transmitted up the chain of command, and eventually someone would make a decision. Charlie had thoroughly enjoyed being an astronaut. He had walked on the moon! Yet here he was, working on a vehicle that was being built by committee. He describes his work on the shuttle with refreshing candor: "It was boring." To his regret, a flight in the shuttle seemed too far off in the future. Charlie resigned from NASA in September 1975.

Even before Apollo-Soyuz, Deke Slayton had been appointed to manage the space shuttle's approach and landing test program. Its series of flights would test the shuttle's aerodynamic capabilities when released from the back of a Boeing 747 carrier aircraft. The key question was whether the world's biggest glider could actually glide to a safe landing.

Deke did not return as the director of flight crew operations, so NASA needed a new "Godfather" to select future crews. Appointed in January 1976, George Abbey was a former U.S. Air Force pilot who had already worked on NASA projects and had wanted to apply to become a Group 5 astronaut in 1966. To his exasperation, although he met NASA's requirements, the air force would not nominate him because he wasn't a test pilot. He later chose to join NASA and participate in Apollo in a backroom capacity. After working his way up with quiet dedication, there was some irony in that Abbey became the man who had the power to select Group 5 astronauts for shuttle missions.

Ron could hardly have expected to be selected for the shuttle's approach and landing test flights carried out at Edwards AFB. NASA wanted test pilots to make the short but crucial gliding flights in the vehicle, which would be named *Enterprise* following a campaign by *Star Trek* fans. Ron certainly would not have faulted Abbey's choice of Fred Haise and Joe Engle to command the test flights, nor would he have begrudged those two Group 5 colleagues the likelihood of early orbital flights. The two copilots Gordon Fullerton, who had been heavily involved in shuttle development since early 1973, and Dick Truly, a future NASA administrator, were also test pilots. If a line was forming to fly the shuttle into orbit, those four at least were ahead of Ron, and other test pilots with spaceflight experience were available. John Young and Ken Mattingly were clearly preparing for early flights, and there were men from Group 5 who had not yet flown.

NASA also sought new astronauts to help fill the seven seats on board future shuttle missions. Wayne Skaggs, one of Ron's fellow Screaming Eagles on board *Ticonderoga*, had left the navy to become a pilot with Delta Airlines. In early 1976 the airline business experienced a slump. Although he hadn't seen anything in writing, Wayne wondered whether an opportunity to fly higher and faster with NASA might arise soon. Who better to ask than his old buddy who had been to the moon?

On a convenient stopover, Wayne visited the Johnson Space Center, hoping to get in touch with Ron. He had lunch before making his move, and looking out the window, he saw a familiar figure walking up the steps to the cafeteria. After exchanging greetings and catching up, Wayne asked about future astronaut selections, and Ron confirmed that NASA would soon be recruiting again. What Ron said next took Wayne by surprise. "He told me he felt he was probably too old at that time to participate as a pilot in the shuttle program. I don't know why he thought that, but that was what he indicated to me and that he was considering going into retirement." Ron was forty-two years old at the time, but perhaps more significantly he had been an astronaut for nearly ten years. "Maybe," Wayne suggests, "he just anticipated it would be a long road in the shuttle program before it would actually fly."

Wayne was sorry they didn't have longer to chat, but Ron had to go to a meeting. The two old friends went their separate ways, and when NASA requested applications for what became Group 8, Wayne obtained the forms and deliberated how to answer the questions about flight time. Ultimately, he decided to stick with Delta.

It's hard to believe Ron really considered himself too old at that time, but as Wayne suggests, Ron may have been thinking of how old he expected to be when the shuttle began flying regularly. Interviewed in 1986, Ron provided some insight into his thinking ten years earlier. Reminiscing about his Apollo days when he had "the best job in the world," he posed a question: "What about afterwards? Each person retires, resigns, quits, for a different reason.

All of them, I think, probably from the standpoint that if you're operational-orientated, and you get shunted off to a different kind of a job that's really not the top of the operation . . . you're really not too enthused with taking on those kind of jobs. So the opportunities look better on the outside. The grass is greener on the other side of the fence!"

Whenever two or three space enthusiasts gather together, it doesn't take long for conspiracy theories to emerge. Mentioning George Abbey's name is enough to provoke suggestions of certain astronauts being favored and others being frozen out. If Ron Evans felt he had been "shunted off" to a dead-end shuttle job, might that suggest he fell out with the new Godfather? However tempting it might be to believe this, there is no evidence that Abbey ever felt that Ron was "surplus to requirements." On the contrary, Abbey planned to fly all the remaining Apollo-era astronauts, including two he didn't get along with personally. Abbey knew Ron very well from the early days of Ron's support crew and capcom responsibilities. He has made it known for this book (through his biographer, Michael Cassutt) that he liked Ron and has no memory of any private conversations with him about shuttle assignments. Abbey believes that Ron—along with other astronauts such as Charlie Duke, Gene Cernan, and Jerry Carr—was simply not willing to spend five, or six, or more years in engineering work while waiting for the shuttle. His recollection was that Ron and Jan wanted to leave Houston to pursue a new life and opportunities elsewhere.

Jerry Carr never discussed the matter with Ron, but he thought they likely approached retirement from NASA in a similar manner. "I had finally gotten an opportunity to fly, but I didn't feel like it made a lot of sense for me to wait seven more years for another flight. And I would suspect that maybe Ron may have felt the same way—that he needed to get on with his life and do something different."

Jack Lousma has no difficulty speculating about why Ron decided to leave NASA. "He had done what he came to do! He had had a flight to the moon, and it's hard to beat that, you know! My guess is he had accomplished his objective, and it was time to move on."

Charlie Duke agrees that, for him and for Ron, having been to the moon made a difference. They each had only one flight, but what flights they were!

Everybody had his reasons for leaving. Fred Haise went to Grumman, the place that had built the lunar module "lifeboat" that had saved his life. He never asked others why they chose to leave NASA, but he points out, "Ron would surely have flown shuttle, had he stayed."

Gerry Griffin was disappointed when Ron decided to leave. He felt it was a loss for NASA as well as for Ron. "I'm sorry he didn't get to fly again, because I think he would have been a good commander. On *anything*, shuttle or anything else, because he had that mentality to keep it calm and get the job done and not waste a lot of effort."

The final word on this choice has to be from Ron Evans himself: "There comes a point in your career where you have to decide: am I going to make a total commitment to staying in the space program, or do I get out and make a dollar? And I wanted to get out and make a dollar."

On 30 April 1976 Capt. Ronald E. Evans retired from the U.S. Navy with just over twenty years' service. On 14 January 1977 he made his last flight for Uncle Sam in a NASA T-38 from Los Angeles, via Las Vegas and El Paso, finally touching down at Ellington AFB. His logbook recorded a total of 5,129 flying hours. But it would not be his last flight in the T-38.

Less than two months later, NASA issued a press release confirming that "astronaut Ronald E. Evans will leave NASA March 15 to become a coal industry executive. Evans, 43 . . . has been named Executive Vice President of Western America Energy Corporation and Director of Marketing for WES-PAC Energy, the coal-producing concern of WAEC."

There is one last entry in Ron's logbook. On 29 April 1977, now a member of the public, he made a seventy-two-minute flight in a NASA T-38. He is recorded as the copilot, but the pilot is not identified. Ron was not a particularly sentimental man by nature, but flying airplanes—and particularly high-performance jets—had been an integral part of his life for two decades. He couldn't just walk away without a backward glance. We can assume an astronaut friend arranged to take him up for one last chance to throw the T-38 across the Texas skies. It was a fitting way to mark the end of both his navy and NASA careers.

27. The Most Alien World

I am the master of my fate:
I am the captain of my soul.

—William Ernest Henley (1849–1903)

Delivering a brief but witty after-dinner speech in London in 2014, moon-walker Alan Bean reflected on the key differences between life in the military world and life in the outside world. When he was in the navy, he could always count on people. If a plane was needed at a certain hour, it would be ready. People you couldn't rely on didn't last in the military. He found that "one of the blessings of being at NASA . . . was being surrounded by people you could count on all the time." If a mission flight plan was needed for discussion the next day, the support staff would have it prepared and ready. When Bean left NASA, he found that life wasn't always like that in "the outside world." He came to learn that many people told him only what they thought he wanted to hear. If someone agreed to do something for him, he just didn't know if they would actually do it. For a man who had known teamwork and reliability for so long, it was a rude awakening.

Alan Bean was certainly not alone in his frustration.

That NASA press release of 8 March 1977 was a little misleading. It reported that Ron had a new job based in Scottsdale, Arizona, and the Evans family would be residing there. The implication was that they were following the job. In fact, once Ron began thinking about leaving NASA, they started planning a move to Scottsdale. Ron and Jan had friends there and had often visited the Phoenix suburb. They loved the desert and the western way of life. It helped that Jan's parents had retired to Sun City, near Scottsdale, a few years earlier.

The job itself came from a chance encounter at a racetrack, probably at Daytona, Florida. Ron and Jan, as well as their children, had always enjoyed watching motor racing, including the NASCAR stock car competitions. They met a Kentuckian named Jim Stacy who loved motor racing enough to buy a NASCAR team in May 1977 and a second team the following year.

When he first met the smiling Apollo astronaut and his family, Stacy had already made a fortune in the coal-mining industry. His mines were mainly in West Virginia, but his coal business was based in Scottsdale, where he and his wife had a home. While the two families were watching the races, the conversation isn't hard to imagine.

"Oh, you live in Scottsdale, Jim? That's where we're going to live!"

"Say, Ron, what does an ex-astronaut do for a living?"

"Oh, nothing concrete planned yet."

"Have you ever thought about the coal industry?"

At first the new job seemed to work well for everybody. Ron had entered the business world as he had planned and was making a good deal more money than he ever did at NASA. For Jim Stacy's Western America Energy Corporation, having a real live Apollo astronaut on the board was great publicity and made the business look good. For Jaime Evans, still a teenager, one of the perks of her dad's new job was getting to visit the racetracks and watch the cars and maybe some of the drivers.

Jan can't remember exactly when the luster began to wear off. It probably started with Ron's visits to the mines. As a marketing executive, he didn't really need to get his hands dirty in the Appalachian coal seams, but Ron Evans never took part in any operation without checking the working conditions, whether in the bowels of an aircraft carrier or where the coal was being dug. The mining region was a long way from home, and he spent at least three or four days on each visit. After one trip, he returned home shaking his head and saying, "Oh my gosh! I don't know how those people do it. It's cold, it's wet, and it's dirty!"

On 8 August, only a few months after Ron's arrival, the U.S. Securities and Exchange Commission announced a ten-day suspension in the trading of WAEC securities because of "the lack of adequate and accurate public information concerning the company's operations and financial condition."

On 1 November Jim Stacy signed a contract to rent racing facilities, a move that spawned a series of lawsuits over sums reportedly totaling $1.2 million.

One of Stacy's coal mines had been purchased from a Mr. Robert Kathary of Pennsylvania, who had also leased coal production areas to WAEC. Kathary alleged that Stacy's companies defaulted on payments for the mines. This gave rise to further litigation until Kathary obtained a court judgment in July 1978 for just over $2.9 million.

Even before that judgment, Ron had seen enough. Whether it was the working conditions of the miners or the steady flow of troubling financial revelations, he realized that his new start had been a false start. The coal-mining business—and specifically *that* coal-mining business—was not for him.

Ron never had to wonder if he made the right decision. After he left, Stacy entered into a contract with Rijn-Schelde-Verolme (RSV), a major Dutch shipbuilding and engineering company, for the supply of thin-seam coal-digging machines for open-cast mining. The scheme was doomed from the start. The RSV-supplied diggers kept breaking down. RSV's New York legal advisers had actually warned it not to go into business with Stacy because his companies were deeply in debt and lacked the necessary skills for such a project. In June 1982 Western America Energy sank beneath the waves of the business ocean. In 1983 the giant RSV also foundered, at least partly because of the coal-mining fiasco, resulting in a public inquiry being set up by the Dutch government.

Ron's first foray into the business world had lasted about a year. His former colleague Wally Schirra had a similar experience. His presidency of a financial services company only lasted six months. Schirra realized his good name and reputation were being exploited as a cover for shady dealings, so he left.

After the coal company, Ron found himself with no job and only a navy pension. While he began making careful inquiries to identify a suitable opening, he attended classes and obtained a real estate license as a fallback. Then he had a meeting with the president of the Phoenix-based Sperry Flight Systems Company. It would be a stretch to call it a match made in heaven, but Ron liked the idea of working for an aerospace company with links to the space shuttle program. Sperry's president liked Ron, liked the idea of hiring an Apollo astronaut, and could see Ron having a bright future in the company. Ron was brought in as the director of space systems marketing.

For Jaime, her father's new job was a disappointment. It was "just a regular old job" compared with the cool stuff, like getting invited to motor races. She didn't know what her father did at Sperry, and she didn't pay much attention to it. For Ron, however, it was a proper entry into the corporate world with an income to match. He was now earning more than twice his previous NASA salary.

When the shuttle finally flew in 1981, Sperry in Phoenix was supplying electronic components, cockpit display instrumentation, and an automated landing system. Ron also became the go-to guy when the Phoenix radio reporters wanted a local astronaut to interview in those early, heady days of the shuttle. Company management certainly wasn't averse to Sperry's name getting mentioned at the top of the interviews. The reporters liked Ron and his easy, relaxed approach, and the public liked to phone in their questions. Inevitably, someone would always ask how you "take a rest" in space without a "rest room," and Ron would always begin his careful description of the delicate matter by saying, "Well, there are two ways . . ."

The experience also afforded glimpses of his old life. A family photograph shows Ron and Jan at Edwards AFB sporting Sperry hats and patches of Ken Mattingly's STS-4 mission, as they awaited the return of shuttle orbiter *Columbia*. Ron and Fred Haise occasionally bumped into each other at industry business meetings and social functions. After Jack Lousma left NASA in late 1983, he recalls attending a space-related event in Phoenix, where he was pleased but not too surprised to run into Ron. The old friends had time for a quick visit, but Jack doesn't remember discussing their respective reasons for leaving NASA.

Ron Evans and Sperry Flight Systems seemed a good fit, and for several years that was true. But for all the positives of working in the aerospace industry, Ron could never get used to certain aspects of corporate life: the pettiness, the backbiting, and the office politics. Sperry was by no means immune to it. One of the straws that broke the camel's back involved a member of Ron's marketing team. Nearing retirement age, the man did his job well and was a skillful golfer who was regularly asked to play in a foursome with senior management figures. That seemed to put a lot of middle-management noses out of joint. The man also suffered from a serious medical condition and occasionally had to take time off for treatment.

The president who had brought Ron into Sperry died suddenly and unexpectedly. Things changed afterward. A combination of jealousy and concerns about carrying an ailing "passenger" resulted in Ron's being ordered to get rid of his marketing golf prodigy. Ron refused, point-blank. The man was still doing his job well, and Ron felt he had no reason or justification for firing him. Ron's act of defiance soured his relations with several of his superiors, to the extent that he concluded that the corporate world was an alien one he just didn't want to inhabit. This realization was all the more disappointing to him because in his previous encounters with NASA contractors, such as Grumman and North American Rockwell, he had seen only teamwork and the dedicated pursuit of a common goal.

Comparing life at NASA and life in the world of business, Ron could have been writing the speech that Alan Bean would later deliver in slightly blander form. "No one [in NASA] is what I would call 'petty' or 'brown-nosing' in order to get ahead. In other words, everybody feels that they have just as much capability as the other guy that's down there. And the way you get ahead is to do the best you can, not to go out and 'dig' somebody else, see? Whereas, you know full well that in the corporate and business world, you've got a bunch of dead-heads who really aren't worth a shit, and the only way they're going to be able to get up to the top is to twist the knife on somebody else."

It's hard to avoid the conclusion that many of those who inhabited the disciplined and principled world exemplified by NASA and the armed forces may not have been cut out for the corporate life or at least may have been unprepared to face the world of the two-faced. Some astronauts—Frank Borman and Bill Anders, for instance—managed to bridge the space between the two worlds successfully and with their integrity intact. Others found the transition challenging.

During his time at Sperry, Ron had been allowed to pursue a sideline that he would now expand into a business run by the person best suited to be his boss—himself. He had found it stimulating and enjoyable to tour the country on speaking engagements, whether as a NASA astronaut or as a Sperry man. Now he went out on his own, trying to build his sideline into a viable source of income.

For the first few years after *Apollo 17*, crowds of people turned up just to meet a man who had been to the moon and hear him speak. That lasted roughly until American astronauts returned to space in the shuttle. Space was big news again, even if no one was going back to the moon, but by the time Ron left Sperry, the shuttle had launched twice as many astronauts as NASA had flown in the first decade of the space race. To some members of the public, NASA was beginning to resemble an airline.

While space enthusiasts still wanted to hear a firsthand account from someone who had flown a thousand times higher than the shuttle, as Ron learned to plow his own furrow in the field of business, he gravitated more toward motivational speaking, consulting, and marketing. He saw a widespread corporate desire to expose workers to someone who had achieved the extraordinary. Bosses wanted their people to be motivated to achieve more. Provided the audiences *wanted* to be motivated, Ron found that his inspiration worked. He met a lot of people who took his message to heart. After all, if Captain Evans got to the moon by working his socks off, surely with a bit of effort they could hit their less lofty targets.

Ron set up his own company, Space Age America, and found the "astronaut" tag still opened doors in the aerospace world. There were plenty of connections to be made, and he developed some unlikely business relationships. For one, Ron established a close and lucrative partnership with a Japanese entrepreneur who was setting up a space-age theme park. He required replica spacecraft, including full-scale command modules and lunar modules for children to play in and around. Ron arranged for manufacturers to build the replicas accurately enough to meet the exacting standards of the young space cadets. He also visited Australia to discuss setting up theme parks and a space museum.

He even became a member of the board of directors of a Beverly Hills–based company called Space WIF Corporation, which was developing and marketing space-themed videos featuring former astronauts and animated characters.

Ron never tired of invitations to make after-dinner speeches, and a recording of one of the talks is available on a VHS video titled *Let's Fly to the Moon with Apollo Astronaut Ron Evans: The Ultimate Personal Experience*. He genuinely loved standing in front of a large audience, reliving the adventure of *Apollo 17*. "It turns me on. You bet it does! It turns me on! There's a tremendous excitement, and that excitement comes back to me when I get the opportunity to speak to people about what it's like. Anytime you can give a talk, get them excited, make them happy . . . everyone is listening intently to your words. That's got to be satisfying! I have yet to give a speech and not have total absorption, total attention. I take that back! I did one out at Sun City one night after a long dinner. I lost two out of a hundred, so that's not too bad!"

As they approached their midfifties, Ron and Jan were living the relaxed western lifestyle they had desired. Their home echoed with the barking of the dogs that were a part of that life. Ron continued to slice balls across the fairways of his local golf course. They had never been wealthy, but Ron's various business interests continued to generate income. When Andy Chaikin interviewed him in 1986 for *A Man on the Moon: The Voyages of the Apollo Astronauts*, Ron described being his own boss. "I'm having a beautiful time. Having a great time. I'm out on my own, doing my thing, bringing in a few dollars here and there! Keeping busy!" Somewhere in the background, Jan and Jaime are chatting. Dogs bark. A bird squawks past the window. Ice cubes clink in glasses, and there is the click of a cigarette lighter.

It was a lifestyle Ron and Jan must have thought would continue indefinitely.

28. Ad Astra

Ad Astra per Aspera.

—Kansas state motto

No one ever seems to have persuaded Ron Evans that his voyage to the moon had changed him. Perhaps it depends on what we mean by "changed." Certainly, others perceived him as a man who had done something extraordinary, and that changed *their* perception of *him*.

Those who try to understand humanity's place, bobbing along in time's ever-rolling stream, might look to Arthur C. Clarke's introduction to *2001: A Space Odyssey*. Clarke calculated that since the dawn of our species, some one hundred billion humans have walked the earth. (The figure must be higher since the novel was published, but let's not quibble.) In all that time, only twenty-four of us have ever visited another world in space. For those of us who are old enough to remember the Apollo years, there ought to be a sense of astonishment that in all the long history of humanity, ours was the generation that witnessed both the small step and the giant leap. Yet how much more remarkable it all was for those twenty-four Americans who were able to make the journey.

Modestly claiming simply to have been in the right place at the right time, all of those astronauts were born between 1923 and 1935; they possessed good health, received good educations, and (with one honorable exception) had outstanding flying skills. "Just doing my job. Just a happy accident really . . ."

However much each of us may yearn to be in some way special, a sobering truth lies at the heart of our little existences: most lives are ordinary unless you get to do something extraordinary. But if you thought of what you did as just the job you trained for and got paid for, might that explain an unwillingness to admit to any change? Can a person really fly to the moon and see every trace of home disappearing behind its cratered horizon without being changed? Ron Evans was certainly awed by his experiences, and he suggested that for many of the Apollo astronauts, the clue was in the voice more than the words. He felt that all of his lunar colleagues shared a sense of awe at what they had seen, even if they didn't wax lyrical about it or even say it out loud. If he sounded excited on his spacewalk, it was because he *was* excited. He earned his paycheck for being able to sound excited while still methodically ticking off the requirements of the flight plan and getting the job done.

After Ron's return to Earth, he was frequently asked how he had been changed by his experience. It was as if the questioners needed to hear that visiting the moon transforms a person. It got to the point where Ron began asking himself if he *should* have had some sort of revelation or epiphany. The reinforcement of his belief in God didn't really count because he had been a believer before the flight. Seeing the precision of the universe for himself simply confirmed what he already believed, so it wasn't an epiphany.

When pressed, even many years later, he laid it on the line. "Maybe people try to make me different, and maybe I'm trying to make myself different, but I'm not! I'm the same as I was before I left. I didn't change. I didn't change!" But he did accept that he had become a better person after the mission by having the experience. That acknowledgment would be something of a contradiction unless we think in terms of a car whose tank has been filled, greatly extending its range. Likewise, *Apollo 17* extended Ron's perspective with a tankful of new experiences but without altering the nature of the man himself. By his own admission, talking to a gathering of people about his flight to the moon allowed him to shed the analytical, engineering, "right stuff" persona and become an enthusiastic storyteller. But when he arrived home, he was once again the same old Ron.

Media commentators at the time of Apollo frequently lamented what they perceived as a character defect in the astronauts who went to the moon. "If only they could have been more expressive!" "Why couldn't they have bared their souls for us?" "Why couldn't we have sent a poet to the moon?" But if pilots don't make good poets, it is (almost) equally true that poets don't make good pilots. (There are exceptions.) The men selected to fly Apollo missions were all test pilots or, apart from Jack Schmitt, highly experienced fighter pilots, and they made the grade because of their analytical personalities. They were chosen for their ability to make the right decisions, coolly and efficiently, in difficult or even emergency situations. Their reactions didn't always make for the most exciting soundtrack, but every Apollo astronaut who flew returned safely. There will come a time when poets can broadcast descriptions of the moon for us in iambic pentameter, but for operational reasons, that time was not during the days of Apollo.

Some of the Apollo astronauts did experience epiphanies. Edgar Mitchell wrote that he felt "a sense of universal connectedness" and spent much of the rest of his life studying human consciousness and a "sense of unity and wholeness with the cosmos." He drew attention to the fact that fellow test pilots Jim Irwin and Charlie Duke later became Christian evangelists and that Alan Bean established a very successful career as an artist. All four were lunar module pilots and thus had "a bit more latitude for contemplation on the way home."

Mitchell didn't quite suggest that commanders and command module pilots lacked introspection or spontaneous expression. But if Ron Evans lacked the poetic gene, it seems reasonable to conclude that he wasn't selected for his ability to philosophize about the wonders of the universe but, self-evidently, for his ability to navigate a spacecraft safely to the moon and back. With Ron Evans, it was very much "what you see is what you get."

How does any man contemplate the life that lies ahead if the events for which he will be remembered took place before he turned forty? How does he emerge from the bright shadow of that past? These questions have dogged most of the lunar pioneers to a greater or lesser extent. Some clearly asked themselves the questions but could never find wholly satisfactory answers. Others moved on. Pete Conrad always claimed to be prouder of his efforts to rescue the damaged *Skylab* than of being the third man on the moon. John Young and Ken Mattingly each went on to command two shuttle missions, and Jack Schmitt represented New Mexico for six years as a U.S. senator.

Ron Ammons always credited his friend Ron Evans with being a well-grounded family man who was able to move on. "That was a part of his life, and he left it behind, and he didn't really dwell on it."

But Ron didn't walk away without a backward glance. "There has to be a certain amount of nostalgia with it, because there's a certain amount of nostalgia with anything you look back on that you've been able to accomplish and felt good about doing. . . . You remember the good times you had while doing it. And I really had a good time, I *really* had a good time! I was totally confident I could handle most anything that happened. I took the risk—and there were risks, and I realized there were risks there: there are certain things you can't handle! But in my mind it was worth taking those risks. And I had a ball! I really did! I had a ball!"

But how do you stride into your future if you are forever defined by your past? In 1986, aged not quite fifty-three, Ron clearly still had the moon on his mind. "I'm not going to allow myself to say that's the greatest thing that's ever going to happen to me in my life. I'm just not going to do that! I don't believe—even though you have a lot of nostalgia about what you've been able to accomplish—I don't believe you can live forever on those particular accomplishments. You've got to have a near-term goal, near-term ideas, newer things so that you can get that same feeling of accomplishment. If you ever get to the position in your life where you cannot feel you're accomplishing something, you know, you might as well forget it! So maybe that's what drives me, see? Keeps me going!

"I don't think there will ever be the same type of experience available for any of us, but that's not to say, at this point in our lives, that we can say that. When I'm sixty-five and looking back, *then* I might say the best thing that ever happened in my life was going to the moon. But I'm not going to tell myself that now, because I have other things I want to accomplish, other things I want to get done.... Don't bring yourself down by saying the best is behind you! That's my philosophy."

In early March 1990, as he approached his own retirement from the U.S. Navy, Rear Adm. Jimmie Taylor was looking for a guest speaker for a Navy League event. He was the chief of naval air training, and as the function would be attended by the chief of naval operations, Taylor wanted someone special. He invited his former squadron mate Ron Evans. Ron and Jan flew from Phoenix to Corpus Christi, and Jimmie saw his old friends settled into their accommodations. Ron gave "an absolutely superb talk" about his life at NASA and his flight to the moon and back, complete with his trademark "there are two ways" account of the things they never dwelled on in the press.

Relaxing later in the officers' club, Ron and Jan enjoyed catching up with Jimmie and his wife, Annette. The memory makes Jimmie laugh. "If I hadn't known that he'd gone through astronaut training and had gone to the moon and back on *Apollo 17*, I couldn't have told it, or seen any indication that his personality or his demeanor had changed at all. Same old Ron!"

About three weeks later, Tom Stafford was at Los Angeles International Airport, waiting for a flight, when he ran into a familiar figure. Ron Evans was also passing through, and the two retired Apollo astronauts had time to sit down, drink a coffee, and reminisce a little. Tom had to smile when Ron rolled up a trouser leg and pulled a cigarette out from the back of his sock. He was keeping alive the old navy tradition. After an all-too-brief encounter, Ron went on his way with a cheery, "See you later, Tom!" Stafford smiled and waved. "Right, see you later, Ron."

In 1986 Ron had become a Rotarian, joining the Rotary Club of Scottsdale. Four years later, he and Jan were preparing for a Rotary excursion to the Grand Canyon. Ron had arranged a surprise for Jan—a preliminary visit to the little town of Williams on the way. He told her on Friday, 6 April, so she had time to make any extra preparations, and they agreed to go to bed early, around 9:30 p.m., to allow an early start on Saturday.

Just before retiring for the night, Ron told Jan, "I kind of feel like I might be catching a cold. My throat feels scratchy." Jan suggested the classic remedy, an aspirin, and Ron swallowed it with a small glass of Seven-Up before climbing into bed.

About 6:30 a.m. on Saturday, Jan woke up and tip-toed to the bathroom so as not to disturb Ron. On returning to bed, she realized that something awful had happened to Ron. She called 911. The call taker kept telling her she had to get him on the floor and start CPR, but Jan knew it was too late. Ron was dead. He was long gone. When the paramedics arrived, they concluded that he had suffered a massive heart attack in his sleep, dying in the early hours of 7 April 1990. Ron was fifty-six years old.

Jaime and her husband, Lamar, arrived within ten minutes of receiving Jan's call. Jon and his family booked a flight from Texas and arrived the next day. Either the paramedics or the police had wanted to notify NASA, and Jan had somehow managed to give them a number. Thus, the news spread, and obituaries appeared from nowhere. Jan later wondered who had written them.

She was surrounded and comforted by family, friends, and neighbors. Pastor Tom Erickson from her church was a tower of strength, and while Jan drew heavily on her faith, even Tom's best efforts could not cushion her grief and sense of disbelief. It was nearly two weeks before the shock subsided, leaving the pain of realization and acceptance.

Ron was cremated, and a memorial service was held at two o'clock on Wednesday, 11 April, at Valley Presbyterian Church in Scottsdale. Meanwhile, in Kansas, porch lights in St. Francis were switched on in memory of the town's most famous son, and a moment of silence was observed during a Lenten service at Ron's boyhood church. The Reverend John Bartlett had stunned his congregation with a sad announcement three days earlier on Palm Sunday.

The service in Scottsdale united mourners from Kansas, Texas, Arizona, and beyond. Ron's brother Jay was there, struggling to understand how the man who could have died in any number of ways over Vietnam or flying to the moon and back had passed away in his bed. Tom Stafford remembered his airport meeting with a smiling Ron only two weeks earlier. Now he was looking at a gold-colored box containing his old friend's ashes. "It was real sad. He was a really great guy, and it's just a tragedy that he died at such a young age."

The first two men on the moon, Neil Armstrong and Buzz Aldrin, attended the service, and Neil was accompanied by his wife Jan. The original Godfather, Deke Slayton, was there, as were Dick Gordon and Paul Weitz, whose memories of Ron stretched back as far as their days at Monterey. Jack Schmitt was there as well as, of course, Gene Cernan with Barbara. Although by now divorced, they came together for Ron and Jan.

Ron's death was a huge and painful shock to Cernan, as intimations of mortality always are, particularly in the case of someone so close. Gene later wrote that delivering a eulogy at the service was "one of the most difficult things I ever had to do." Although no record of the complete text exists, Cernan was at his best, talking about "a special person with whom I went on the journey of a lifetime." He also found comfort in the poem "High Flight" by John Gillespie Magee Jr., one man who did combine the skills of a pilot and a poet while treading "the high untrespassed sanctity of space."

Ron Ammons delivered an even more personal eulogy, talking about his friend's familiar grin, his sense of wonder, and his yearning to explore. He reflected on how the old El Lago neighborhood family had largely moved away to new pastures. Then he added, "I guess we never faced the reality that this type of reunion would someday take place. But it has, and as usual 'the family' is here."

Ron's death certificate confirmed that he died of a heart attack. Marv Miller remembers taking a call from Jay Evans shortly after they heard the news. Both found it hard to believe that a man whose fitness had been a necessary prerequisite for navy flying and astronaut selection was actually gone. Like all retired Apollo astronauts, Ron was invited back to NASA annually around the time of his birthday for a full medical examination. He always went and had done so in November 1989. As far as Jan is aware, nothing unusual had emerged. Apart from the odd cold or flu, Ron's only medical issue had been his needing back surgery a few years earlier. She wondered if repeated hard landings on carrier decks had contributed to that. The post-op therapy had been to take long daily walks with the dogs around the golf course, and he had thoroughly enjoyed the exercise.

While Ron's death only months after a full NASA medical examination may seem surprising, it is worth noting that his Group 5 colleague Jim Irwin had suffered a serious heart attack barely a week after his annual NASA medical in March 1973. He later survived a second one, before dying of a third in 1991 at the age of sixty-one.

Right up until the night of 6 April 1990, Ron seemed fit and well to those who saw him regularly, and his sudden death came completely out of the blue. No one will ever know for certain, but one possible culprit had dogged him throughout his life since he first shared that pack of Wings cigarettes as a boy in St. Francis. Jan—a former smoker herself in the days when most people did—acknowledges the possible link and that Jaime and Jon had often said, "Well, if Daddy hadn't smoked . . ." But Jan recalls many non-smoking friends and acquaintances who also died prematurely. She is inclined to believe that in life, bad things just sometimes happen to good people.

Jan Evans eventually picked up the pieces and carried on. Everyone who knows her recognizes her as a strong person, and she continues to be a leading light of the sisterhood of astronaut wives and widows. They were often called the Astronaut Wives Club, although they never had an organized club, as such. They all had similar experiences, and they treasured coming together as old friends and sharing their chatter, laughter, and sometimes tears.

Jan also kept in touch with Gene and Jack, and as *Apollo 17* anniversaries rolled past, the three came together for reunion celebrations, sometimes joined by other familiar faces from the "glory years." When Gene Cernan passed away on 16 January 2017, that left Jack Schmitt as the sole surviving crewmember of *Apollo 17*.

There has never been a day when Ron is absent from Jan's thoughts. How could there be? He was the love of her life, and she had enjoyed every day of their thirty-two years together, even the days when they were separated by thousands—and sometimes hundreds of thousands—of miles. They say time heals all wounds, but some pain never goes away. "It's with you every day," she says wistfully. "You never forget it. It's a part of life. I can't change it, so I accept it."

Ron could have been buried at Arlington National Cemetery among his fellow astronauts and the fallen of many wars and conflicts, but Jan didn't want him "out there," far from those who had known him best. She wanted him near her and his home. Thus, the mortal remains of Ron Evans rest amid the tranquil greenery of Valley Presbyterian Church's garden of remembrance. There in the shade of a tree, a simple plaque displays the name by which everyone knew him—Ron Evans. On every anniversary of his birth, his death, his flight to the moon, and their wedding, visitors to the garden will see the single red rose that Jan has placed on Ron's plaque.

What are we to make of the life, career, and legacy of Ron Evans? When a man lives long enough to hold his baby granddaughter in his arms, he knows that the legacy that really matters is secure. His two children are thriving. Jon owns a successful construction equipment and tool rental business in Austin, Texas. He shared the raising of his wife's three boys—Jeremy, James, and Joshua—and they had twins, Jace and Jannah. On 13 November 2019 Jannah bore Ron and Jan's first great-grandchild, Savannah Mae. Jaime and Lamar Harner, who work together in their flourishing accountancy and fiduciary practice in Scottsdale, have a daughter, Wesley, and a son, Cade.

Ron Evans was raised in the agricultural heartland of the Midwest. He wanted to expand his horizons, and very few have fulfilled such a common ambition in such a spectacularly uncommon way. His navy service led him to participate in one of the great tragedies of the twentieth century and then the greatest scientific triumph of any century.

As we enter the third decade of the twenty-first century, a new era of lunar exploration is dawning, but even if landing and working on the moon become commonplace in the future, our hearts and our memories should still reserve a place for the two dozen men who were first to explore it. Neil Armstrong was first to get his boots dirty, but they all made their contributions, along with hundreds of thousands of other Americans, most of whose names we don't recall. The last of the pioneers—Gene Cernan, Ron Evans, and Jack Schmitt—deserve to be remembered for bringing the first golden age of lunar exploration to a spectacular and successful conclusion. Let it not be forgotten that for nearly half a century, Ron Evans has held the record for the longest time anyone has ever orbited another celestial body.

Every time a big Arizona moon rises over Scottsdale, Jan Evans looks up at it with nostalgia and remembers how she met and married one of the men who flew there a long time ago. Ron Evans was a happy soul who lived his life to the fullest, but when all is said and done, the saddest thing about the life of Ron Evans is that he and Jan did not have the chance to grow old together.

Sources

Books

Aldrin, Buzz, with Ken Abraham. *Magnificent Desolation: The Long Journey Home from the Moon*. London: Bloomsbury, 2009.

Armstrong, Neil, Edwin Aldrin, and Michael Collins. *First on the Moon*. Toronto: Little, Brown, 1970.

Baker, David. *The History of Manned Spaceflight*. London: New Cavendish Books, 1981.

Bilstein, Roger E. *Stages to Saturn: A Technological History of the Apollo/Saturn Launch Vehicles*. NASA History Series publication SP-4206. Washington DC: NASA, 1996.

Brun, Nancy. *Astronautics and Aeronautics, 1974: A Chronology*. NASA History Series publication SP-4019. Washington DC: NASA, 1977. https://history.nasa .gov/AAchronologies/1974.pdf.

Brun, Nancy, and Eleanor Ritchie. *Astronautics and Aeronautics, 1975: A Chronology*. NASA History Series publication SP-4020. Washington DC: NASA, 1979. https://history.nasa.gov/AAchronologies/1975.pdf.

Cassutt, Michael. *The Astronaut Maker*. Chicago: Chicago Review Press, 2018.

Cernan, Eugene, with Don Davis. *The Last Man on the Moon*. New York: St. Martin's Press, 1999.

Chaikin, Andrew. *A Man on the Moon: The Voyages of the Apollo Astronauts*. London: Michael Joseph, 1994.

Clarke, Arthur C. *2001: A Space Odyssey*. London: Arrow Books, 1990.

Collins, Michael. *Carrying the Fire*. London: W. H. Allen, 1975.

Compton, David. *Where No Man Has Gone Before*. NASA History Series publication SP-4214. Washington DC: NASA, 1989.

Cooper, Gordon, with Bruce Henderson. *Leap of Faith: An Astronaut's Journey into the Unknown*. New York: HarperCollins, 2000.

Cooper, Paul L. *Weekend Warriors*. Manhattan KS: Sunflower University Press, 1997.

Crotts, Arlin. *The New Moon: Water, Exploration and Future Habitation*. New York: Cambridge University Press, 2014.

Cunningham, Walter, with Mickey Herskowitz. *The All-American Boys.* New York: Macmillan, 1977.

Debenham, Frank. *In the Antarctic: Stories of Scott's Last Expedition.* London: John Murray, 1952.

Duke, Charlie, and Dotty Duke. *Moonwalker.* Nashville: Oliver Nelson Books, 1990.

Ertel, Ivan, Roland Newkirk, and Courtney Brooks. *The Apollo Spacecraft: A Chronology.* Vol. IV. NASA History Series publication SP-4009. Washington DC: NASA, 1978.

Ezell, Edward, and Linda Ezell. *The Partnership: A History of the Apollo-Soyuz Test Project.* Mineola NY: Dover Books, 2010. Reprint of NASA History Series publication SP-4209.

French, Francis, and Colin Burgess. *In the Shadow of the Moon.* Lincoln: University of Nebraska Press, 2007.

Furniss, Tim. *Manned Spaceflight Log: New Edition.* London: Jane's, 1986.

Ginter, Steve. *Navy Fighter Squadrons: Part Four: Vought's F-8 Crusader.* Naval Fighters Series Number 19. Forest Lake MN: Specialty Press, 1990. Printed on demand: Lightning Source UK, Milton Keynes.

Glenn, John, with Nick Taylor. *John Glenn: A Memoir.* New York: Bantam Books, 1999.

Grissom, Virgil "Gus." *Gemini: A Personal Account of Man's Venture into Space.* New York: Macmillan, 1968.

Hansen, James R. *First Man: The Life of Neil A. Armstrong.* New York: Simon & Schuster, 2005.

Harland, David M. *Exploring the Moon: The Apollo Expeditions.* Chichester: Praxis, 1999.

Highland Park High School. *The Highlander.* Topeka KS: Highland Park High School, 1951. Yearbook.

Irwin, James B., and William Emerson Jr. *To Rule the Night: The Discovery Voyage of Astronaut Jim Irwin.* New York: A. J. Holman, 1973.

Johnston, Richard S., Lawrence F. Dietlein, and Charles A. Berry, eds. *Biomedical Results of Apollo.* NASA History Series publication SP-368. Washington DC: NASA, 1975.

Kelly, Thomas J. *Moon Lander: How We Developed the Apollo Lunar Module.* Washington DC: Smithsonian Institution Press, 2001.

Kranz, Gene. *Failure Is Not an Option.* New York: Simon & Schuster, 2000.

Logsdon, John M. *After Apollo? Richard Nixon and the American Space Program.* New York: Palgrave Macmillan, 2015.

———. *John F. Kennedy and the Race to the Moon.* New York: Palgrave Macmillan, 2010.

Lovell, Jim, and Jeffrey Kluger. *Lost Moon: The Perilous Voyage of Apollo 13.* New York: Houghton Mifflin, 1994.

Lucchitta, B. K., and H. H. Schmitt. "Orange Material in the Sulpicius Gallus Formation at the Southwestern Edge of Mare Serenitatis." In *Proceedings of the Fifth Lunar Conference* (Lunar and Planetary Institute). Vol. 1. New York: Pergamon Press, 1974. http://adsabs.harvard.edu/full /1974LPSC. . . . 5..223L.

Lyndon B. Johnson Space Center. *Apollo 17 Preliminary Science Report.* NASA SP-330. Washington DC: NASA, 1973. https://www.hq.nasa.gov/alsj/a17/as17psr.pdf.

Masursky, Harold, G. W. Colton, and Farouk El-Baz, eds. *Apollo over the Moon: A View from Orbit.* NASA SP-362. Washington DC: NASA, 1978.

Mitchell, Edgar D., with Dwight Williams. *The Way of the Explorer: An Apollo Astronaut's Journey through the Material and Mystical Worlds.* New York: G. P. Putnam's Sons, 1996.

Morrison, Phil. *Pivotal Moments '64.* 2nd ed. Mustang OK: Tate Publishing, 2014.

Orloff, Richard W., and David M. Harland. *Apollo: The Definitive Sourcebook.* Chichester: Praxis, 2006.

Platzer, Max F., Richard W. Bell, Louis V. Schmidt, and Conrad F. Newberry. "Naval Officer Graduate Education in Aerospace Engineering at the Naval Postgraduate School." In *Aerospace Engineering Education during the First Century of Flight,* edited by Barnes McCormick, Conrad Newberry, and Eric Jumper, 800–818. Reston VA: American Institute of Aeronautics and Astronautics, 2003. https:// calhoun.nps.edu/handle/10945/55365.

Ryan, Peter. *The Invasion of the Moon, 1969: The Story of Apollo 11.* Harmondsworth: Penguin, 1969.

Schirra, Walter M., with Richard N. Billings. *Schirra's Space.* Boston: Quinlan Press, 1988.

Scott, David, and Alexei Leonov. *Two Sides of the Moon.* New York: Simon & Schuster, 2004.

Scott, Robert Falcon. *Tragedy and Triumph: The Journals of Captain R.F. Scott's Last Polar Expedition.* New York: Konecky & Konecky, 1998.

Seaver, George. *Edward Wilson of the Antarctic.* London: John Murray, 1959.

Shayler, David. *Around the World in 84 Days: The Authorized Biography of Skylab Astronaut Jerry Carr.* Burlington ON: Apogee Books, 2008.

Shayler, David, and Colin Burgess. *The Last of NASA's Original Pilot Astronauts.* Chichester: Praxis, 2017.

Shulimson, Jack, and Charles Johnson. *The U.S. Marines in Vietnam: The Landing and the Buildup, 1965.* Washington DC: History and Museums Division, U.S. Marine Corps, 1978.

Slayton, Donald K., and Michael Cassutt. *Deke! U.S. Manned Space: From Mercury to the Shuttle.* New York: Forge Books, 1994.

Solomon, Susan. *The Coldest March: Scott's Fatal Antarctic Expedition*. New Haven CT: Yale University Press, 2001.

Stafford, Thomas P., and Michael Cassutt. *We Have Capture*. Washington DC: Smithsonian Institution Press, 2002.

Stockdale, Jim, and Sybil Stockdale. *In Love and War*. New York: Bantam Books, 1985.

Wilhelms, Don E. *To a Rocky Moon: A Geologist's History of Lunar Exploration*. Tucson: University of Arizona Press, 1993.

Wolfe, Tom. *The Right Stuff*. New York: Farrar, Straus, Giroux, 1979, 1983.

Woods, W. David. *How Apollo Flew to the Moon*. Chichester: Praxis, 2008.

Worden, Al, with Francis French. *Falling to Earth*. Washington DC: Smithsonian Books, 2011.

Periodicals and Online Articles

All Hands. "Navy Wings Issue." No. 509, June 1959. https://media.defense.gov/2019/Jul/25/2002162167/-1/-1/1/AH195906.pdf.

Bjorkman, Eileen. "Have Gun, Will Dogfight." *Air & Space Magazine*, October 2015. https://www.airspacemag.com/history-of-flight/11_on 2015-f8-crusader-at-60-180956611/.

Cooper, Gary M. "Pilot Gives Hope Tough Act to Follow." *Stars and Stripes*, 29 December 1965. https://www.stripes.com/news/us/pilot-gives-hope-tough-act-to-follow-1.258100.

Daily Mail. "A Joint Step for Mankind." 25 May 1972.

———. "TV Row Threat to Apollo 17." 4 December 1972.

Department of the Navy, Naval History and Heritage Command. "SCB-27 Modernization of *Essex/Ticonderoga* Class Aircraft Carriers." 8 October 2001. ibiblio.org/hyperwar/OnlineLibrary/photos/usnshtp/cv/scb27cl.htm.

Encyclopedia.com. "Great Depression in the Mid-West." Updated 18 August 2020. https://www.encyclopedia.com/economics/encyclopedias-almanacs-transcripts-and-maps/Midwest-great-depression.

Grosvenor, Gilbert M. "Summing Up Mankind's Greatest Adventure." *National Geographic* 144, no. 3 (September 1973).

Guilmartin, John F. "B-17 Aircraft." *Britannica*. https://www.britannica.com/technology/B17.

Hauri, Erik H., Alberto E. Saal, Malcolm J. Rutherford, and James A. Van Orman. "Water in the Moon's Interior: Truth and Consequences." *Earth and Planetary Science Letters* 409 (20 November 2014): 252–64. http://dx.doi.org/10.1016/j.epsl.2014.10.053.

Hauri, Erik H., Thomas Weinreich, Alberto E. Saal, Malcolm C. Rutherford, and James A. Van Orman. "High Pre-Eruptive Water Contents Preserved in Lunar Melt Inclusions." *Science* 33, no. 6039 (8 July 2011): 231–15.

Jones, Eric M., and Ken Glover, eds. *Apollo Lunar Surface Journal (Apollo 17)*. Last revised 5 June 2018. https://www.hq.nasa.gov/alsj/.

Kansas Engineer Magazine. "Meet the Authors." University of Kansas. Undated.

Morning Call (Mansfield OH). "U.S. Navy Flier Is Killed in S. Pacific." 22 December 1965.

Muson, Howard. "Comedown from the Moon—What Has Happened to the Astronauts." *New York Times*, 3 December 1972. https://www.nytimes.com/1972/12/03/archives/comedown-from-the-moonwhat-has-happened-to-the-astronauts.html.

Naval Aviation News. "Ranger Men Try Their New Suits." May 1959. navymats.com/NavalAviationMay59.pdf.

Nevin, David. "Collins Has Cool to Cope with Space and the Easter Bunny." *Life*, 4 July 1969.

Norris, Robert S., and Hans M. Kristensen. "Declassified: U.S. Nuclear Weapons at Sea during the Cold War." *Bulletin of the Atomic Scientists* 72 (8 January 2016): 58–61. https://doi.org/10.1080/00963402.2016.1124664.

NRC Handelsblad (Amsterdam, Netherlands). "RSV Was Warned about Partner in U.S." [In Dutch.] 4 October 1982. https://www.delpher.nl/nl/kranten/view?identifier=kbnrc01:000028101:mpeg21:p001.

Pinson, Jerald. "Glossopteris and the Moving Continents." *All Is Leaf*, 28 October 2017. https://allisleaf.com/2017/10/28/glossopteris-moving-continents/.

Porter, Gareth. "The Real Tonkin Gulf Deception Wasn't by Lyndon Johnson." *Truthout*, 5 August 2014. https://truthout.org/articles/the-real-tonkin-gulf-deception-wasn't-by-lyndon-johnston/.

Prados, John. "Essay: 40th Anniversary of the Gulf of Tonkin Incident." *National Security Archive*, 4 August 2004. https://nsarchive2.gwu.edu/NSAEBB/NSAEBB132/essay.htm.

Quanah (TX) Tribune-Chief. "Glenn's Heroics Call to Mind Astronaut-trainee Givens." 19 November 1998.

Rufu, Raluca, Oded Aharonson, and Hagai B. Perets. "A Multiple-Impact Origin for the Moon." *Nature Geoscience* 10 (9 January 2017): 89–94. https://doi.org/10.1038/ngeo2866.

Saal, Alberto E., Erik H. Hauri, Mauro L. Cascio, James A. Van Orman, Malcolm J. Rutherford, and Reid F. Cooper. "Volatile Content of Lunar Volcanic Glasses and the Presence of Water in the Moon's Interior." *Nature* 454 (10 July 2008): 192–95. https://www.nature.com/articles/nature07047.

Saint Francis (KS) Herald (Special Souvenir Edition). "Welcome Home, Ron Evans." 1 February 1973.

Sanger, David E. "U.S. Confirms It Lost an H-Bomb off Japan in '65." *New York Times*, 9 May 1989. https://nyti.ms/29ylau8.

Santa Cruz Sentinel. "Admiral Rickover Speaks Out." 17 April 1964.

Schlom, David. "Target America 1972: When Terrorists Threatened Apollo." *Ad Astra*, November–December 2001. https://space.nss.org/wp-content/uploads /Ad-Astra-Magazine-v13n6f5.pdf.

Schmitt, Harrison H. "Apollo 17: Diary of the 12th Man." America's Uncommon Sense (website). https://www.americasuncommonsense.com/1-apollo-17-diary -of-the-12th-man/.

Science. "Lunar Water Signs Found in Apollo Mission Rocks." 10 July 2008. https:// www.pbs.org/newshour/science/science-july-dec08-moonwater_07-10.

SEC *News Digest.* "Trading Suspended in Western American Energy Corporation." Issue 77-152 (8 August 1977). https://www.sec.gov/news/digest/1977/dig080877.pdf.

Settles, Bob. "Seven Years a Sailor." *Stories from the Funny Farm.* Undated. http:// www.bobandlois.com.

Stars and Stripes. "Flying Brothers Hailed, May Be Held Again." 18 May 1959.

Time. "Apollo 17: Farewell Mission to the Moon." 11 December 1972.

——. "Portfolio from Apollo." 8 January 1973.

Times (UK). "Awesome Launch Puts Apollo on Course After Last-Minute Delay." 8 December 1972.

——. "Home Thoughts from the Moon." 6 December 1972.

Unofficial U.S. Navy Site. "USS Oriskany (CV-34)." https://www.navysite.de/cv /cv34.htm.

——. "USS Oriskany (CVA 34) WestPac Cruise Book 1960." https://www.navysite .de/cruisebooks/cv34-60/index.html.

——. "USS Ranger (CV-61)." https://www.navysite.de/cvn/cv61.htm.

——. "USS Ranger (CVA 61) WestPac Cruise Book 1959." https://www.navysite .de/cruisebooks/cv61-59/index.html.

——. "USS Ticonderoga (CV-14)." https://www.navysite.de/cv/cv14.htm.

——. "USS Ticonderoga (CVA 14) WestPac Cruise Book 1964." https://www .navysite.de/cruisebooks/cv14-64/index.html.

——. "USS Ticonderoga (CVA 14) WestPac Cruise Book 1965–66." https://www .navysite.de/cruisebooks/cv14-65/index.html.

——. "USS Wisconsin (BB 64) Cruise Book 1952–54." https://www.navysite.de /cruisebooks/bb64-53/index.html.

UPI. "Former Astronaut Dies." 8 April 1990. https://www.upi.com/Archives/1990 /04/08/Former-astronaut-dies/9412639547200/.

Virtual Aircraft Museum. "Vought F8U/F-8 Crusader 1955." 24 October 2010. aviastar .org/air/usa/chance_crusader.php.

Weaver, Michael. "An Examination of the F-8 Crusader through Archival Sources." *Journal of Aeronautical History* 8 (2018): 63–85. Paper no. 2018/02. https://

www.aerosociety.com/media/8037/an-examination-of-the-f-8-crusader-through -archival-sources.pdf.

Wiki.org. "Military: Vought F-8 Crusader." https://military.wikia.org/wiki/Vought _F-8_Crusader.

Wolpert, Stuart, University of California–Los Angeles. "Moon Was Produced by a Head-on Collision between Earth and a Forming Planet." *ScienceDaily*, 29 January 2016. www.sciencedaily.com/releases/2016/01 /160129090451.htm.

Woods, David, and Ben Feist, eds. *Apollo 17 Flight Journal*, 16 February 2018. NASA History Division, Apollo Flight Journal series (to Mission Day 5). https://history .nasa.gov/afj/ap17fj/index.html.

Woods, David, Ben Feist, Ronald Hansen, and Lennox Waugh, eds. *Apollo 14 Flight Journal*, 3 January 2017. NASA History Division. https://history.nasa.gov/afj /ap14fj/index.html.

Interviews and Personal Communications

All interviews were conducted by the author, by telephone, unless otherwise indicated.

Allen, Jack. 21 March 2018.

Allen, John. 5 March 2018.

———. 30 October 2018.

Ammons, Ron. 13 June 2018.

———. 12 July 2018.

Armstrong, Neil. Interview by Alex Malley, then the chief executive officer of CPA Australia, 2011.

———. Interview by Gay Byrne for CSL Associates, Dublin, 17 November 2003.

Bailey, Gail R. Email exchange with the author, 5 March 2018.

Berry, Charles A. Johnson Space Center (JSC) Oral History interview by Carol Butler, Houston, 29 April 1999.

Brand, Vance. 28 February 2019.

Brown, Howard. 4 April 2018.

———. 2 October 2018.

Butler, Barbara. 4 December 2018.

———. 29 May 2019.

———. 27 June 2019.

Carr, Jerry. 19 March 2018.

Cavicke, Dick. Email exchanges with the author, 11–15 December 2018 to 24–25 February 2020.

Dale, Fred. 2 April 2018.

Denney, Emily. 13 January 2018.

Duke, Charles M. Conversation with the author at Spacefest in Tucson, 9 August 2019.

——. JSC Oral History interview by Doug Ward, Houston, 12 March 1999.

El-Baz, Farouk. 9 May 2018.

——. JSC Oral History interview by Rebecca Wright, Boston, 2 November 2009.

Engle, Joe H. 27 March 2018.

——. JSC Oral History interview by Rebecca Wright, Houston, 22 April 2004.

Evans, Jan. 21 August 2017.

——. 25 August 2017.

——. 7 September 2017.

——. 21 September 2017.

——. 26 September 2017.

——. 5 October 2017.

——. 16 October 2017.

——. 24 October 2017.

——. 31 October 2017.

——. 7 November 2017.

——. 6 December 2017.

——. 4 January 2018.

——. 11 January 2018.

——. 22 January 2018.

——. 8 February 2018.

——. 20 February 2018.

——. 6 March 2018.

——. 27 April 2018.

——. 7 May 2018.

——. 31 May 2018.

——. 19 July 2018.

——. 7 August 2018.

——. 18 September 2018.

——. 17 October 2018.

——. 25 October 2018.

——. 18 December 2018.

——. 16 January 2019.

——. 13 March 2019.

——. 25 March 2019.

——. 21 May 2019.

——. 13 November 2019.

——. JSC Oral History interview by Jennifer Ross-Nazzal, Scottsdale AZ, 7 August 2003.

Evans, Jay Dale. 9 October 2017.

Evans, Ron. Interview conducted by Andy Chaikin, Scottsdale AZ, October 1986.

Geronime, Gene. 12 March 2018.

Griffin, Gerry. 14 November 2018.

———. JSC Oral History interview by Doug Ward, Houston, 12 March 1999.

Haise, Fred. 10 May 2018.

Harner, Jaime. 10 July 2018.

———. 31 July 2018.

Holland, Marlo. 5 March 2018.

Holm, John. 12 March 2018.

James, Ralph. 8 March 2019.

Kintzel, Craig. 25 March 2018.

Klein, Tom. 14 March 2018.

Lucchitta, Baerbel. Email exchanges with the author, 10–12 September 2019.

Lousma, Jack. 7 February 2018.

———. JSC Oral History interview by Carol Butler, Houston, 7 March 2001.

———. JSC Oral History interview by Jennifer Ross-Nazzal, Houston, 15 March 2010.

Mattingly, T. K. Email exchange with the author, 21 March 2018.

Miller, Marvin. 14 October 2017.

———. 30 January 2018.

Morrison, Phil. 29 November 2017.

O'Hara, Dee. 28 February 2018.

———. JSC Oral History interview by Rebecca Wright, Mountain View CA, 23 April 2002.

Schmitt, Harrison H. Interview by the author, Washington DC, 18 March 2003.

———. 18 April 2018.

Scott, David R. Email exchanges with the author, 13 December 2017–20 January 2018; and 9 June–1 July 2018.

Sears, Kelly Evans. Email exchanges with the author, 9–11 October 2017; and 8–15 July 2020.

Settles, Bob. Email exchanges with the author, 31 January–1 February 2018.

Skaggs, Wayne. 7 March 2018.

Stafford, Thomas P. 17 November 2018.

Taylor, Jimmie. 28 March 2018.

Weitz, Paul. JSC Oral History interview by Rebecca Wright, Flagstaff AZ, 26 March 2000.

Worden, Al. 12 March 2019.

Other Sources

Alfred Wegener Institute. "The Copernicus of Geosciences: Alfred Wegener Presented His Revolutionary Theory of Continental Drift 100 Years Ago." Press release. Bremerhaven, 21 December 2011.

Author's scrapbook, personal diary, and personal records of *Apollo 17*.

BBC. Sound recordings of live *Apollo 17* mission events and highlights.

Bean, Alan. After-dinner speech at Autographica 20. London, 21 March 2014.

Cernan, Gene. Public talk at Spacefest VI. Pasadena CA, 13 May 2014.

Cernan, Gene, and Harrison Schmitt, accompanied by Jan Evans. Werner von Braun Lecture. National Air and Space Museum, Washington DC, 18 March 2003.

City of St. Francis KS (website). www.stfranciskansas.com.

Discovery Channel. *Sea Wings*. Vol. 2, "The Last of the Gunfighters." Narrated by Edward Easton. Documentary. Aired 31 December 1995; uploaded to YouTube on 23 October 2011. https://www.youtube.com/watch?v=5ic-EOpLxto.

Drew, Dennis M. "Rolling Thunder 1965: Anatomy of a Failure." Report no. AU-ARI-CP-86-3. Maxwell AFB AL: Air University Press, October 1986. https://apps.dtic.mil/dtic/tr/fulltext/u2/a215903.pdf.

Evans, Jan. *The Beginning of It All*. 1984. Personal record of events surrounding Ron's selection as an astronaut.

Evans, Marie. Letter sent to the Miller family, St. Francis KS, May 1969.

Evans, Ron. Astronaut application papers (partial), July 1963.

———. Astronaut application papers (partial), December 1965.

———. Aviation logbooks, 1956–77.

———. Correspondence received following astronaut selection, 1966.

———. Personal correspondence to Jan Pollom, 1957.

Feist, Ben, comp. *Apollo 17 in Real Time: A Real-Time Journey through the Last Landing on the Moon*. Multimedia project. Forty-Fifth Anniversary Edition, 2017. https://apolloinrealtime.org/17/.

Herdrich, David J. *Apollo 17 and American Samoa* (website). News bulletin with photographs, 14 December 1972. https://members.tripod.com/~Tavita_Herdrich/apollo17.html.

Kansas Historical Society (website). Topeka. www.kshs.org.

KTAR Radio, Scottsdale AZ. Ron Evans interview and phone-in (undated but ca. 1983).

McDivitt, James. Public talk at Autographica 11. Birmingham, UK, 19 April 2008.

Miller, Rear Adm. H. L. Declassified memo from Commander, Carrier Division Three to Commander, Seventh Fleet, dated 29 March 1965. In History of Task Force 77, 2 September 1964–17 March 1965. Glenn Helm Collection, box 08, folder 18. The Vietnam Center and Sam Johnson Vietnam Archive, Texas Tech University, Lubbock. https://vva.vietnam.ttu.edu/repositories/2/digital_objects/250957.

NASA. "Apollo 14 Crew Training Summary." 1 September 1969. https://www.hq
.nasa.gov/alsj/a14/A14CrewTraining.pdf.

———. *Apollo 14 Press Kit*. NASA Release no. 71-3K. Washington DC, 21 January
1971. https://www.nasa.gov/specials/apollo50th/pdf/A14_PressKit.pdf.

———. "Apollo 17 Crew Training Summary." 15 September 1971. https://www.hq
.nasa.gov/alsj/a17/A17CrewTraining.pdf.

———. *Apollo 17 Press Kit*. NASA Release no. 72-220K. Washington DC, 14 November 1972. For release 26 November 1972. https://www.hq.nasa.gov/alsj/a17/A17
_PressKit.pdf.

———. "Apollo 17 Site Selection." NASA Press Release MSC 72-44. 16 February 1972.
https://www.nasa.gov/centers/johnson/pdf/83124main_1972.pdf.

———. *Apollo-Soyuz Test Project Press Kit*. NASA Release no. 75-118. Washington DC, 10 June 1975. https://history.nasa.gov/astp/documents/astp%20press
%20kit%20(us).pdf.

———. "Astronaut Evans to Leave NASA." NASA Press Release 77-17. 8 March 1977.

———. "Flight and Support Crews for Apollo 15." Press Release MSC 70-34. 26
March 1970. nasa.gov/centers/Johnson/news/releases/1969_1971/index.html.

———. "Flight Crews for Apollos 13 and 14." Press Release MSC 69-56. 6 August
1969. nasa.gov/centers/Johnson/news/releases/1969_1971/index.html.

———. *The Mission Transcript Collection: U.S. Human Spaceflight Missions from
Mercury Redstone 3 to Apollo 17*. NASA History Office CD-ROM Release SP-
2000-4602. 1 January 2000.

———. "NASA's SOFIA Discovers Water on Sunlit Surface of Moon." Press release.
26 October 2020; last updated 4 January 2021. https://www.nasa.gov/press
-release/nasa-s-sofia-discovers-water-on-sunlit-surface-of-moon/.

———. "Report of Panel 20: In-Flight Fire Emergency Provisions Panel, Appendix
D-20 to Final Report of Apollo 204 Review Board." In *Report of Apollo 204
Review Board to the Administrator, National Aeronautics and Space Administration*, D-20-1–D-20-25. Washington DC: U.S. Government Printing Office,
1967. https://www.history.nasa.gov/Apollo/204appendices/AppendixD19-21
.pdf.

NASA, Flight Planning Branch, Crew Procedures Division. *Apollo 17 Final Flight
Plan*. Houston: Manned Spacecraft Center, 23 October 1972.

NASA, Mission Evaluation Team. *Apollo 14 Mission Report*. NASA Release MSC-
04112. Houston: Manned Spacecraft Center, May 1971. https://www.hq.nasa
.gov/alsj/a14/A14_MissionReport.pdf.

———. *Apollo 17 Mission Report*. NASA Release JSC-07904. Houston: Lyndon
B. Johnson Space Center, March 1973. https://history.nasa.gov/alsj/a17/A17
_MissionReport.pdf.

Naval History and Heritage Command. "Vietnam War Highlights: Naval Operations, January–April 1966." Uploaded 29 April–6 May 2015. https://history.navy .mil/research/archives/digital-exhibits-highlights/Vietnam-war/naval-historical -summary-january-1966-highlights.html.

Richard Nixon Presidential Library and Museum. Correspondence, memos, and audio recordings relating to *Apollo 17*. Yorba Linda CA. www.nixonlibrary.gov.

Saturn Flight Evaluation Working Group. *Saturn V Launch Vehicle Flight Evaluation Report-AS-512: Apollo 17 Mission*. NASA Release MPR-SAT-FE-73-1. Huntsville AL: George C. Marshall SFC, 28 February 1973. https://www.ibiblio.org /apollo/Documents/lvfea-AS512-Apollo17.pdf.

Schmitt, Harrison H. "2011 AGI Medal in Memory of Ian Campbell." Address to the American Geosciences Institute. Minneapolis, 9 October 2011. https://www .geosociety.org/awards/11speeches/IanCampbell.htm.

Slayton, Deke. Letter to Lt. Ron Evans concerning astronaut selection evaluations, 20 December 1965.

Training Office, Crew Training and Simulation Division, NASA. "Apollo 17 Technical Crew Debriefing." NASA Release MSC-07631. Houston: Manned Spacecraft Center, 4 January 1973. https://www.hq.nasa.gov/alsj/a17/a17tecdbrf.pdf.

TRW Systems for Systems Engineering Division, Apollo Spacecraft Program Office. *Mission Requirements SA-512/CSM-114/LM-12, J-3 Type Mission (Apollo 17): Lunar Landing*. NASA Release MSC-05180. Houston: Manned Spacecraft Center, March 1972. https://history.nasa.gov/afj/ap17fj/pdf/a17_mission-requirements.pdf.

University of Kansas. Exchange of emails regarding Ron Evans's 1956 graduation with Kathy Lafferty, Spencer Research Laboratory, and Stefanie Shackelford, Alumni Office, 2019.

U.S. Navy. Communications to Ron Evans, 1955–73.

———. *Welcome Aboard: Guide to USS* Wisconsin *BB-64*. Pamphlet. June 1952.

Index

In the Outward Odyssey: A People's History of Spaceflight series

A Long Voyage to the Moon: The Life of Naval Aviator and Apollo 17
Astronaut Ron Evans
Geoffrey Bowman
Foreword by Jack Lousma

The Light of Earth: Reflections on a Life in Space
Al Worden with Francis French
Foreword by Dee O'Hara

To order or obtain more information on these or other University of Nebraska Press titles, visit nebraskapress.unl.edu.